INTRODUCTION TO ALGEBRAIC GEOMETRY

INTRODUCTION TO ALGEBRAIC GEOMETRY

BY

W. GORDON WELCHMAN

Massachusetts Institute of Technology, formerly Fellow of Sidney Sussex College, Cambridge

CAMBRIDGE
AT THE UNIVERSITY PRESS
1950

CAMBRIDGE
UNIVERSITY PRESS

University Printing House, Cambridge CB2 8BS, United Kingdom

Cambridge University Press is part of the University of Cambridge.

It furthers the University's mission by disseminating knowledge in the pursuit of education, learning and research at the highest international levels of excellence.

www.cambridge.org
Information on this title: www.cambridge.org/9781316601808

First published 1950
First paperback edition 2015

A catalogue record for this publication is available from the British Library

ISBN 978-1-316-60180-8 Paperback

CONTENTS

Chapter IV. The Conic

Chapter V. Configurations

Chapter VI. Metrical Geometry

Chapter VII. Homographic Ranges on a Conic

Chapter VIII. Two Conics, Reciprocation of one Conic into Another. Particular Cases

Chapter IX. Two Conics. Apolarity

Chapter X. Two-two Correspondences

Chapter XI. Application of Matrix Algebra

CHAPTER XII. INVARIANTS AND COVARIANTS

PREFACE

When a book is called an introduction, it is natural to inquire what further fields of study the author has in mind. This book contains a treatment of the theory of conics which illustrates techniques that have a wider range of application. The aim of the book is to lead a student as rapidly as possible to the study of configurations, loci and transformations in space of three, four and five dimensions. A university course in Geometry leading to the threshold of postgraduate study might well include a discussion of linear series of sets of points on rational curves during which it could be shown for example that a one to one correspondence between a rational cubic curve in space of three dimensions and a rational quartic curve in space of four dimensions determines a representation of the lines of the former space by the points of the latter. The mapping of a quadric surface in three dimensions on a plane could be used to derive properties of cubic and quartic curves in a plane and could be associated with the elementary theory of higher plane curves. The representation of the lines of a space of three dimensions by the points of a quadric primal in space of five dimensions could also be introduced. These and similar topics provide a field of study that is full of interest and by no means beyond the scope of an able university student. In order to facilitate progress, however, it is desirable to apply to the more elementary problems the types of reasoning that are used in advanced work, instead of employing methods that are not capable of extension.

Such are the considerations that led to the general plan of this book, in which the first three chapters establish the foundations that are necessary for the more advanced work to which the book is an introduction. The remaining chapters are mainly concerned with the theory of conics because it is helpful to clarify ideas by two dimensional examples before starting to consider space of three or more dimensions. The book was nearly ready for the press ten years ago, but the war intervened and the manuscript spent the next six years in various cellars, whence it emerged in a slightly mildewed condition. Its publication is due to the kindness of Dr D. W. Babbage of Magdalene College, who has revised the proofs and made valuable contributions. In the five years during which I was writing the book I enjoyed the great advantage of collaboration with A. Robson, who had aroused my interest in Geometry when I was a boy at Marlborough College. He was writing an Introduction to Analytical Geometry for use in schools, with which I hope most readers of this book will be already acquainted.

Many students for whom a study of mathematics is a preparation for a career outside the academic world will not have time to reach the topics that have been mentioned above. Many will find that at some stage in the development of the subject the concepts are beyond their grasp. Geometry, however, is remarkable in that for all levels of ability and at all stages of progress it can provide interesting fields of study in which a student can exercise ingenuity and originality. I hope that in some small way this book may help to maintain and spread the study of Geometry, that it may help others to derive as much pleasure from this study as I have myself, and that it may help some students to appreciate the volumes on *Principles of Geometry* by Professor H. F. Baker, under whose guidance so many of my contemporaries at Cambridge University learned the fascination of the subject.

W. GORDON WELCHMAN

CAMBRIDGE, MASS.

March 1950

CHAPTER I

INTRODUCTION AND DEFINITIONS

1·1. The development of Geometry

1·11. Geometry is a vast subject which is still growing. Just as a game is governed by rules, so each branch of geometry is governed by the axiomatic assumptions on which it is founded and by the restrictions which are imposed on the course of development. The rules of a game develop with the game and are changed from time to time in accordance with the experience of players. Similarly, in geometry the nature of the foundations and the attitude of students to the subject are changed from time to time in accordance with the researches of geometers. Whenever the foundations are changed, a new type of geometry is introduced, and some confusion is inevitably caused by the use of the same language in all types of geometry.

A thorough knowledge of the rules of a game and an appreciation of the objects of the rules are most easily acquired by the actual experience of playing the game. Similarly an understanding of the attitude to the subject which has suggested the form of the foundations of Algebraic Geometry is most easily acquired by a study of the subject itself. A reader should not be discouraged if he finds difficulty in understanding the first three chapters of this volume, which are mainly concerned with foundations, but should proceed with the development of the theory in the rest of the book.

The nature of the foundations of Algebraic Geometry is justified by the great developments of the last sixty years. It will be impossible to give any idea of these developments in this book. They are largely due to the investigation of spaces of more than three dimensions and to the theory of algebraic functions. But an attempt is made to introduce some of the types of reasoning which have proved fruitful, so that the book may be a useful introduction to more advanced work. In fact the book is just as much concerned with the development of technique as with the proving of theorems, and for this reason a theorem will often be proved by several different methods.

1·12. *Different types of geometry in a plane*

The Algebraic Geometry of two dimensions, with which this volume is chiefly concerned, may be regarded historically as the logical successor to a sequence of seven different types of geometry in a plane, which we shall denote by $G_1, G_2, \ldots, G_7$.* Euclidean Geometry, G_1, is a theory of figures suggested by physical considerations and based on a complicated set of axioms. The introduction of co-ordinates by Descartes led to G_2, and the combination of G_1 and G_2 is still called Modern Geometry by some writers. It is these two geometries which are of direct use in various branches of Natural Science and in Applied Mathematics, for in them a plane is regarded as existing in nature and the co-ordinates of G_2 are only introduced to represent the points of a plane.

$G_3, G_4, \ldots, G_7$ are abstract geometries. The real Cartesian Geometry G_3 differs from G_2 in that the points of a plane are no longer regarded as physical elements, but are defined to be an ordered pair of real numbers. The language of G_2 is retained in G_3, but lines, circles, conics, perpendicularity, etc., are defined analytically. The introduction of G_3 is important as a stepping-stone to the other geometries.

In the complex Cartesian Geometry G_4, a point of a plane is defined to be an ordered pair of complex numbers instead of real numbers. It includes G_3, but widens its scope by obviating the necessity for considering many special cases; thus in G_3, for example, we must make a distinction, unnecessary in G_4, between a line which does not meet a given circle and one which meets it in two distinct points.

In G_3 and G_4 two lines may or may not meet. This distinction can be avoided by the introduction of the homogeneous Cartesian Geometries G_5 and G_6. In G_5 a point is defined to be an ordered triplet of real numbers (x, y, z) not all zero, with the convention that $(\lambda x, \lambda y, \lambda z)$ is the same point as (x, y, z) provided $\lambda \neq 0$; G_6 is similar to G_5 except that x, y, z are complex numbers. In both geometries a line of the plane is determined by a homogeneous linear equation in the co-ordinates. If $z \neq 0$ the point (x, y, z) is the point whose ordinary Cartesian co-ordinates are $(x/z, y/z)$. This last is meaning-

* This notation is borrowed from A. Robson, *An Introduction to Analytical Geometry*, I and II (Cambridge, 1940 and 1947), which we shall in future refer to as $R(1)$ and $R(2)$. For fuller details the reader is referred to Chapter VIII of $R(1)$, of which the present paragraph is a summary.

less if z vanishes, but the convention is made that a point of the form $(\xi, \eta, 0)$ represents the common 'intersection' of the system of parallel lines whose equation, in non-homogeneous Cartesian Geometry, is $\eta x - \xi y + \lambda = 0$, with λ variable. The aggregate of such points constitutes the line at infinity $z = 0$; two lines in G_5 or G_6 always have just one common point, which, if the lines are parallel, lies on $z = 0$. It further appears that in the complex homogeneous Cartesian Geometry G_6 there are two points of special significance lying on the line at infinity. These are the so-called circular points at infinity; they are determined by the equations $z = 0$, $x^2 + y^2 = 0$, and their co-ordinates are therefore $(1, \pm i, 0)$. These points take their name from the fact that all circles of the plane pass through them; in G_6 a circle differs from a general conic only in that it passes through these two special points. The significance of the circular points is, however, wider than this property suggests. In fact all properties of direction and distance which may be associated with a geometrical figure, properties which are inherited from the Euclidean geometry G_1, can be interpreted in terms of the relationships of the figure to the circular points; in these relationships the notions of direction and distance play no part.

The transition from G_6 to G_7 is obvious. The definitions of a point and a line are the same as in G_6, but no special significance is attached to the line $z = 0$ or to the points $(1, \pm i, 0)$. In G_7, direction and distance are meaningless terms.

This chain of abstract geometries $G_3, G_4, \ldots, G_7$ may easily be extended to space of three dimensions, or indeed to space of any number of dimensions, by increasing the number of co-ordinates.

1·2. Metrical, projective, and algebraic geometries

1·21. In $R(1)$ a distinction is made between projective and metrical properties in G_6. It is shown* that, if x, y, z and x', y', z' are homogeneous co-ordinates in two planes π and π', the one-to-one correspondence between the points of π and π' which is obtained by projection from a point O not lying in either plane is given by equations of the form

$$\begin{aligned} x &= a_1 x' + b_1 y' + c_1 z', \\ y &= a_2 x' + b_2 y' + c_2 z', \\ z &= a_3 x' + b_3 y' + c_3 z', \end{aligned}$$

* $R(1)$ Chapter IX.

where the determinant

$$\begin{vmatrix} a_1 & b_1 & c_1 \\ a_2 & b_2 & c_2 \\ a_3 & b_3 & c_3 \end{vmatrix}$$

does not vanish, i.e. by a non-singular linear transformation. Conversely it can be proved that any non-singular linear transformation between x, y, z and x', y', z' is equivalent to a series of successive projections. In consequence, linear transformations are often called *projective transformations*.

A property of a configuration of points and lines in π is said to be a *projective property* if the corresponding configuration in π' always has the same property however O and π' are chosen. Since the line at infinity and the circular points in π do not generally correspond to the line at infinity and circular points in π', properties which involve these elements are not projective, and are said to be *metrical*. For instance, the property of being a circle is not a projective property of a conic. For the discussion of projective properties it is more convenient to use G_7 than G_6.

1·22. It is difficult to give intelligible definitions of Projective Geometry and of Algebraic Geometry at this stage, for the reasons indicated in 1·11. In both these geometries, the points and lines of a plane are defined as in G_7,* but the field of study is strictly limited by certain restrictions, and only loci of certain definite types are considered. In Algebraic Geometry of a plane for instance the only curves that are considered are those which are given by equations

$$\phi(x, y, z) = 0,$$

where $\phi(x, y, z)$ is a homogeneous polynomial in the co-ordinates x, y, z. In Projective Geometry of a plane the only curves that are discussed are those which may be defined from the points and lines by means of linear transformations. The class of curves discussed in Projective Geometry is contained in the class of curves discussed in Algebraic Geometry, and is far larger than might at first be supposed. A similar distinction is made in space of three or more dimensions. Before giving a more precise definition of Algebraic Geometry the use of the word 'algebraic' will be explained.

* See 1·12, p. 3.

1·23. A polynomial in n variables $x_1, x_2, \ldots, x_n$ is the sum of a number of terms of the form

$$c\, x_1^{m_1} x_2^{m_2} \ldots x_n^{m_n},$$

in which $m_1, m_2, \ldots, m_n$ are positive integers or zero, and c is a number which is called a coefficient of the polynomial. If the highest value of m_i which occurs in the terms of the polynomial is k_i, the polynomial is said to be of degree k_i in the variable x_i. If the sum $m_1 + m_2 + \ldots + m_n$ has the same value M for all the terms, the polynomial is said to be homogeneous and of degree M.

An *algebraic equation* in $x_1, x_2, \ldots, x_n$ is an equation of the form

$$\phi(x_1, x_2, \ldots, x_n) = 0,$$

in which $\phi(x_1, x_2, \ldots, x_n)$ is a polynomial in $x_1, x_2, \ldots, x_n$. A process is said to be algebraic when it depends only on algebraic equations and their solution.

It should be noted that, when the powers of the variables which occur in the terms of the polynomial ϕ are assigned, the equation $\phi = 0$ is determined by the ratios of the coefficients, and the ratios of the coefficients are determined by the equation. This process of passing from the ratios of the coefficients to the equation and vice versa is an algebraic process.

1·24. The Algebraic Geometry which is to be studied in this volume is only one of many Algebraic Geometries, though it is the one that has received most attention. It will be denoted by G_a, and its full title should be 'Algebraic Geometry of the Field of Complex Numbers'. The elements of G_a in space of any number of dimensions are defined as in G_7 by ordered sets of complex numbers, but the only curves, surfaces and other constructs that are discussed are those which may be defined by algebraic equations in which the coefficients are complex numbers and the variables are complex variables.

Every theorem in the subject must be algebraic, i.e. it must be capable of statement as a theorem about algebraic equations. An isolated theorem in G_a may be proved by non-algebraic methods, just as theorems about a real variable may legitimately be proved by the use of a complex variable, but the theorem itself must be algebraic. Apart from this exception, every step in an argument must be an appeal to an algebraic theorem and therefore expressible in terms of algebraic equations. Consequently any construct, correspondence, or theorem that may be derived from an argument in G_a is itself algebraic and may be determined or expressed by algebraic equations. It is seldom necessary to write down the equations which would justify each step of an argument, but it is essential to know that at any stage algebraic equations can be introduced to represent what has been obtained. In fact the algebra which underlies Algebraic Geometry may be kept below the surface until it is needed, and the attractive synthetic character which is usually associated with Projective Geometry is not destroyed.

The numbers which are used to define the elements of G_a are those of the field of complex numbers. Other types of Algebraic Geometry may be obtained by using some other field of numbers, for instance, real numbers, rational numbers, or integers. It is even of interest to consider 'finite geometries', in which there are only a finite number of points, but this book is only concerned with G_a.

1·25. If Projective Geometry is regarded as that part of Algebraic Geometry which may be deduced from the study of linear transformations, nearly all the geometry of this book is projective, though the underlying attitude to the subject is algebraic. The real extension of the subject comes at a later stage, when non-linear transformations are considered. The transformation

$$x:y:z = y'z':z'x':x'y',$$

$$x':y':z' = yz:zx:xy$$

is a simple example of a non-linear *birational* transformation, and Algebraic Geometry in its more advanced stages is a study of properties which are unaffected by birational transformations.†

Projective Geometry of space of any number of dimensions may also be regarded as an independent subject founded on purely geometrical axioms with no appeal to algebra or analysis. From this point of view the subject consists of all deductions that may be made from these axioms. A symbolism is introduced which is shown to be equivalent to ordinary algebra, and the subject is thereby connected with Algebraic Geometry. This line of approach is discussed in the first volume of *Principles of Geometry*, by H. F. Baker.*

* Cambridge University Press, 1922.

† See footnote to 2·25, p. 47.

1·26. The restriction to algebraic processes which is introduced in the definition of Algebraic Geometry does not exclude the projective theorems of G_6 which are proved in $R(1)$. For it is shown that all the projective arguments in G_6 hold good in G_7, and since the arguments are actually algebraic they also hold good in G_a. The deduction of metrical algebraic theorems in G_6 from G_a will be discussed in Chapter VI of this volume.

1·27. In this book the following abbreviations are used:

IF = if and only if,

CONDITION = necessary and sufficient condition,

ONE = one and only one,

TWO = two and only two.

It should be noticed that, if A and B denote two statements or theorems, then the statement 'A is true IF B is true' is equivalent to the statement 'B is true IF A is true'. Other equivalent statements are 'The truth of A is a CONDITION for the truth of B', and 'The truth of B is a CONDITION for the truth of A'.

Of course there is no unique CONDITION for the truth of a given statement, any equivalent statement being a CONDITION. But sometimes there is a particularly obvious CONDITION, which may be referred to as '*the* CONDITION' without ambiguity.

It will be convenient to use bars to denote pairs of elements of the same kind when the order of the elements is immaterial. Thus $|P_1, P_2|$ will mean the unordered pair of points P_1, P_2, and $|x_1, x_2|$ will mean the unordered pair of values x_1, x_2 of the variable x. Brackets of various kinds are used to denote ordered sets of elements.

1·3. Complex numbers. Linear dependence

1·31. *Analysis*

For an introduction to the abstract foundations of Analysis the reader is referred to *Pure Mathematics*, by G. H. Hardy.* The subject is founded on the conception of real numbers, and every theorem in Analysis must ultimately be a deduction from the principle of continuity, which states that every Dedekind section of the real numbers defines ONE real number. In the theory of real numbers the word 'infinity' is used in certain phrases, such as

* Cambridge University Press, 1938.

'tends to infinity', which have precise meanings, but infinity is not a real number. Similarly in real non-homogeneous geometry, G_3, there is no such thing as a point at infinity, though the idea of a point tending to infinity is defined. But in real homogeneous geometry, G_5, there are points at infinity in a plane, namely the points (x, y, z) for which $z = 0$. Thus the use of the word 'infinity', which is denoted by the symbol ∞, is different in different branches of mathematics, and even in different types of geometry.

1·32. *Complex variable*

A complex number is defined to be an ordered pair (a, b) of real numbers a and b, so the complex numbers may be represented by the points of a plane in G_3, as in the Argand diagram. This plane is called the complex plane and has no points at infinity, but it is convenient to regard the complex plane as being completed or closed by the addition of one element, which is called the point at infinity. This point at infinity does not correspond to any complex number, but such expressions as 'a neighbourhood of infinity' and 'tends to infinity' have precise meanings. This notion of completing the complex plane by the addition of one point at infinity is in agreement with G_6 but in contrast with G_5. For if the points of a line in G_6 are represented by the values of a complex variable, as is possible in many ways,* ONE point of the line has no representation, and this point corresponds to the point at infinity in the closed complex plane. On the other hand, the transition from G_3 to G_5 may be regarded as a process of completing the plane of G_3 by the addition of a whole line of points at infinity. The use of the symbol ∞ in G_a and the notion of freedom will be discussed later.

1·33. *Linear dependence*

A good introduction to the theory of homogeneous linear equations will be found in Chapter XVII of *Advanced Algebra*, vol. III, by Durell and Robson.† The dummy suffix summation convention, which is explained in Chapter XVI of the same book, will be used, but the convention will not be restricted to Greek suffixes.

* E.g. if the equation of the line is $ax+by+cz = 0$, where at least one, say a, of the coefficients does not vanish, we can represent its points by the values of the complex number y/z; a general point of the line corresponds to ONE such value but the particular point for which $z = 0$ has no representation.

† London: G. Bell and Sons, Ltd., 1937.

Consider m sets of n numbers each

$$(a_{11}, a_{12}, \ldots, a_{1n}),\ (a_{21}, a_{22}, \ldots, a_{2n}),\ \ldots,\ (a_{m1}, a_{m2}, \ldots, a_{mn}), \qquad (\cdot 331)$$

and suppose that in each set not all the numbers are zero. A set $(b_1, b_2, \ldots, b_n)$ is said to be *linearly dependent on the sets* ($\cdot$331) IF there exist m numbers $\rho_1, \rho_2, \ldots, \rho_m$ not all zero such that

$$b_1 = \rho_i a_{i1}, \quad b_2 = \rho_i a_{i2}, \quad \ldots, \quad b_n = \rho_i a_{in}, \qquad (\cdot 332)$$

where the summations are for $i = 1, 2, \ldots, m$. Also the m sets ($\cdot$331) are said to be *linearly dependent* IF there exist m numbers

$$\kappa_1, \kappa_2, \ldots, \kappa_m$$

not all zero such that

$$\kappa_i a_{i1} = \kappa_i a_{i2} = \ldots = \kappa_i a_{in} = 0. \qquad (\cdot 333)$$

Thus, if the m sets ($\cdot$331) are linearly dependent, and if the number κ_s is not zero, then the set $(a_{s1}, a_{s2}, \ldots, a_{sn})$ is linearly dependent on the remaining $m-1$ sets of ($\cdot$331).

The sets ($\cdot$331) are said to be *linearly independent* IF no numbers $\kappa_1, \kappa_2, \ldots, \kappa_m$ exist which satisfy ($\cdot$333).

Let r be the rank of the matrix (a_{ij}) whose rows are the m sets ($\cdot$331). Then $r \leqslant m$ and $r \leqslant n$. The sets are linearly independent IF $r = m \leqslant n$. In all other cases $r < m$ and there are r sets of ($\cdot$331) on which the remaining sets are linearly dependent.

If $x_1, x_2, \ldots, x_n$ are variables, the m linear forms

$$a_{1j}x_j, \quad a_{2j}x_j, \quad \ldots, \quad a_{mj}x_j, \qquad (\cdot 334)$$

in which the summations are for $j = 1, 2, \ldots, n$, are said to be linearly dependent or independent according as the sets ($\cdot$331) are linearly dependent or independent. When a set $(b_1, b_2, \ldots, b_n)$ is linearly dependent on the sets ($\cdot$331), the form $b_j x_j$ is said to be linearly dependent on the forms ($\cdot$334), and the linear equation

$$b_j x_j = 0 \qquad (\cdot 335)$$

is implied by the linear equations

$$a_{1j}x_j = 0, \quad a_{2j}x_j = 0, \quad \ldots, \quad a_{mj}x_j = 0. \qquad (\cdot 336)$$

Finally the linear equations ($\cdot$336) are also said to be linearly dependent or independent according as the sets ($\cdot$331) are linearly dependent or independent, and the equation ($\cdot$335) is said to be linearly dependent on the equations ($\cdot$336) IF the set $(b_1, b_2, \ldots, b_n)$ is linearly dependent on the sets ($\cdot$331).

As only linear dependence is considered in this volume, the word 'linear' will in future be omitted.

1·34. *A convention*

Two sets $(a_1, a_2, \dots, a_n)$ and $(b_1, b_2, \dots, b_n)$, neither of which is $(0, 0, \dots, 0)$, are dependent IF there are non-zero numbers ρ and σ such that

$$\rho a_1 + \sigma b_1 = \rho a_2 + \sigma b_2 = \dots = \rho a_n + \sigma b_n = 0. \qquad (\cdot 341)$$

If none of the numbers a and b are zero, this is equivalent to saying that

$$a_1 : a_2 : \dots : a_n = b_1 : b_2 : \dots : b_n. \qquad (\cdot 342)$$

When some of the numbers a and b are zero it is convenient to regard (·342) as meaning that there are non-zero numbers ρ and σ for which (·341) is true. For example, if $a_1 = 0$, (·342) means that $b_1 = 0$ and that

$$a_2 : a_3 : = : a_n = b_2 : b_3 : \dots : b_n.$$

Thus, with this convention, the two sets are dependent IF (·342) is true.

1·35. *Solution of linear equations*

Consider the m homogeneous linear equations (·336), and let r be the rank of the matrix (a_{ij}). Then $r \leqslant m$ and $r \leqslant n$. If $r = n \leqslant m$, the equations are only true if $x_1 = x_2 = \dots = x_n = 0$. *This will not be regarded as a solution.*

If $r < n$ there are r independent equations among the set (·336) on which the remaining equations, if any, are dependent. It is possible to choose $n - r$ independent solutions, and, if $(x_{\rho 1}, x_{\rho 2}, \dots, x_{\rho n})$, with $\rho = 1, 2, \dots, n - r$, is any set of $n - r$ independent solutions, then every solution is dependent on these and can be expressed in the form

$$(\kappa_\rho x_{\rho 1}, \kappa_\rho x_{\rho 2}, \dots, \kappa_\rho x_{\rho n}), \qquad (\cdot 351)$$

the set of numbers $(\kappa_1, \kappa_2, \dots, \kappa_n)$ being uniquely determined. Conversely for any set of values of $\kappa_1, \kappa_2, \dots, \kappa_n$ not all zero the set (·351) is a solution of the equations (·336).

Thus the equations certainly have a solution if $m \leqslant n - 1$, but if $m \geqslant n$ there is a solution IF all the n-row determinants of the matrix (a_{ij}) vanish. This is expressed by writing

$$\| a_{ij} \| = 0 \qquad (\cdot 352)$$

and the process of obtaining these equations, which are CONDITIONS for the existence of a solution, is called *elimination* of the variables $x_1, x_2, \dots, x_n$ from the equations (·336). From (·336) equations may also be derived which involve $x_2, x_3, \dots, x_n$ but not x_1, and which have a solution IF the original equations have a solution; this process is called elimination of the variable x_1.

1·36. *Linear systems*

If $(y_{\lambda 1}, y_{\lambda 2}, \ldots, y_{\lambda n})$, with $\lambda = 1, 2, \ldots, r$, are independent sets, the system of sets

$$(\kappa_\lambda y_{\lambda 1}, \kappa_\lambda y_{\lambda 2}, \ldots, \kappa_\lambda y_{\lambda n}) \qquad (\cdot 361)$$

obtained by giving the numbers $\kappa_1, \kappa_2, \ldots, \kappa_r$ all sets of values except $0, 0, \ldots, 0$ is called a *linear system of rank* r. Each set of the system is determined by ONE set of values of $\kappa_1, \kappa_2, \ldots, \kappa_r$.

Similarly, if

$$b_{\mu j} x_j = 0 \quad (\mu = 1, 2, \ldots, r;\ j = 1, 2, \ldots, n) \qquad (\cdot 362)$$

are r independent equations, the system of equations

$$\kappa_\mu b_{\mu j} x_j = 0 \qquad (\cdot 363)$$

is called a *linear system of rank* r, and is satisfied by all the solutions of (·362).

If m sets $(x_{i1}, x_{i2}, \ldots, x_{in})$, $i = 1, 2, \ldots, m$, are not independent, let r be the rank of the matrix (x_{ij}). Then $r < m$, and from the system of sets

$$(k_i x_{i1}, k_i x_{i2}, \ldots, k_i x_{in}) \qquad (\cdot 364)$$

it is possible in infinitely many ways to choose r independent sets

$$(y_{\lambda 1}, y_{\lambda 2}, \ldots, y_{\lambda n}), \quad \lambda = 1, 2, \ldots, r. \qquad (\cdot 365)$$

However these r independent sets are chosen, the linear system

$$(\kappa_\lambda y_{\lambda 1}, \kappa_\lambda y_{\lambda 2}, \ldots, \kappa_\lambda y_{\lambda n}) \qquad (\cdot 366)$$

of rank r is the same as the system (·364), each set of the system being determined by infinitely many sets of values of $k_1, k_2, \ldots, k_m$.

Similarly the m equations (·336) determine a system of equations

$$k_i a_{ij} x_j = 0 \quad (i = 1, 2, \ldots, m;\ j = 1, 2, \ldots, n) \qquad (\cdot 367)$$

If the equations (·336) are not independent, the rank r of the matrix (a_{ij}) is less than m and it is possible to choose r independent equations

$$b_{\mu j} x_j = 0 \quad (\mu = 1, 2, \ldots, r;\ j = 1, 2, \ldots, n),$$

from the system (·367). However these r independent equations are chosen, the system (·367) is the same as the linear system

$$\kappa_\mu b_{\mu j} x_j = 0$$

of rank r.

Finally, if $r < n$, the solutions of a linear system of rank r of linear homogeneous equations in n variables form a linear system of sets of rank $n-r$. Also, if $s<n$, a linear system of rank s of sets of n numbers is the complete system of solutions of a linear system of rank $n-s$ of linear homogeneous equations in n variables.

For convenience a linear system of rank r will be denoted by L_r.

1·4. Ratio sets. Space of *n* dimensions

1·41. *Ratio sets*

If $x_1 = x_{11}, x_2 = x_{12}, \ldots, x_n = x_{1n}$ is a solution of a set of homogeneous equations in the variables $x_1, x_2, \ldots, x_n$, then

$$x_1 = \kappa x_{11}, x_2 = \kappa x_{12}, \ldots, x_n = \kappa x_{1n}$$

is also a solution, where κ is any number other than zero. In homogeneous geometry all these solutions are regarded as identical, and it is convenient to have a name for a whole system of identical solutions. As the variables which occur in G_a are complex variables, the following definition of a *ratio set* of *index n* is introduced.

A ratio set is an ordered set of complex numbers $(c_1, c_2, \ldots, c_n)$, not all of which are zero, with the provision that, if κ is any complex number other than zero, the ratio set $(\kappa c_1, \kappa c_2, \ldots, \kappa c_n)$ is the same as the ratio set $(c_1, c_2, \ldots, c_n)$. The complex numbers $c_1, c_2, \ldots, c_n$ are called coefficients, and n is called the index of the ratio set. The set $(0, 0, \ldots, 0)$ is not a ratio set.

A homogeneous linear equation

$$a_1x_1 + a_2x_2 + \ldots + a_nx_n = 0$$

may be regarded as an equation in the variable ratio set

$$(x_1, x_2, \ldots, x_n),$$

and the equation is uniquely determined by the ratio set

$$(a_1, a_2, \ldots, a_n).$$

Conversely when the equation is given, or rather when the system of solutions of the equation is given, the ratio set $(a_1, a_2, \ldots, a_n)$ is uniquely determined, though the actual values of the coefficients are not uniquely determined.

The definitions of dependence and independence of sets apply equally well to ratio sets. In fact the word 'set' may be replaced by 'ratio set' throughout 1·33. The only difference is that two dependent ratio sets are identical, whereas two dependent sets are proportional. The contents of 1·35 and 1·36 may also be restated in terms of ratio sets. Each equation of the system (·363) is determined by ONE ratio set $(\kappa_1, \kappa_2, \ldots, \kappa_r)$. There is ONE ratio set of index one, and a linear system of rank one of ratio sets of index n consists of ONE ratio set. The theory of homogeneous linear equations will now be summarized in the language of ratio sets.

1·42. *Ratio sets of index n*

The ratio set $(x_1, x_2, \ldots, x_n)$ whose coefficients $x_1, x_2, \ldots, x_n$ are complex variables, is denoted by $(x)_n$, and a particular ratio set obtained by giving particular values to the coefficients is said to be a *value* of the variable ratio set $(x)_n$.

(i) All the ratio sets $(x)_n$ form an L_n since they are the ratio sets dependent on the n independent ratio sets

$$(1,0,0,\ldots,0,0),\ (0,1,0\ldots,0,0),\ \ldots,\ (0,0,0,\ldots,0,1).$$

This system will be denoted by σ.

(ii) An L_r formed by the ratio sets of σ which are dependent on r independent ratio sets of σ is denoted by σ_r.

(iii) Any $r+1$ ratio sets of σ_r are dependent, but it is possible in infinitely many ways to choose r independent ratio sets

$$(x_{11}, x_{12}, \ldots, x_{1n}),\ (x_{21}, x_{22}, \ldots, x_{2n}),\ \ldots,\ (x_{r1}, x_{r2}, \ldots, x_{rn})$$

belonging to σ_r. However this choice is made the linear system

$$(\kappa_i x_{i1}, \kappa_i x_{i2}, \ldots, \kappa_i x_{in}) \quad (i = 1, 2, \ldots, r)$$

is the system σ_r, and each ratio set of σ_r corresponds to ONE value of $(\kappa_1, \kappa_2, \ldots, \kappa_r)$.

(iv) Thus the system σ_r of ratio sets $(x)_n$ is equivalent to the system formed by all the ratio sets of index r.

(v) Since any $n+1$ sets of σ are dependent it follows that $1 \leqslant r \leqslant n$. Also from (iii) it follows that any σ_n is identical with σ. If $r < n$, σ_r is said to be a linear subsystem of σ.

(vi) It also follows from (iii) that any r independent ratio sets of σ belong to ONE σ_r.

(vii) All the linear equations

$$X_1 x_1 + X_2 x_2 + \ldots + X_n x_n = 0$$

in the variable ratio set $(x)_n$ form a linear system of rank n, and each equation is determined by ONE ratio set $[X_1, X_2, \ldots, X_n]$, square brackets being used to distinguish sets of coefficients in equations from sets of variables.

(viii) All the ratio sets $[X_1, X_2, \ldots, X_n]$ or $[X]_n$ form an L_n which will be denoted by Σ. Linear systems Σ_r contained in Σ are defined as above.

(ix) The ratio sets $(x)_n$ of a linear system σ_r are the solutions of a linear system of equations of rank $n-r$ and the corresponding ratio sets $[X]_n$ form a Σ_{n-r}.

(x) Conversely a linear system Σ_s contained in Σ determines a linear system of equations of rank s whose solutions form a σ_{n-s}.

(xi) Thus s independent sets $[X]_n$ determine a Σ_s and a σ_{n-s} just as r independent sets $(x)_n$ determine a σ_r and a Σ_{n-r}.

(xii) If h ratio sets

$$(x_{11}, x_{12}, \ldots, x_{1n}), \quad (x_{21}, x_{22}, \ldots, x_{2n}), \quad \ldots, \quad (x_{h1}, x_{h2}, \ldots, x_{hn})$$

are not independent, and if r $(<h)$ is the rank of the matrix (x_{ij}) formed by these ratio sets, then the system

$$(\kappa_i x_{i1}, \kappa_i x_{i2}, \ldots, \kappa_i x_{in}) \quad (i = 1, 2, \ldots, h)$$

is a σ_r, each ratio set of σ_r corresponding to infinitely many values of $(\kappa_1, \kappa_2, \ldots, \kappa_n)$. Thus any set of ratio sets $(x)_n$ determines ONE subsystem of σ. Similarly any set of ratio sets $[X]_n$ determines ONE subsystem of Σ.

1·43. *Space of n dimensions*

A space of n dimensions may now be defined as follows. A *point* is defined to be a ratio set of index $n+1$, and all points $(x)_{n+1}$ are said to form a space of n dimensions, which is denoted by the symbol $[n]$. Thus the points of $[n]$ are the values of the variable ratio set $(x)_{n+1}$, but as soon as they have been so defined the points of $[n]$ are regarded as geometrical elements, and a point of $[n]$ is said to be determined or represented by a value of $(x)_{n+1}$.

The points of $[n]$ which satisfy a linear equation

$$X_1 x_1 + X_2 x_2 + \ldots + X_{n+1} x_{n+1} = 0$$

are said to form a *prime* of $[n]$, and the primes of $[n]$ are determined or represented by the ratio sets $[X]_{n+1}$ just as the points of $[n]$ are determined or represented by the ratio sets $(x)_{n+1}$. The definitions of dependence, independence and linear systems of ratio sets are also applied to points and primes of $[n]$.

It follows from (v) that any L_{n+1} of points is identical with $[n]$. If $r<n$ an L_{r+1} of points of $[n]$ is called a *subspace* of r dimensions.

By (iv) a subspace of r dimensions is equivalent to an $[r]$, and it may therefore be said to be an $[r]$ contained in $[n]$.

By (vi) any $r+1$ independent points of $[n]$ lie in ONE $[r]$. By (ix) the points of an $[r]$ in $[n]$ are the common points of an L_{n-r} of primes of $[n]$, and in particular an $[n-1]$ of $[n]$ is a prime of $[n]$. By (x) the common points of the primes of an L_s are the points of a $[n-s]$. Thus any $r+1$ independent points of $[n]$ lie in ONE $[r]$ which is formed by the common points of $n-r$ independent primes, while the common points of any $n-r$ independent primes form an $[r]$. A $[0]$ is a point, a $[1]$ is called a *line*, a $[2]$ is called a *plane*, a $[3]$ is called a *solid*, and an $[n-2]$ is called a *secundum* of $[n]$.

1·44. *The incidence theorem*

Before proceeding to discuss spaces of one, two and three dimensions in greater detail, it is of interest to mention here without proof an important general theorem which may be called *the incidence theorem*.

If S_r and S_s are two subspaces $[r]$ and $[s]$ of $[n]$, the points of S_r are the common points of $n-r$ independent primes $\pi_1, \pi_2, \ldots, \pi_{n-r}$, and the points of S_s are the common points of $n-s$ independent primes $\pi'_1, \pi'_2, \ldots, \pi'_{n-s}$. The common points of S_r and S_s are the common points of all the $2n-r-s$ primes π and π', and therefore form a space $[i]$, with the convention that $i=-1$ if S_r and S_s have no common point. If the $2n-r-s$ primes are independent, $i=r+s-n$, but this is not necessarily the case, so $i \geqslant r+s-n$. The space $[i]$ is called the *meet* or the *intersection* of S_r and S_s, and is denoted by (S_r, S_s).

Let $P_1, P_2, \ldots, P_{r+1}$ be $r+1$ independent points of S_r and let $P'_1, P'_2, \ldots, P'_{s+1}$ be $s+1$ independent points of S_s. Then any subspace of $[n]$ which contains the $r+s+2$ points P and P' must contain S_r and S_s. But the $r+s+2$ points P and P' determine a linear system of points forming a subspace $[j]$, where $j \leqslant r+s+1$. This $[j]$ is the subspace of lowest dimension which contains both S_r and S_s; it is called the *join* of S_r and S_s and is denoted by $[S_r, S_s]$.

The incidence theorem states that

$$i+j=r+s.$$

The theorem is of particular interest because it is used as an axiom in the foundation of Projective Geometry on purely geometrical axioms mentioned in 1·25. From the point of view of Algebraic Geometry the theorem is a deduction from the theory of linear equations.

1·5. Space of one dimension. Ratio variables

1·51. *Points*

A point is defined to be a ratio set (x, y) of index two. The points obtained by allowing the complex numbers x and y to take all values other than $x = y = 0$ are said to form a *space of one dimension* or a *line*. The numbers x and y are called homogeneous co-ordinates of the point (x, y), the points $(1, 0)$ and $(0, 1)$ are called the *reference points*, and the point $(1, 1)$ is called the *unit point*. Capital letters will be used to denote points, and P_i will denote the point (x_i, y_i). P_1 and P_2 are identical IF $x_1y_2 - x_2y_1 = 0$ and distinct IF

$$x_1y_2 - x_2y_1 \neq 0.$$

The set $(0, 0)$ is not a ratio set and does not represent a point.

1·52. *Parametric representation*

The points of the line other than $(1, 0)$ may be represented by the values of $\lambda = x/y$, for each value of λ determines ONE point and each point other than $(1, 0)$ determines ONE value of λ. It is convenient to use the symbol ∞ and to say that the point $(1, 0)$ corresponds to $\lambda = \infty$, although ∞ is not a complex number. In fact the statements '$\lambda = \infty$' and 'λ has the value ∞' are conventional ways of saying that the point (x, y) is $(1, 0)$.* The variable λ is called a *non-homogeneous co-ordinate* or a *parameter* of the points of the line.

1·53. *Ratio variables*

If ξ, η are any two complex variables, and if ξ_1, η_1 are any two values of ξ, η which are not both zero, the ratio set (ξ, η) is called a *ratio variable* and the ratio set (ξ_1, η_1) is called a *value* of the ratio variable (ξ, η). The values of a ratio variable may be identified with the points of a line, but it is convenient to have another name. A ratio set (α, β) is called a *ratio number* when it is regarded as a value of a ratio variable.

As in 1·52 the values of a ratio variable (ξ, η) may be represented by the values of $\rho = \xi/\eta$, with the convention that $\rho = \infty$ means that

* Cf. 1·32, p. 8.

(ξ, η) is $(1, 0)$. Conversely any complex variable ρ may be replaced by a ratio variable (ξ, η) by the substitution $\rho = \xi/\eta$, and in practice we shall pass freely from one to the other.

1·54. *Choice of reference points and unit point*

The points of a line are identified with all the ratio sets (x, y). If P_1 and P_2 are two distinct points, and if (κ_1, κ_2) is a ratio variable, then by 1·42 (iii) all ratio sets (x, y) may be expressed in the form $(\kappa_1 x_1 + \kappa_2 x_2, \kappa_1 y_1 + \kappa_2 y_2)$, each ratio set (x, y) being given by ONE value of (κ_1, κ_2). The numbers κ_1, κ_2 may therefore be used as homogeneous co-ordinates of the points of the line instead of x, y, and in the new system of co-ordinates the reference points are P_1, P_2.

Any point P_3 corresponds to a value (b_1, b_2) of (κ_1, κ_2), and if P_3 is distinct from P_1, P_2 then $b_1 b_2 \neq 0$. If x', y' are now defined by

$$b_1 x' : b_2 y' = \kappa_1 : \kappa_2,$$

then x', y' may be used as homogeneous co-ordinates of the points of the line instead of κ_1, κ_2. In this third system of co-ordinates the reference points are P_1, P_2 and the unit point is P_3. Thus:

If P_1, P_2, P_3 are any three distinct points of the line, a co-ordinate system may be chosen in which P_1, P_2 are the reference points and P_3 is the unit point.

1·55. *Linear transformations*

Consider the linear transformation

$$\left.\begin{aligned} x &= a_1 x' + b_1 y', \\ y &= a_2 x' + b_2 y' \end{aligned}\right\} \qquad (\cdot 551)$$

between the complex variables x, y and x', y'. Suppose that $a_1 b_2 - a_2 b_1 \neq 0$, so that the equations may be solved for x', y', giving

$$\left.\begin{aligned} (a_1 b_2 - a_2 b_1) x' &= b_2 x - b_1 y, \\ (a_1 b_2 - a_2 b_1) y' &= -a_2 x + a_1 y. \end{aligned}\right\} \qquad (\cdot 552)$$

If x_1 and y_1 are the value of x and y obtained by substituting $x' = x_1'$, $y' = y_1'$ in (·551), then the substitution $x' = \rho x_1'$, $y' = \rho y_1'$ gives $x = \rho x_1$, $y = \rho y_1$, while $x' = y' = 0$ gives $x = y = 0$. Thus each value of the ratio variable (x', y') determines ONE value of the ratio variable (x, y). Similarly the equations (·552) show that each value of (x, y) determines ONE value of (x', y'). In fact the transformation may be regarded as a relation between the ratio variables (x, y) and (x', y').

If $a_1b_2 - a_2b_1 = 0$, the transformation is said to be *singular*. All values of (x', y') determine the same value of (x, y), except the value given by $a_1x' + b_1y' = a_2x' + b_2y' = 0$, which gives $x = y = 0$.

1·56. *Change of co-ordinates*

Consider the line whose points are the values of the ratio variable (x, y). It follows from 1·55 that, provided $a_1b_2 - a_2b_1 \neq 0$, the points of the line are equally well determined in accordance with the definition of 1·51 by the values of the ratio variable (x', y') which is related to (x, y) by (·551). In fact x', y' may be regarded as homogeneous co-ordinates of the points of the line, and the transformation (·551) is said to determine a transformation of co-ordinates. The complex variable $\lambda' = x'/y'$ is therefore another parameter of the points of the line, and is related to the parameter λ by the equation

$$\lambda = \frac{a_1\lambda' + b_1}{a_2\lambda' + b_2} \tag{·561}$$

or

$$a_2\lambda\lambda' + b_2\lambda - a_1\lambda' - b_1 = 0. \tag{·562}$$

A singular transformation does not give a transformation of co-ordinates.

The system of co-ordinates mentioned at the beginning of 1·54, in which the reference points are P_1 and P_2, is given by

$$x = x_1x' + x_2y', \quad y = y_1x' + y_2y'.$$

The most general system in which the reference points are P_1 and P_2 is given by

$$x = \rho x_1x' + \sigma x_2y', \quad y = \rho y_1x' + \sigma y_2y'.$$

1·57. *Choice of co-ordinate system*

If the points of a line are (x, y), it is possible to choose a transformation of co-ordinates of the form (·551) so that any three distinct assigned points P_1, P_2, P_3 are determined by any three distinct assigned values (x_1', y_1'), (x_2', y_2'), (x_3', y_3') of the ratio variable (x', y'). For the coefficients a_1, b_1, a_2, b_2 can be chosen to satisfy the three equations

$$x_i(a_2x_i' + b_2y_i') - y_i(a_1x_i' + b_1y_i') = 0, \tag{·571}$$

where $i = 1, 2, 3$ and no summation is implied, and in the transformation so obtained (x_1, y_1), (x_2, y_2), (x_3, y_3) correspond to

(x'_1, y'_1), (x'_2, y'_2), (x'_3, y'_3). Moreover this transformation cannot be singular, since three distinct values of (x', y') determine three distinct values of (x, y), so a transformation of co-ordinates is obtained having the required property.

In particular, as we have already seen in 1·54, a homogeneous co-ordinate system can be chosen in which any three distinct assigned points of the line are $(1, 0)$, $(0, 1)$ and $(1, 1)$. Thus it is possible to choose a parameter λ so that any three distinct assigned points of the line correspond to $\lambda = \infty$, $\lambda = 0$, and $\lambda = 1$.

1·6. Space of two dimensions

1·61. *Points*

A *point* is defined to be a ratio set (x, y, z) of index three. The points obtained by allowing x, y, and z to take all values other than $x = y = z = 0$ are said to form a *space of two dimensions* or a *plane*.

The points of a plane are all the ratio sets of index three, which form an L_3. The definitions of dependence, independence and linear systems of ratio sets of index three are also applied to the points with which they are identified. Two points are distinct IF they are independent. In fact the points of the plane are the L_3 formed by all the ratio sets (x, y, z). The numbers x, y, z are called homogeneous co-ordinates of the point (x, y, z), the points $(1, 0, 0)$, $(0, 1, 0)$, $(0, 0, 1)$ are called the reference points, and the point $(1, 1, 1)$ is called the unit point. Capital letters will be used to denote points, and P_i will denote the point (x_i, y_i, z_i). Thus P_1 and P_2 are distinct IF $x_1 : y_1 : z_1 \neq x_2 : y_2 : z_2$ and identical IF $x_1 : y_1 : z_1 = x_2 : y_2 : z_2$. The set $(0, 0, 0)$ is not a ratio set and does not represent a point.

1·62. *Non-homogeneous co-ordinates*

If $\lambda = x/z$ and $\mu = y/z$, when λ and μ are given any assigned values ONE point (x, y, z) is determined. Conversely each point (x, y, z) for which $z \neq 0$ determines ONE value of λ and ONE value of μ. The numbers λ, μ are called non-homogeneous co-ordinates of the points of the plane. A point $(a, b, 0)$ does not correspond to any values of λ and μ. By analogy with 1·52 it might be said that $(a, b, 0)$ corresponds to $\lambda = \mu = \infty$, but the points of the form $(a, b, 0)$ are not all identical. For this reason it is not so easy to pass from homogeneous to non-homogeneous co-ordinates in space of two dimensions as it was in space of one dimension.

1·63. *Lines*

A *line* in the plane is defined to be an L_2 of points. On putting $n = 3$ in 1·42 it follows that

(i) If P_1, P_2 are any two distinct points of a line, the line is the linear system of points

$$(\kappa_1 x_1 + \kappa_2 x_2,\ \kappa_1 y_1 + \kappa_2 y_2,\ \kappa_1 z_1 + \kappa_2 z_2).$$

(ii) The points of the line may be identified with the values of the ratio variable (κ_1, κ_2), so this definition of a line in a plane is in agreement with the definition of a line as a space of one dimension.

(iii) The points of a line are the solutions of a linear equation

$$Xx + Yy + Zz = 0,$$

which is called the *point equation* of the line.

(iv) The ratio set $[X, Y, Z]$ formed by the coefficients in this point equation is uniquely determined by the line, and conversely a given ratio set $[X, Y, Z]$ determines ONE line. Thus the lines of the plane may be identified with the ratio sets $[X, Y, Z]$ just as the points are identified with the ratio sets (x, y, z). The numbers X, Y, Z are called *line co-ordinates*; capital letters and square brackets are used to distinguish them from x, y, z, which are now called *point co-ordinates*. Small letters are used to denote lines, and p_i will denote the line $[X_i, Y_i, Z_i]$, whose point equation is

$$p_i \equiv X_i x + Y_i y + Z_i z = 0.$$

(v) The lines $[X, Y, Z]$ which contain a point P_i are the solutions of the equation

$$P_i \equiv x_i X + y_i Y + z_i Z = 0,$$

which is called the *line equation* of the point P_i.

(vi) Two distinct points P_1 and P_2 lie on ONE line,

$$[y_1 z_2 - y_2 z_1,\ \ z_1 x_2 - z_2 x_1,\ \ x_1 y_2 - x_2 y_1],$$

which is called the *join* of P_1 and P_2 and is denoted by $P_1 P_2$.

(vii) Two distinct lines p_1 and p_2 have ONE common point,

$$(Y_1 Z_2 - Y_2 Z_1, Z_1 X_2 - Z_2 X_1, X_1 Y_2 - X_2 Y_1),$$

which is called the *meet* of p_1 and p_2 and is denoted by $p_1 p_2$.

(viii) A *pencil* of lines is defined to be an L_2 of lines. If p_1, p_2 are any two distinct lines of a pencil, the lines of the pencil are the linear system

$$[\kappa_1 X_1 + \kappa_2 X_2, \kappa_1 Y_1 + \kappa_2 Y_2, \kappa_1 Z_1 + \kappa_2 Z_2].$$

(ix) The lines of the pencil may be identified with the values of the ratio variable (κ_1, κ_2).

(x) The lines of a pencil are the solutions of a linear equation

$$xX+yY+zZ = 0,$$

and are therefore the lines which contain a point, called the *vertex* of the pencil.

(xi) The lines of the pencil determined by two distinct lines p_1, p_2 have point equations

$$\kappa_1 p_1 + \kappa_2 p_2 = 0$$

or $$\kappa_1(X_1 x + Y_1 y + Z_1 z) + \kappa_2(X_2 x + Y_2 y + Z_2 z) = 0,$$

since they are the lines

$$[\kappa_1 X_1 + \kappa_2 X_2, \kappa_1 Y_1 + \kappa_2 Y_2, \kappa_1 Z_1 + \kappa_2 Z_2].$$

The vertex of the pencil is the point $p_1 p_2$, or

$$(Y_1 Z_2 - Y_2 Z_1, Z_1 X_2 - Z_2 X_1, X_1 Y_2 - X_2 Y_1).$$

(xii) The points of the line determined by two distinct points P_1, P_2 have line equations

$$\kappa_1 P_1 + \kappa_2 P_2 = 0,$$

or $$\kappa_1(x_1 X + y_1 Y + z_1 Z) + \kappa_2(x_2 X + y_2 Y + z_2 Z) = 0.$$

The line $P_1 P_2$ is

$$[y_1 z_2 - y_2 z_1, z_1 x_2 - z_2 x_1, x_1 y_2 - x_2 y_1].$$

(xiii) Points which lie on the same line are said to be *collinear*. Three points P_1, P_2, P_3 are collinear IF they are dependent, i.e. IF

$$\begin{vmatrix} x_1 & y_1 & z_1 \\ x_2 & y_2 & z_2 \\ x_3 & y_3 & z_3 \end{vmatrix} = 0.$$

(xiv) Lines which contain or pass through the same point are said to be *concurrent*. Three lines p_1, p_2, p_3 are concurrent IF they are dependent, i.e. IF

$$\begin{vmatrix} X_1 & Y_1 & Z_1 \\ X_2 & Y_2 & Z_2 \\ X_3 & Y_3 & Z_3 \end{vmatrix} = 0.$$

(xv) More generally $m(>3)$ points are collinear IF the matrix formed by their co-ordinates is of rank two, and m lines are concurrent IF the matrix formed by their co-ordinates is of rank two.

1·64. *Choice of triangle of reference and unit point*

The points $(1, 0, 0)$, $(0, 1, 0)$ and $(0, 0, 1)$ were called the reference points, and their line equations are $X = 0$, $Y = 0$, $Z = 0$. The triangle whose vertices are the reference points is called the *triangle of reference*, and its sides, which are the joins of pairs of vertices, are called the *reference lines*. The reference lines are thus [1, 0, 0], [0, 1, 0], [0, 0, 1] and their point equations are $x = 0$, $y = 0$, $z = 0$.

Any four points are dependent, and if P_1, P_2, P_3 are independent the system of points

$$(\kappa_1 x_1 + \kappa_2 x_2 + \kappa_3 x_3, \quad \kappa_1 y_1 + \kappa_2 y_2 + \kappa_3 y_3, \quad \kappa_1 z_1 + \kappa_2 z_2 + \kappa_3 z_3)$$

is the whole plane, and each point of the plane corresponds to ONE value of the ratio set $(\kappa_1, \kappa_2, \kappa_3)$. The numbers $\kappa_1, \kappa_2, \kappa_3$ may therefore be used as homogeneous co-ordinates of the points of the plane instead of x, y, z, and in this new system of co-ordinates the reference points are P_1, P_2, P_3.

Any point P_4 corresponds to a value (b_1, b_2, b_3) of $(\kappa_1, \kappa_2, \kappa_3)$, and if P_4 does not lie on any of the lines P_2P_3, P_3P_1, P_1P_2, then $b_1 b_2 b_3 \neq 0$. If x', y', z' are now defined by

$$b_1 x' : b_2 y' : b_3 z' = \kappa_1 : \kappa_2 : \kappa_3,$$

then x', y', z' may be used as homogeneous co-ordinates of the points of the plane instead of $\kappa_1, \kappa_2, \kappa_3$. In this third system of co-ordinates the reference points are P_1, P_2, P_3 and the unit point is P_4. Thus

If P_1, P_2, P_3, P_4 are any four points of the plane, no three of which are collinear, a co-ordinate system may be chosen in which P_1, P_2, P_3 are the reference points and P_4 is the unit point.

1·65. *Linear transformations*

Consider the linear transformation

$$\left.\begin{aligned} x &= a_1 x' + b_1 y' + c_1 z', \\ y &= a_2 x' + b_2 y' + c_2 z', \\ z &= a_3 x' + b_3 y' + c_3 z', \end{aligned}\right\} \qquad (\cdot 651)$$

which is said to be singular IF

$$m \equiv \begin{vmatrix} a_1 & b_1 & c_1 \\ a_2 & b_2 & c_2 \\ a_3 & b_3 & c_3 \end{vmatrix} = 0.$$

Suppose $m \neq 0$, so that the equations (·651) may be solved for x', y', z', giving

$$\left.\begin{aligned} mx' &= A_1 x + A_2 y + A_3 z, \\ my' &= B_1 x + B_2 y + B_3 z, \\ mz' &= C_1 x + C_2 y + C_3 z, \end{aligned}\right\} \qquad (\cdot 652)$$

where A_i, B_i, C_i are the cofactors of a_i, b_i, c_i in the determinant m. Then x', y', z' may be taken as homogeneous co-ordinates instead of x, y, z.

The points of a line $[X, Y, Z]$ satisfy the equation

$$Xx + Yy + Zz = 0,$$

so, by substituting from (·651), the new co-ordinates of the points of the line satisfy

$$(a_1 X + a_2 Y + a_3 Z)\, x' + (b_1 X + b_2 Y + b_3 Z)\, y' + (c_1 X + c_2 Y + c_3 Z)\, z' = 0.$$

The new co-ordinates X', Y', Z' of the line are therefore given by

$$\left.\begin{aligned} X' &= a_1 X + a_2 Y + a_3 Z, \\ Y' &= b_1 X + b_2 Y + b_3 Z, \\ Z' &= c_1 X + c_2 Y + c_3 Z. \end{aligned}\right\} \qquad (\cdot 653)$$

Since $m \neq 0$, these equations may be solved for X, Y, Z, giving

$$\left.\begin{aligned} mX &= A_1 X' + B_1 Y' + C_1 Z', \\ mY &= A_2 X' + B_2 Y' + C_2 Z', \\ mZ &= A_3 X' + B_3 Y' + C_3 Z'. \end{aligned}\right\} \qquad (\cdot 654)$$

Thus, when the point co-ordinates are transformed by a non-singular transformation (·651), the line co-ordinates are transformed by (·654).

It can be shown that, if P_1, P_2, P_3, P_4 are any four points, no three of which are collinear, and if (x_1', y_1', z_1'), (x_2', y_2', z_2'), (x_3', y_3', z_3'), (x_4', y_4', z_4') are any four ratio sets, no three of which are dependent, then the coefficients in (·651) may be so chosen that in the new system of co-ordinates (x', y', z') the four given points have the four given sets of co-ordinates. We can in fact, by 1·64, introduce an intermediate co-ordinate system (x'', y'', z'') referred to which the points P_1, P_2, P_3, P_4 have co-ordinates $(1, 0, 0)$, $(0, 1, 0)$, $(0, 0, 1)$, $(1, 1, 1)$. The required co-ordinates (x', y', z') are then defined by

$$\begin{aligned} x' &= \lambda x_1' x'' + \mu x_2' y'' + \nu x_3' z'', \\ y' &= \lambda y_1' x'' + \mu y_2' y'' + \nu y_3' z'', \\ z' &= \lambda z_1' x'' + \mu z_2' y'' + \nu z_3' z'', \end{aligned}$$

where λ, μ, ν are the solutions of the equations

$$\begin{aligned} x_4' &= \lambda x_1' + \mu x_2' + \nu x_3', \\ y_4' &= \lambda y_1' + \mu y_2' + \nu y_3', \\ z_4' &= \lambda z_1' + \mu z_2' + \nu z_3'. \end{aligned}$$

A singular transformation does not give a transformation of co-ordinates, for if $m = 0$ the equations (·651) cannot be solved for x', y', z'. There are sets of values of x', y', z' not all zero which give $x = y = z = 0$, and all other values of x', y', z' give points (x, y, z) which either lie on a certain line or coincide, the latter case arising IF the matrix

$$\begin{pmatrix} a_1 & b_1 & c_1 \\ a_2 & b_2 & c_2 \\ a_3 & b_3 & c_3 \end{pmatrix}$$

is of rank one.

It should be noted in the general case that, by (·651) and (·654),

$$xX + yY + zZ \equiv x'X' + y'Y' + z'Z',$$

where (x, y, z) and $[X, Y, Z]$ are the original co-ordinates of any point and line, and (x', y', z') and $[X', Y', Z']$ their new co-ordinates. This is of course not a new result but is implicit in the way the new line co-ordinates were chosen.

We have here regarded a linear transformation as effecting a change in the co-ordinates of a point P but not in the actual position of P. We may, however, interpret the equations (·651) in a different sense and say that they transform a point $P(x, y, z)$ into another point P' whose co-ordinates, *referred to the original scheme*, are (x', y', z'). When a linear transformation is regarded thus as a transformation of the plane, we can say that there exists a non-singular linear transformation carrying any four points P_1, P_2, P_3, P_4, no three of which are collinear, respectively into any other four points P'_1, P'_2, P'_3, P'_4, no three of which are collinear. Similar considerations apply to linear transformations in space of any number of dimensions.

1·66. *Duality*

The points of a plane were defined to be ratio sets (x, y, z), and a line was defined to be an L_2 of points. It then appeared that the lines of the plane are also essentially ratio sets $[X, Y, Z]$. The equation

$$xX + yY + zZ = 0,$$

which is symmetrical in x, y, z and X, Y, Z is both the CONDITION for the point (x, y, z) to lie on the line $[X, Y, Z]$ and the CONDITION for the line $[X, Y, Z]$ to contain the point (x, y, z). Thus the points and lines of the plane form exactly similar systems and there is a complete symmetry between them, as is borne out by 1·63.

Instead of regarding a line as composed of a linear system of points we may regard the lines as the fundamental elements and a point as determined by or composed of an L_2 of lines, namely the lines through the point. These two attitudes to the elements of the plane are called *dual* attitudes, and elements which correspond in them are called dual elements.*

The following pairs of elements are dual:

A point P_1, (x_1, y_1, z_1).	*A line* p_1, $[X_1, Y_1, Z_1]$.
Lines $[X, Y, Z]$ *through* P_1, *given by* $x_1 X + y_1 Y + z_1 Z = 0.$	*Points* (x, y, z) *on* p_1, *given by* $X_1 x + Y_1 y + Z_1 z = 0.$
The join P_1P_2 *of* P_1 *and* P_2, *i.e. the line* $[y_1z_2 - y_2z_1, z_1x_2 - z_2x_1, x_1y_2 - x_2y_1]$.	*The meet* p_1p_2 *of* p_1 *and* p_2, *i.e. the point* $(Y_1Z_2 - Y_2Z_1, Z_1X_2 - Z_2X_1, X_1Y_2 - X_2Y_1)$.
The points of P_1P_2, *i.e.* $(\kappa_1 x_1 + \kappa_2 x_2, \kappa_1 y_1 + \kappa_2 y_2, \kappa_1 z_1 + \kappa_2 z_2)$.	*The lines through* p_1p_2, *i.e.* $[\kappa_1 X_1 + \kappa_2 X_2, \kappa_1 Y_1 + \kappa_2 Y_2, \kappa_1 Z_1 + \kappa_2 Z_2]$.

Any statement in plane geometry is a statement about points and lines. It is essentially a statement about two classes of ratio sets (x, y, z) and $[X, Y, Z]$, and the geometrical nature of the statement is obtained by calling the first class of ratio sets points and the second class lines. In view of the symmetry between points and lines it is equally legitimate to call the second class of ratio sets points and the first class lines. By so doing another geometrical statement is obtained which is called the dual of the original statement. Thus with every theorem of plane geometry there is associated a dual theorem, and the dual of the dual theorem is the original theorem. The proof of a theorem in plane geometry is essentially an argument about the two classes of ratio sets (x, y, z) and $[X, Y, Z]$, in which the classes are called points and lines respectively. If in the same argument the two classes are called lines and points respectively, instead of points and lines, a proof of the dual theorem is obtained. *Thus the truth of a theorem implies the truth of the dual theorem.* This is the *principle of duality* in a plane.

The theorems of 1·63 (vi) and (vii), that two distinct points lie on a line and that two distinct lines meet in a point, are dual theorems

* Cf. $R(1)$, p. 61.

and are known as the *incidence theorems* in a plane. They are particular cases of the incidence theorem of 1·44. For this theorem states that if $r = s = 0$ and $i = -1$, then $j = 1$, and that if $r = s = 1$ and $j = 2$, then $i = 0$.

1·67. The two dual attitudes to plane geometry, which give rise to what are really two different statements of the same algebraic theorem, should be equally easy to grasp. In practice this is not so, because we attempt to form a mental picture of a plane, and this picture is suggested by the Euclidean plane. We draw figures to help us to understand geometrical theorems, though these figures cannot represent complex geometry. In consequence it is much easier to grasp the first attitude, in which the points are fundamental, because it is more easy to picture a line as composed of points than to picture a point as composed of lines. We can draw something that gives the impression of being a line, but we cannot draw a pencil of lines.

In the development of plane geometry we shall tend to avoid complicated algebra by the use of descriptive arguments, and it is sometimes quite difficult to pass directly from a descriptive argument to the dual argument. For this reason when a theorem has been proved by a descriptive argument we shall sometimes go through the dual proof, though this is not necessary to establish the truth of the dual theorem. When this is not done in the text the reader should do it for himself as an exercise.

1·7. Space of three dimensions

1·71. *Points*

A *point* is defined to be a ratio set (x, y, z, t) of index four. The L_4 formed by all the points (x, y, z, t) is called a *space of three dimensions* or a *solid*. Until space of more than three dimensions is again discussed 'space' will mean space of three dimensions. The numbers x, y, z, t are called homogeneous co-ordinates of the points, and P_i will denote the point (x_i, y_i, z_i, t_i). The points for which $t \neq 0$ may be represented by non-homogeneous co-ordinates λ, μ, ν defined by $\lambda = x/t$, $\mu = y/t$, $\nu = z/t$.

1·72. *Planes*

A *plane* is defined to be an L_3 of points. On putting $n = 4$ in 1·42 it follows that

(i) If P_1, P_2, P_3 are any three independent points of a plane, the plane is the linear system of points

$$(\kappa_1 x_1+\kappa_2 x_2+\kappa_3 x_3, \kappa_1 y_1+\kappa_2 y_2+\kappa_3 y_3, \kappa_1 z_1+\kappa_2 z_2+\kappa_3 z_3, \kappa_1 t_1+\kappa_2 t_2+\kappa_3 t_3).$$

(ii) The points of the plane may be identified with the ratio sets $(\kappa_1, \kappa_2, \kappa_3)$.

(iii) The points of any plane are the solutions of a linear equation

$$Xz+Yy+Zz+Tt = 0,$$

which is called the *point equation* of the plane.

(iv) The planes of space may be identified with the ratio sets $[X, Y, Z, T]$. The numbers X, Y, Z, T are called *plane co-ordinates*; capital letters and square brackets are used to distinguish them from x, y, z, t which are now called *point co-ordinates*. Small Greek letters are used to denote planes, and π_i will denote the plane $[X_i, Y_i, Z_i, T_i]$, whose point equation is

$$\pi_i \equiv X_i x+Y_i y+Z_i z+T_i t = 0.$$

(v) The planes $[X, Y, Z, T]$ which contain a point P_i are the solutions of the equation

$$P_i \equiv x_i X+y_i Y+z_i Z+t_i T = 0,$$

which is called the *plane equation* of the point P_i.

(vi) Three independent points P_1, P_2, P_3 lie in ONE plane, denoted by $P_1P_2P_3$, whose point equation is

$$\begin{vmatrix} x & y & z & t \\ x_1 & y_1 & z_1 & t_1 \\ x_2 & y_2 & z_2 & t_2 \\ x_3 & y_3 & z_3 & t_3 \end{vmatrix} = 0.$$

(vii) Three independent planes π_1, π_2, π_3 have ONE common point, whose plane equation is

$$\begin{vmatrix} X & Y & Z & T \\ X_1 & Y_1 & Z_1 & T_1 \\ X_2 & Y_2 & Z_2 & T_2 \\ X_3 & Y_3 & Z_3 & T_3 \end{vmatrix} = 0.$$

(viii) A *sheaf* of planes is defined to be an L_3 of planes. If π_1, π_2, π_3 are any three independent planes of a sheaf, the planes of the sheaf are the linear system

$$[\kappa_1 X_1+\kappa_2 X_2+\kappa_3 X_3, \kappa_1 Y_1+\kappa_2 Y_2+\kappa_3 Y_3, \kappa_1 Z_1+\kappa_2 Z_2+\kappa_3 Z_3, \kappa_1 T_1+\kappa_2 T_2+\kappa_3 T_3].$$

(ix) The planes of the sheaf may be identified with the ratio sets $(\kappa_1, \kappa_2, \kappa_3)$.

(x) The planes of a sheaf are the solutions of a linear equation

$$xX+yY+zZ+tT = 0,$$

and are therefore the planes which contain a point, called the *vertex* of the sheaf.

(xi) The planes of the sheaf determined by three independent planes π_1, π_2, π_3 have point equations

$$\kappa_1\pi_1+\kappa_2\pi_2+\kappa_3\pi_3 = 0.$$

The vertex of the sheaf is the point $\pi_1\pi_2\pi_3$ whose plane equation is given in (vii).

(xii) The points of the plane determined by three independent points P_1, P_2, P_3 have plane equations

$$\kappa_1 P_1+\kappa_2 P_2+\kappa_3 P_3 = 0.$$

The equation of the plane $P_1P_2P_3$ is given in (vi).

(xiii) Points which lie on the same plane are said to be *coplanar* Four points P_1, P_2, P_3, P_4 are coplanar IF they are dependent, i.e. IF

$$\begin{vmatrix} x_1 & y_1 & z_1 & t_1 \\ x_2 & y_2 & z_2 & t_2 \\ x_3 & y_3 & z_3 & t_3 \\ x_4 & y_4 & z_4 & t_4 \end{vmatrix} = 0.$$

(xiv) Planes which contain or pass through the same point are said to be *concurrent*. Four planes $\pi_1, \pi_2, \pi_3, \pi_4$ are concurrent IF they are dependent, i.e. IF

$$\begin{vmatrix} X_1 & Y_1 & Z_1 & T_1 \\ X_2 & Y_2 & Z_2 & T_2 \\ X_3 & Y_3 & Z_3 & T_3 \\ X_4 & Y_4 & Z_4 & T_4 \end{vmatrix} = 0.$$

(xv) More generally m (>4) points are coplanar IF the matrix of the points is of rank three, and m planes are concurrent IF the matrix of the planes is of rank three.

1·73. *Lines*

A *line* is defined to be an L_2 of points. Again, it follows from 1·42 that

(i) If P_1, P_2 are any two distinct points of a line, the line is the linear system of points

$$(\kappa_1 x_1+\kappa_2 x_2, \kappa_1 y_1+\kappa_2 y_2, \kappa_1 z_1+\kappa_2 z_2, \kappa_1 t_1+\kappa_2 t_2).$$

(ii) The points of the line may be identified with the ratio sets (κ_1, κ_2).

(iii) Two distinct points P_1, P_2 lie on ONE line which is called the *join* of the points and is denoted by P_1P_2.

(iv) The points of a line are the solutions of a linear system of linear equations, $Xx+Yy+Zz+Tt=0$, of rank two. The planes given by the equations of this system are the planes which contain or pass through the line.

(v) A *pencil* of planes is defined to be an L_2 of planes. Thus the planes through a line are the planes of a pencil.

(vi) If π_1, π_2 are two distinct planes of a pencil, the pencil is the system of planes

$$[\kappa_1 X_1+\kappa_2 X_2, \kappa_1 Y_1+\kappa_2 Y_2, \kappa_1 Z_1+\kappa_2 Z_2, \kappa_1 T_1+\kappa_2 T_2].$$

(vii) The planes of the pencil may be identified with the ratio sets (κ_1, κ_2).

(viii) Two distinct planes belong to ONE pencil.

(ix) The common points of the planes of a pencil are the solutions of a linear system of equations of rank two and therefore form an L_2 of points, which is a line. This line is called the *vertex* of the pencil.

(x) The common points of two distinct planes π_1, π_2 are the points of a line, which is called the *meet* or *intersection* of the planes and is denoted by $\pi_1\pi_2$.

(xi) The common planes of two sheaves with distinct vertices P_1, P_2 are the planes of the pencil whose vertex is the line P_1P_2.

(xii) The points of a plane π form a linear system

$$(\kappa_1 x_1+\kappa_2 x_2+\kappa_3 x_3, \kappa_1 y_1+\kappa_2 y_2+\kappa_3 y_3, \kappa_1 z_1+\kappa_2 z_2+\kappa_3 z_3, \kappa_1 t_1+\kappa_2 t_2+\kappa_3 t_3),$$

and may be represented by the ratio sets $(\kappa_1, \kappa_2, \kappa_3)$. A line in π is represented by an L_2 of ratio sets $(\kappa_1, \kappa_2, \kappa_3)$, and two such systems have ONE ratio set in common. Thus two distinct lines in a plane meet in ONE point.

(xiii) The planes of a sheaf with vertex P form a linear system

$$[\kappa_1 X_1+\kappa_2 X_2+\kappa_3 X_3, \kappa_1 Y_1+\kappa_2 Y_2+\kappa_3 Y_3, \kappa_1 Z_1+\kappa_2 Z_2+\kappa_3 Z_3, \kappa_1 T_1+\kappa_2 T_2+\kappa_3 T_3],$$

and may be represented by the ratio sets $(\kappa_1, \kappa_2, \kappa_3)$. A pencil of planes contained in the sheaf is represented by an L_2 of ratio sets $(\kappa_1, \kappa_2, \kappa_3)$, and two such systems have ONE ratio set in common.

Any line through P is the vertex of ONE pencil of planes, so two distinct lines through a point lie in ONE plane.

(xiv) Points which lie on the same line are said to be *collinear*. Thus three points P_1, P_2, P_3 are collinear IF they are dependent. More generally $m(\geq 3)$ points are collinear IF the matrix of the points is of rank two.

(xv) Planes which pass through the same line are also said to be *collinear*. Thus three planes π_1, π_2, π_3 are collinear IF they are dependent. More generally $m(\geq 3)$ planes are collinear IF the matrix of the planes is of rank two.

1·74. *Duality*

The points (x, y, z, t) have been regarded as the fundamental elements of space and the lines and planes have been defined to be linear systems of points of rank two and three. But it has been shown that the planes are also essentially ratio sets $[X, Y, Z, T]$, that any line is the vertex of a pencil of planes, and that any point is the vertex of a sheaf of planes. Another attitude to the geometry of space is therefore possible, in which the planes are regarded as the fundamental elements, and the lines and points are regarded as linear systems of planes of rank two and three. These two attitudes are called *dual* attitudes, and the following pairs of elements are dual:

A point P_1.	*A plane* π_1.
A line regarded as an L_2 *of points.*	*A line regarded as an* L_2 *of planes.*
A plane regarded as an L_3 *of points.*	*A point regarded as an* L_3 *of planes.*
Points of the line P_1P_2.	*Planes through the line* $\pi_1\pi_2$.
Planes through P_1P_2.	*Points on* $\pi_1\pi_2$.
Lines through P_1.	*Lines in* π_1.
Lines meeting P_1P_2, *i.e. lines containing some point of* P_1P_2.	*Lines meeting* $\pi_1\pi_2$, *i.e. lines in some plane through* $\pi_1\pi_2$.
Planes through P_1.	*Points of* π_1.
The plane $P_1P_2P_3$.	*The point* $\pi_1\pi_2\pi_3$.
Lines in a plane π *and containing a point* P *of* π, *i.e. a pencil of lines* in π *with vertex* P.	*Lines through a point* P *and in a plane* π *through* P, *i.e. a pencil of lines with vertex* P *lying in* π.

The reasoning of 1·66 may be extended to space, showing that to every statement or theorem there corresponds a dual statement or theorem, and that the truth of a theorem implies the truth of the dual theorem. The notion of duality depends on the dimension of the space that is being considered, duality in a plane being different from duality in space of three dimensions. For example, in a plane the dual of a line is a point, whereas in space a line is self-dual.

1·75. *Incidence theorems*

The following six theorems, which follow from 1·72 and 1·73, and which occur in dual pairs, are particular cases of the incidence theorem mentioned in 1·44 and are known as the incidence theorems of space of three dimensions.

(*i*) *Two distinct points P_1 and P_2 lie on* ONE *line, which is called the* join *of P_1 and P_2 and is denoted by P_1P_2.*

Two distinct planes π_1 and π_2 have ONE *line in common, which is called the* meet *of π_1 and π_2 and is denoted by $\pi_1\pi_2$.*

(*ii*) *Two distinct lines p_1 and p_2 in a plane π have* ONE *common point, which is called the* meet *of p_1 and p_2 and is denoted by (p_1p_2).*

Two distinct lines p_1 and p_2 through a point P have ONE *common plane, which is called the* join *of p_1 and p_2 and is denoted by $[p_1p_2]$.*

(*iii*) *A line p and a point P which does not lie on p lie in* ONE *plane, which is called the* join *of p and P and is denoted by pP or Pp.*

A line p and a plane π which does not contain p have ONE *common point, which is called the meet of p and π and is denoted by $p\pi$ or πp.*

It should be noticed that the incidence theorems in a plane* are included in these but that they are no longer dual. Two lines in space, which have no common point and consequently do not lie in a plane, are said to be *skew*. In plane geometry the symbol p_1p_2 means the meet of the lines p_1 and p_2, but in space the symbol is meaningless if the two lines are skew and ambiguous if the two lines intersect. For this reason the round and square brackets were introduced in (ii) to distinguish between the meet and the join of p_1 and p_2. This convention will be used throughout this book. Thus, if P_1, P_2, P_3, P_4 and p_1, p_2, p_3, p_4 are four points and four lines lying in

* See 1·66, p. 25.

the same plane, (P_1P_2, P_3P_4) is the meet of the lines P_1P_2 and P_3P_4, while $[p_1p_2, p_3p_4]$ is the join of the points p_1p_2 and p_3p_4. Four more trivial theorems which also occur in dual pairs are the following:

(iv) *Three independent points P_1, P_2, P_3 lie in* ONE *plane which is denoted by $P_1P_2P_3$. This plane contains the lines P_2P_3, P_3P_1, P_1P_2 and is therefore $[P_1, P_2P_3]$.*

Three independent planes π_1, π_2, π_3 have ONE *common point which is denoted by $\pi_1\pi_2\pi_3$. This point lies on the lines $\pi_2\pi_3, \pi_3\pi_1, \pi_1\pi_2$ and is therefore $(\pi_1, \pi_2\pi_3)$.*

(v) *If a plane π does not contain two lines p_1 and p_2 and meets them in distinct points, there is* ONE *line in π which meets p_1 and p_2, namely $[\pi p_1, \pi p_2]$.*

If a point P does not lie on two lines p_1 and p_2, and if the planes Pp_1 and Pp_2 are distinct, there is ONE *line through P which meets p_1 and p_2, namely (Pp_1, Pp_2).*

1·76. *Tetrahedron of reference*

A *tetrahedron* is the figure determined by four independent points, which are called the *vertices*. The four planes which contain three vertices are called *faces* and the six lines which contain two vertices are called *edges*. Each edge lies on two faces and each vertex on three faces. The figure is self-dual, and may be determined by its four faces which are four independent planes.

The four independent points $(1, 0, 0, 0)$, $(0, 1, 0, 0)$, $(0, 0, 1, 0)$, $(0, 0, 0, 1)$ are called the *reference points*, and the tetrahedron determined by them is called the *tetrahedron of reference*. The faces of this tetrahedron are the planes $[1, 0, 0, 0]$, $[0, 1, 0, 0]$, $[0, 0, 1, 0]$, $[0, 0, 0, 1]$, and are called the *reference planes*. The point $(1, 1, 1, 1)$ and the plane $[1, 1, 1, 1]$ are called the unit point and unit plane.

Any five points are dependent and, if P_1, P_2, P_3, P_4 are four independent points, the L_4 of points

$$(\kappa_1x_1+\kappa_2x_2+\kappa_3x_3+\kappa_4x_4, \kappa_1y_1+\kappa_2y_2+\kappa_3y_3+\kappa_4y_4,$$
$$\kappa_1z_1+\kappa_2z_2+\kappa_3z_3+\kappa_4z_4, \kappa_1t_1+\kappa_2t_2+\kappa_3t_3+\kappa_4t_4)$$

is the whole space, each point of space corresponding to ONE ratio set $(\kappa_1, \kappa_2, \kappa_3, \kappa_4)$. The numbers $\kappa_1, \kappa_2, \kappa_3, \kappa_4$ may therefore be used as homogeneous point co-ordinates instead of x, y, z, t, and when this is done the reference points are P_1, P_2, P_3, P_4.

Any point P_5 corresponds to a value (b_1, b_2, b_3, b_4) of $(\kappa_1, \kappa_2, \kappa_3, \kappa_4)$, and if P_5 does not lie on any of the faces of the tetrahedron $P_1P_2P_3P_4$, then $b_1b_2b_3b_4 \neq 0$. If x', y', z', t' are now defined by

$$b_1x' : b_2y' : b_3z' : b_4t' = \kappa_1 : \kappa_2 : \kappa_3 : \kappa_4,$$

then x', y', z', t' may be used as homogeneous point co-ordinates, and in this third system of co-ordinates the reference points are P_1, P_2, P_3, P_4 and P_5 is the unit point. Thus

If P_1, P_2, P_3, P_4, P_5 are any five points no four of which are dependent, a co-ordinate system may be chosen in which the reference points are P_1, P_2, P_3, P_4 and the unit point is P_5.

1·77. The reasoning of 1·65 may also be extended to space. A linear transformation

$$\begin{aligned} x &= a_1x' + b_1y' + c_1z' + d_1t', \\ y &= a_2x' + b_2y' + c_2z' + d_2t', \\ z &= a_3x' + b_3y' + c_3z' + d_3t', \\ t &= a_4x' + b_4y' + c_4z' + d_4t', \end{aligned}$$

determines a transformation of point co-ordinates IF the determinant of the transformation is not zero, i.e. IF the transformation is non-singular. If P_1, P_2, P_3, P_4, P_5 are any five points no four of which are dependent, and if (x_i', y_i', z_i', t_i') $(i = 1, 2, 3, 4, 5)$, are any five ratio sets no four of which are dependent, then it is possible to choose the coefficients of the transformation so that in the co-ordinate system x', y', z', t' the five given points have the five given sets of co-ordinates.

1·78. *Line co-ordinates*

It is frequently convenient to consider a line not as a set of points or as the base of a pencil of planes but as an element in itself, and for this purpose it is necessary to define a set of numbers which shall determine the line uniquely and which we may call its co-ordinates. We could, for example, define the co-ordinates of a *general* line as the set of co-ordinates of the points in which it meets the planes $x = 0, y = 0$; but to this definition there are the objections that it is unsymmetrical and that it breaks down in the case of a line which intersects the line $x = 0, y = 0$. The system of line co-ordinates now to be described is symmetrical and valid for all lines of [3]; these co-ordinates are usually known as Plücker co-ordinates, although they were first introduced by Cayley.* The system is a special case of a general co-ordinate system for $[r]$'s in $[n]$.

If P_1 and P_2 are two points of a line p, its line co-ordinates are defined as the ratio set $\{l, m, n, l', m', n'\}$ whose elements are proportional to the two-rowed determinants of the matrix

$$\begin{pmatrix} x_1 & y_1 & z_1 & t_1 \\ x_2 & y_2 & z_2 & t_2 \end{pmatrix};$$

thus

$$l:m:n:l':m':n' = \begin{vmatrix} x_1 & t_1 \\ x_2 & t_2 \end{vmatrix} : \begin{vmatrix} y_1 & t_1 \\ y_2 & t_2 \end{vmatrix} : \begin{vmatrix} z_1 & t_1 \\ z_2 & t_2 \end{vmatrix} : \begin{vmatrix} y_1 & z_1 \\ y_2 & z_2 \end{vmatrix} : \begin{vmatrix} z_1 & x_1 \\ z_2 & x_2 \end{vmatrix} : \begin{vmatrix} x_1 & y_1 \\ x_2 & y_2 \end{vmatrix}. \qquad (\cdot 781)$$

For this definition to be useful it is necessary that it should be independent of the particular choice of the pair of points on p, and it is easily seen that this is the case. If instead of P_1 and P_2 we had started with two other distinct points of p, say

$$(\rho_1 x_1 + \rho_2 x_2, \rho_1 y_1 + \rho_2 y_2, \rho_1 z_1 + \rho_2 z_2, \rho_1 t_1 + \rho_2 t_2)$$

and $$(\sigma_1 x_1 + \sigma_2 x_2, \sigma_1 y_1 + \sigma_2 y_2, \sigma_1 z_1 + \sigma_2 z_2, \sigma_1 t_1 + \sigma_2 t_2),$$

where $\rho_1 \sigma_2 \neq \rho_2 \sigma_1$, we should obtain the same line-co-ordinates as before, each multiplied by the non-zero determinant $\begin{vmatrix} \rho_1 \rho_2 \\ \sigma_1 \sigma_2 \end{vmatrix}$; thus, for example,

$$\begin{vmatrix} \rho_1 x_1 + \rho_2 x_2 & \rho_1 t_1 + \rho_2 t_2 \\ \sigma_1 x_1 + \sigma_2 x_2 & \sigma_1 t_1 + \sigma_2 t_2 \end{vmatrix} = \begin{vmatrix} \rho_1 & \rho_2 \\ \sigma_1 & \sigma_2 \end{vmatrix} \begin{vmatrix} x_1 & t_1 \\ x_2 & t_2 \end{vmatrix}.$$

The ratio set is the same in each case.

The Plücker co-ordinates of a line are not independent. In fact, by expanding the zero determinant

$$\begin{vmatrix} x_1 & y_1 & z_1 & t_1 \\ x_2 & y_2 & z_2 & t_2 \\ x_1 & y_1 & z_1 & t_1 \\ x_2 & y_2 & z_2 & t_2 \end{vmatrix}$$

in terms of the two-row minors of the first two rows, we have

$$\Omega \equiv ll' + mm' + nn' = 0;$$

conversely we shall show that any ratio set $\{l, m, n, l', m', n'\}$ the six coefficients of which satisfy $\Omega = 0$ determines ONE line p.

* Cayley, *Collected Mathematical Papers*, IV, p. 447 (1891), and VII, p. 66 (1894).

The points of the line P_1P_2 are

$$(\kappa_1 x_1 + \kappa_2 x_2, \kappa_1 y_1 + \kappa_2 y_2, \kappa_1 z_1 + \kappa_2 z_2, \kappa_1 t_1 + \kappa_2 t_2).$$

Its meets with the reference planes are obtained by putting $\kappa_2 = x_1$, $\kappa_1 = -x_2$, etc., and are therefore the points

$$(0, n', -m', l), \; (-n', 0, l', m), \; (m', -l', 0, n), \; (-l, -m, -n, 0). \quad (\cdot 782)$$

Suppose then we have a ratio set $\{l, m, n, l', m', n'\}$ satisfying $\Omega = 0$. Since not all of l, m, n, l', m', n' are zero, at least two of the four points whose co-ordinates are given by ($\cdot$782) are distinct. Thus, if $l \neq 0$, the points $(0, n', -m', l)$ and $(-l, -m, -n, 0)$ are distinct. The co-ordinates of the line joining them can then, by ($\cdot$781), be written down as

$$\{l^2, lm, ln, -mm' - nn', lm', ln'\},$$

or, by virtue of the relation $\Omega = 0$, as $\{l, m, n, l', m', n'\}$. Thus a ratio set $\{l, m, n, l', m', n'\}$ satisfying $\Omega = 0$ determines ONE line p whose meets with the reference planes are the points ($\cdot$782). The planes joining p to the reference points are

$$[0, n, -m, l'], \; [-n, 0, l, m'], \; [m, -l, 0, n'], \; [-l', -m', -n', 0], \quad (\cdot 783)$$

since these planes contain the points $\cdot$782 by virtue of $\Omega = 0$.

If π_1 and π_2 are any two distinct planes through p, the planes through p are

$$[\kappa_1 X_1 + \kappa_2 X_2, \kappa_1 Y_1 + \kappa_2 Y_2, \kappa_1 Z_1 + \kappa_2 Z_2, \kappa_1 T_1 + \kappa_2 T_2],$$

and the planes joining p to the reference points are obtained by putting $\kappa_2 = X_1$, $\kappa_1 = -X_2$, etc. Comparison with ($\cdot$783) shows that

$l : m : n : l' : m' : n' =$

$$\begin{vmatrix} Y_1 Z_1 \\ Y_2 Z_2 \end{vmatrix} : \begin{vmatrix} Z_1 X_1 \\ Z_2 X_2 \end{vmatrix} : \begin{vmatrix} X_1 Y_1 \\ X_2 Y_2 \end{vmatrix} : \begin{vmatrix} X_1 T_1 \\ X_2 T_2 \end{vmatrix} : \begin{vmatrix} Y_1 T_1 \\ Y_2 T_2 \end{vmatrix} : \begin{vmatrix} Z_1 T_1 \\ Z_2 T_2 \end{vmatrix} \quad (\cdot 784)$$

Thus

The lines of space may be identified with the ratio sets

$$\{l, m, n, l', m', n'\}$$

which satisfy $\Omega = 0$. The numbers l, m, n, l', m', n' are called line co-ordinates, or Plücker co-ordinates. The line co-ordinates of a line p may be written down either from any two distinct points P_1, P_2 of p by ($\cdot$781), or from any two distinct planes π_1, π_2 through p by ($\cdot$784). Curly brackets are used for line co-ordinates, and the line

$$\{l_i, m_i, n_i, l'_i, m'_i, n'_i\}$$

is denoted by p_i.

In fact the lines of space of three dimensions may be represented by the points of a space of five dimensions, [5], which lie on the locus given by the equation

$$x_1x_4 + x_2x_5 + x_3x_6 = 0, \qquad (\cdot 785)$$

whose $x_1, x_2, \ldots, x_6$ are homogeneous point co-ordinates in the [5]. This representation will not be discussed in this volume.

Two lines p_1 and p_2 intersect IF

$$l_1l_2' + l_1'l_2 + m_1m_2' + m_1'm_2 + n_1n_2' + n_1'n_2 = 0. \qquad (\cdot 786)$$

For if P_1, P_1' are two distinct points of p_1 and P_2, P_2' are two distinct points of p_2, the lines p_1 and p_2 intersect IF P_1, P_1', P_2, P_2' are coplanar, i.e. IF

$$\begin{vmatrix} x_1 & y_1 & z_1 & t_1 \\ x_1' & y_1' & z_1' & t_1' \\ x_2 & y_2 & z_2 & t_2 \\ x_2' & y_2' & z_2' & t_2' \end{vmatrix} = 0.$$

On expanding this determinant in terms of the two-row minors of the first two rows, which are the line co-ordinates of p_1, the equation ·786 is obtained.

1·79. *Co-ordinates of an [r] in [n]*

The line co-ordinates of a line in [3] furnish a special case of a method, due to Grassmann, of assigning co-ordinates to the subspaces of given dimension of any linear space. If $(x_1, x_2, \ldots, x_{n+1})$ are homogeneous point co-ordinates in a space $[n]$, and if

$$(x_1^{(i)}, x_2^{(i)}, \ldots, x_{n+1}^{(i)}) \; (i = 1, 2, \ldots, r+1)$$

are $r+1$ independent points of a given r-dimensional subspace S_r of $[n]$, the Grassmannian co-ordinates of S_r may be defined as the ratio set whose elements are the $\binom{n+1}{r+1}$ determinants of the matrices obtained by selecting, in a pre-determined order, groups of $r+1$ columns from the matrix

$$\begin{pmatrix} x_1^{(1)} & x_2^{(1)} & \ldots & x_{n+1}^{(1)} \\ x_1^{(2)} & x_2^{(2)} & \ldots & x_{n+1}^{(2)} \\ \ldots & \ldots & \ldots & \ldots \\ x_1^{(r+1)} & x_2^{(r+1)} & \ldots & x_{n+1}^{(r+1)} \end{pmatrix}.$$

It is easily verified, as in 1·78, that we obtain the same ratio set if we start with any other $r+1$ independent points of S_r.

Dually, if $[X_1, X_2, \ldots, X_{n+1}]$ are homogeneous prime co-ordinates in $[n]$, and if $[X_1^{(i)}, X_2^{(i)}, \ldots, X_{n+1}^{(i)}]$ $(i = 1, 2, \ldots, n-r)$ are $n-r$ independent primes through S_r, we may define the co-ordinates of S_r as the ratio set whose elements are the $\binom{n+1}{n-r}$ or $\binom{n+1}{r+1}$ determinants of $n-r$ rows and columns obtained from the matrix

$$\begin{pmatrix} X_1^{(1)} & X_2^{(1)} & \ldots & X_{n+1}^{(1)} \\ X_1^{(2)} & X_2^{(2)} & \ldots & X_{n+1}^{(2)} \\ \ldots & \ldots & \ldots & \ldots \\ X_1^{(n-r)} & X_2^{(n-r)} & \ldots & X_{n+1}^{(n-r)} \end{pmatrix}.$$

It may be proved that this ratio set is the same as the first, except for the order of the elements.

If we interpret the Grassmannian co-ordinates as point co-ordinates in a space of $\binom{n+1}{r+1} - 1$ dimensions we obtain in this space an algebraic locus of points corresponding to the $[r]$'s of $[n]$, which is known as the Grassmannian of $[r]$'s of $[n]$. Thus (·785) is the Grassmannian of lines of [3]. In the general case, however, the Grassmannian is not given by a single equation.

1·8. Projection

1·81. The theorems which have been proved about spaces of one, two and three dimensions are all capable of algebraic expression, and consequently any geometrical construction which may be deduced from them is an algebraic construction. For instance in 1·75 (v) the construction of the unique line through P which meets p_1 and p_2 is an algebraic construction. *Projection* is an important construction, which depends only on the theorems of incidence and is therefore algebraic.

1·82. *Projection in a plane*

Consider first projection in a plane. If a line a and a point V which does not lie on a are chosen, any point P other than V determines ONE line VP which meets a in ONE point P'. This point P' of a is called the projection of P on to a from V, and V is called the vertex of projection. Each point P' of a is thus the projection of every point of the line VP' except V.

Let b be another line which does not contain V. Then each point P' of a is the projection of ONE point P of b, and each point P of b projects into ONE point P' of a. This one to one correspondence between points of b and points of a is called a *perspectivity*, and V is called the *centre of perspective*.

1·83. *Projection in space from a point vertex*

In space, if a plane α and a point V not on α are chosen, any point P other than V determines ONE line VP which meets α in ONE point P'. This point P' is called the projection of P on to α from V, and V is called the vertex of projection. Each point P' of α is the projection of every point of the line VP' except V. The points of any line p which does not contain V project into the points of the line (α, Vp), and a perspectivity is established between p and (α, Vp).

If β is another plane which does not contain V, the one to one correspondence between points of β and points of α obtained by projection from V is also called a perspectivity.

1·84. *Projection in space from a line vertex*

If two skew lines a and v are chosen, any point P which does not lie on v determines ONE plane Pv which meets a in ONE point P'. This point P' is called the projection of P on to a from v, and v is called the vertex of projection. Each point P' of a is thus the projection of every point of the plane $P'v$ except the points of v. If b is another line skew to v each point P' of a is the projection of ONE point P of b, and each point P of b projects into ONE point of a. This one to one correspondence between the points of b and the points of a is called a *line perspectivity* and v is called the *line centre of perspective*.

Projection from a line vertex v on to a line a which is skew to v is equivalent to two successive projections from point vertices. For, if V_1 and V_2 are any two distinct points of v, projection from V_1 on to the plane V_2a followed by projection in this plane on to a from V_2 is equivalent to projection on to a from v.

1·9. Symbolic notation. Syzygies

1·91. We have already introduced the notation P_i for a point whose homogeneous co-ordinates in a given linear space are

$(x_i, y_i, z_i, \ldots)$, and we have defined the line P_1P_2 as the aggregate of points P with co-ordinates

$$(\kappa_1 x_1 + \kappa_2 x_2, \kappa_1 y_1 + \kappa_2 y_2, \kappa_1 z_1 + \kappa_2 z_2, \ldots),$$

where κ_1 and κ_2 are not both zero. It is convenient to denote the above point by $\kappa_1 P_1 + \kappa_2 P_2$ and to write

$$P \equiv \kappa_1 P_1 + \kappa_2 P_2.$$

A symbolic identity of this kind, which is merely a shortened way of writing the equations $x = \kappa_1 x_1 + \kappa_2 x_2$, $y = \kappa_1 y_1 + \kappa_2 y_2$, etc., is known as a *syzygy*. More generally we can say that if the three distinct points P_1, P_2, P_3 are collinear they are connected by a syzygy of the form

$$\kappa_1 P_1 + \kappa_2 P_2 + \kappa_3 P_3 \equiv 0,*$$

where not all of $\kappa_1, \kappa_2, \kappa_3$ vanish. If four distinct points P_1, P_2, P_3, P_4 are collinear then they are connected by two distinct syzygies, e.g.

$$\kappa_1 P_1 + \kappa_2 P_2 + \kappa_3 P_3 \equiv 0, \quad \lambda_1 P_1 + \lambda_2 P_2 + \lambda_4 P_4 \equiv 0,$$

from which any other syzygy connecting the points can be obtained by ordinary linear operations.

Similarly, if P_1, P_2, P_3 are three non-collinear points, the plane $P_1P_2P_3$ is the aggregate of points

$$P \equiv \kappa_1 P_1 + \kappa_2 P_2 + \kappa_3 P_3,$$

and four coplanar points P_1, P_2, P_3, P_4 are connected by a syzygy of the form

$$\kappa_1 P_1 + \kappa_2 P_2 + \kappa_3 P_3 + \kappa_4 P_4 \equiv 0.$$

The extension of the notation is obvious. Thus, if X, Y, Z, T are reference points in [3], the point (x, y, z, t) is the point whose symbolic expression is $xX + yY + zZ + tT$. As in 1·76 the co-ordinates can be so chosen that any one particular point not lying on any face of the tetrahedron of reference can be taken as unit point, i.e. as the point $X + Y + Z + T$.

1·92. Dually, if π_i denotes the prime whose homogeneous prime co-ordinates are $[X_i, Y_i, Z_i, \ldots]$, i.e., whose equation in point co-ordinates is

$$X_i x + Y_i y + Z_i z + \ldots = 0,$$

the pencil or L_2 of primes determined by π_1 and π_2 is the aggregate of primes

$$\pi \equiv \kappa_1 \pi_1 + \kappa_2 \pi_2.$$

* We do not, of course, distinguish between this and the syzygy

$$\rho\kappa_1 P_1 + \rho\kappa_2 P_2 + \rho\kappa_3 P_3 \equiv 0 \quad (\rho \neq 0).$$

Three primes π_1, π_2, π_3 of a pencil are connected by a syzygy of the form

$$\kappa_1\pi_1+\kappa_2\pi_2+\kappa_3\pi_3 \equiv 0,$$

and so on.

If x, y, z, t are reference primes in [3], the prime $[X, Y, Z, T]$ has the symbolic expression $Xx+Yy+Zz+Tt$. The duality of points and planes in [3] is implicit in the two different interpretations that can be given to this symbolic expression. A similar statement applies to the duality of points and primes in any linear space.

1·93. The symbolic notation is a convenient one for establishing simple intersection relations between linear spaces. Take, for example, the incidence theorem in [3] that a line p meets a plane π, which does not contain p, in ONE point. If A, B are two distinct points of p, and C, D, E are three distinct non-collinear points of π, these five points, since they lie in [3], are connected by a syzygy of the form

$$aA+bB+cC+dD+eE \equiv 0. \qquad (\cdot 931)$$

They cannot also be connected by another syzygy

$$a'A+b'B+c'C+d'D+e'E \equiv 0,$$

because from these two syzygies it would be possible to deduce two others connecting three of the points with each of the remaining two, and this would imply that the five points are coplanar. We can write (·931) in the form

$$(aA+bB) \equiv -(cC+dD+eE) \equiv P,$$

and the two expressions for P show at once that it is a point both of p and π; further, P is unique, for the existence of another point of intersection would imply another syzygy connecting A, B, C, D, E.

This is a trivial example. A rather less obvious one is the theorem that *there is* ONE *line which meets three given general lines* a, b, c *in* [4]. Let $A_1, A_2, B_1, B_2, C_1, C_2$ be three pairs of distinct points on the given lines. They are general points of [4] and are therefore connected by ONE syzygy

$$a_1A_1+a_2A_2+b_1B_1+b_2B_2+c_1C_1+c_2C_2 \equiv 0.$$

This at once expresses the fact that the points

$$A \equiv a_1A_1+a_2A_2, \quad B \equiv b_1B_1+b_2B_2, \quad C \equiv c_1C_1+c_2C_2$$

of a, b, c are connected by the syzygy

$$A+B+C \equiv 0,$$

and are therefore collinear. If there were another transversal line of a, b, c, meeting them, say, in the points

$$A' \equiv a_1' A_1 + a_2' A_2, \quad B' \equiv b_1' B_1 + b_2' B_2, \quad C' \equiv c_1' C_1 + c_2' C_2,$$

then A', B', C', being collinear, are connected by a syzygy

$$\lambda A' + \mu B' + \nu C' \equiv 0,$$

or $$\lambda a_1' A_1 + \lambda a_2' A_2 + \mu b_1' B_1 + \mu b_2' B_2 + \nu c_1' C_1 + \nu c_2' C_2 \equiv 0,$$

contrary to the fact that there is ONE syzygy connecting the six points $A_1, A_2, \ldots, C_2$.

CHAPTER II

ALGEBRAIC SYSTEMS AND CORRESPONDENCES

2·1. Polynomial equations

2·11. *Roots of a polynomial equation in one variable*

A little of the theory of non-linear algebraic equations is needed. Consider first a polynomial $\phi(\lambda)$ in one variable λ given by

$$\phi(\lambda) \equiv a_n \lambda^n + a_{n-1}\lambda^{n-1} + \ldots + a_1\lambda + a_0.$$

Suppose that $a_n \neq 0$, so that $\phi(\lambda)$ is of degree n. We shall assume that $\phi(\lambda)$ can be expressed in the form

$$\phi(\lambda) \equiv a_n(\lambda - b_1)(\lambda - b_2) \ldots (\lambda - b_n).$$

The only solutions of the equation $\phi(\lambda) = 0$ are

$$\lambda = b_1, \lambda = b_2, \ldots, \lambda = b_n,$$

and the numbers $b_1, b_2, \ldots, b_n$ are called the *roots* of $\phi(\lambda) = 0$.

If $\lambda = b$ is a solution of $\phi(\lambda) = 0$, $\lambda - b$ is a factor of $\phi(\lambda)$ and $\phi(\lambda)$ may be expressed in the form

$$\phi(\lambda) \equiv (\lambda - b)^r \psi(\lambda),$$

where $r \geqslant 1$ and $\psi(\lambda)$ is a polynomial of degree $n - r$ of which $\lambda - b$ is not a factor. If $r = 1$, the number b is said to be a *simple* root of $\phi(\lambda) = 0$. If $r > 1$, b is said to be a *multiple root of order* r, or a *root of multiplicity* r, or an r*-ple root* of $\phi(\lambda) = 0$, and the remaining roots are the roots of $\psi(\lambda) = 0$. Thus, with the convention that an r-ple root is to be counted r times, *a polynomial equation* $\phi(\lambda) = 0$ *of degree* n *has* n *roots.*

The number b is an r-ple root of $\phi(\lambda) = 0$ IF the equations

$$\phi(\lambda) = \frac{d}{d\lambda}\phi(\lambda) = \ldots = \frac{d^{r-1}}{d\lambda^{r-1}}\phi(\lambda) = 0$$

are satisfied by $\lambda = b$ and $\frac{d^r}{d\lambda^r}\phi(\lambda) \neq 0$ for $\lambda = b$.

2·12. *Algebraic processes*

When a set of n numbers $b_1, b_2, \ldots, b_n$ is given, ONE ratio set $(a_n, a_{n-1}, \ldots, a_0)$ is algebraically determined such that $b_1, b_2, \ldots, b_n$ are the roots of $\phi(\lambda) = 0$. Conversely, when a ratio set

$$(a_n, a_{n-1}, \ldots, a_0)$$

is given, the roots of $\phi(\lambda) = 0$ are algebraically determined. It is not usually possible to distinguish between the roots of $\phi(\lambda) = 0$ algebraically, but if it is known that the roots of another polynomial equation $\chi(\lambda) = 0$ are roots of $\phi(\lambda) = 0$, then $\phi(\lambda)$ is of the form $\chi(\lambda)\psi(\lambda)$ and the roots of $\psi(\lambda) = 0$ are algebraically determined.

2·13. *Systems of polynomial equations in one variable*

Suppose that the coefficients $a_n, a_{n-1}, \ldots, a_0$ of $\phi(\lambda)$ are functions of other variables $\xi_1, \xi_2, \ldots, \xi_s$. By giving different sets of values to $\xi_1, \xi_2, \ldots, \xi_s$ a system of polynomial equations $\phi(\lambda) = 0$ is obtained. It may happen that in a particular polynomial $\phi_0(\lambda)$ of the system all the coefficients vanish, so that the equation $\phi_0(\lambda) = 0$ is satisfied by all values of λ. The equation $\phi_0(\lambda) = 0$ is then said to be an identity, and this is expressed by writing $\phi_0(\lambda) \equiv 0$. If it is known that m distinct numbers $b_1, b_2, \ldots, b_m$ are roots of multiplicity at least $r_1, r_2, \ldots, r_m$ of an equation $\phi_0(\lambda)$ of the system, and if

$$r_1 + r_2 + \ldots + r_m > n,$$

then $\phi_0(\lambda) \equiv 0$. In particular, if an equation of the system is known to have more than n distinct roots, that equation is an identity.

Let $a_n = f(\xi_1, \xi_2, \ldots, \xi_s)$ and suppose that $f(\xi_1, \xi_2, \ldots, \xi_s)$ does not vanish for all values of $\xi_1, \xi_2, \ldots, \xi_s$. Then all the equations $\phi(\lambda) = 0$ of the system are of degree n except those given by values of $\xi_1, \xi_2, \ldots, \xi_n$ for which $f(\xi_1, \xi_2, \ldots, \xi_n) = 0$. But in a particular equation $\phi_1(\lambda) = 0$ of the system it may happen that

$$a_n = a_{n-1} = \ldots = a_{n-h+1} = 0 \quad \text{and} \quad a_{n-h} \neq 0,$$

so that the polynomial $\phi_1(\lambda)$ is only of degree $n-h$. With the convention that ∞ is an h-ple root of $\phi_1(\lambda) = 0$ it is possible to say that every equation $\phi(\lambda) = 0$ of the system, except identities, has n roots.

2·14. *Polynomial equations in a ratio variable*

This convention is justifiable because in G_a a single variable λ is only used to represent a ratio variable (x, y), and when the representation is determined by $\lambda = x/y$ the statement $\lambda = \infty$ means that (x, y) is $(1, 0)$. The polynomial equation $\phi(\lambda) = 0$ represents the homogeneous polynomial equation

$$\phi(x, y) \equiv a_n x^n + a_{n-1} x y^{n-1} + \ldots + a_1 x y^{n-1} + a_0 y^n = 0,$$

and provided the coefficients are not all zero $\phi(x, y)$ can be expressed in the form

$$\phi(x,y) \equiv (d_1 x - c_1 y)(d_2 x - c_2 y) \dots (d_n x - c_n y).$$

The roots of $\phi(x, y) = 0$ are the ratio numbers

$$(c_1, d_1), (c_2, d_2), \dots, (c_n, d_n).$$

Conversely if (c, d) is a root of $\phi(x, y) = 0$, then

$$\phi(x,y) \equiv (dx - cy)^r \psi(x,y),$$

where $r \geqslant 1$ and $dx - cy$ is not a factor of the homogeneous polynomial $\psi(x, y)$. If $r = 1$, (c, d) is said to be a simple root of $\phi(x, y) = 0$. If $r > 1$, (c, d) is said to be an r-ple root of $\phi(x, y) = 0$. Thus, with the convention that an r-ple root is to be counted r times, a homogeneous polynomial equation of order n has n roots. If

$$a_n = a_{n-1} = \dots = a_{n-h-1} = 0$$

and $a_{n-h} \neq 0$, then $(1, 0)$ is an h-ple root.

The ratio number (c, d) is an r-ple root of $\phi(x, y) = 0$ IF the r equations

$$\frac{\partial^{r-1}}{\partial x^{r-1}}\phi(x,y) = \frac{\partial^{r-1}}{\partial x^{r-2}\,\partial y}\phi(x,y) = \dots = \frac{\partial^{r-1}}{\partial y^{r-1}}\phi(x,y) = 0$$

are satisfied by (c, d) and not all the rth partial derivatives vanish. This is a more convenient form of the CONDITIONS for a multiple root than the corresponding ones for $\phi(\lambda) = 0$ given in 2·11. For example (c, d) is a triple root of the cubic equation

$$a_3 x^3 + 3a_2 x^2 y + 3a_1 x y^2 + a_0 y^3 = 0,$$

IF the equations

$$a_3 x + a_2 y = a_2 x + a_1 y = a_1 x + a_0 y = 0$$

are satisfied by (c, d).

2·15. *Algebraic functions*

If $\phi(y, x_1, x_2, \dots, x_s)$ is a polynomial in $y, x_1, x_2, \dots, x_s$ of degree n in y, the equation

$$\phi(y, x_1, x_2, \dots, x_s) = 0$$

determines y as a function of the variables $x_1, x_2, \dots, x_s$. A function obtained in this way is called an *algebraic function*. If $n > 1$ each set of values of $x_1, x_2, \dots, x_s$ determines n values of y, so y is a many-valued function. If $n = 1$, y is said to be a *rational function* of $x_1, x_2, \dots, x_s$ and can be expressed in the form

$$y = \frac{\psi(x_1, x_2, \dots, x_s)}{\chi(x_1, x_2, \dots, x_s)},$$

where ψ and χ are polynomials. In this book there is no need to study algebraic functions of more than two variables, but a polynomial equation in two variables, which determines each as an algebraic function of the other, is discussed in 2·5 and 2·6.

2·2. Freedom of algebraic systems

2·21. *Freedom in different types of geometry*

The notion of freedom is used to compare systems which are composed of infinitely many members, and it is essential to choose some standard system to which others may be compared. Consider first the various types of numbers, e.g. integers, rational numbers, real numbers, and complex numbers. There are infinitely many numbers of each type, but these infinities are not all comparable. It is in fact possible to establish a one to one correspondence between the integers and the rational numbers, but not between the rational numbers and the real numbers.

The abstract idea of real numbers is not mentioned in $R(1)$, but the co-ordinates which determine the points of G_3 are supposed to be real numbers. Thus in G_3 it is natural to take the infinity of real numbers as a standard and to say that the real numbers have freedom one. The points of a line, being determined by the values of one real variable, are also said to have freedom one. The points of a plane, being determined by the values of two real variables, are said to have freedom two, and similarly the points of space are said to have freedom three. This is also expressed by saying that a line contains ∞^1 points, a plane contains ∞^2 points, and space contains ∞^3 points. Since a complex number is determined by two real numbers, there are ∞^2 complex numbers, when the infinity of real numbers is taken as a standard.

In G_4, on the other hand, the points are determined by sets of complex numbers, so it is natural to take the infinity of complex numbers as a standard, and to say that the complex numbers have freedom one. When this is done there are again ∞^1 points on a line, ∞^2 points in a plane, and ∞^3 points in space. Thus the notion of freedom in geometry depends on the type of geometry that is being considered. The difference between the standards in G_3 and G_4 is an advantage when we wish to pass from one geometry to the other. For in G_4 the co-ordinates x, y of the points of a plane π are complex numbers. If $x = x'+ix''$, $y = y'+iy''$, where x', x'', y', y'' are real numbers, then the points of π for which $x'' = y'' = 0$ form a system exactly equivalent to the system of points (x', y') of a plane π' in G_3. Thus all the points of the plane π in G_4 form an ∞^2 system when the infinity of complex numbers is taken as a standard, while the points of the derived plane π' in G_3 also form an ∞^2 system when the infinity of real numbers is taken as a standard.

2·22. *Freedom in G_a*

In G_a the points are determined by ratio sets, so the infinities of ratio sets are taken as standards. It is agreed that the ratio sets of index $n+1$ have freedom n. This is expressed by saying that there are ∞^n ratio sets of index $n+1$, or that the ratio sets of index $n+1$ form an ∞^n system. Thus in G_a, as in G_3 and G_4, there are ∞^1 points on a line, ∞^2 points in a plane, and ∞^3 points in space. There is only one ratio set of index one, namely (1), and it is convenient to say that it has freedom zero. A single point is said to form a space of zero dimensions.

An algebraic system is said to have freedom n if by some algebraic process it is possible to establish an (h, k) correspondence between the elements of the system and the ratio sets of index $n+1$, i.e. a correspondence in which each ratio set of index $n+1$ determines h elements of the system and each element of the system determines k ratio sets of index $n+1$, h and k being positive integers. It can be proved that it is impossible to establish an algebraic (h, k) correspondence between the ratio sets of index $n+1$ and those of index $n'+1$, when $n' \neq n$, so the freedom of an algebraic system is unique. An algebraic system of freedom n, or an algebraic ∞^n system, is denoted by $A\infty^n$. An $A\infty^0$ consists of a finite number of elements.

2·23. *Rational systems*

When it is possible to establish an algebraic (1, 1) correspondence between the elements of an algebraic system and the ratio sets of index $n+1$, the system is said to be a *rational ∞^n system*, and is denoted by $R\infty^n$. An $R\infty^0$ consists of ONE element.

The ratio sets which correspond to the elements of an $R\infty^n$ may be regarded as representing the elements of the $R\infty^n$, so the elements of an $R\infty^n$ may be represented by the points of a space of n dimensions.

Several rational systems have already been mentioned. The points of a line, the lines of a pencil, and the planes of a pencil are rational ∞^1 systems, since their elements may be identified with the values of a ratio variable (κ_1, κ_2). Similarly the points of a plane, the lines of a plane, and the planes of a sheaf are rational ∞^2 systems, while the points of space and the planes of space are rational ∞^3 systems. An L_r is an $R\,\infty^{r-1}$.

2·24. *Rational ∞^1 systems*

The elements of an $R\,\infty^1$ are said to form a *range* and a set of elements of an $R\,\infty^1$ is called a *subrange*. By definition the elements of an $R\,\infty^1$ may be represented by the values of a ratio variable (x, y), which is called a *ratio parameter* of the $R\,\infty^1$; the $(1, 1)$ correspondence between the elements of an $R\,\infty^1$ and the values of a ratio parameter is called a *ratio parametric representation* of the $R\,\infty^1$. If (x, y) is a ratio parameter of an $R\,\infty^1$, the variable λ given by $\lambda = x/y$ may also be used to represent the elements of the $R\,\infty^1$ and is called a *parameter* of the $R\,\infty^1$; the $(1, 1)$ correspondence between the elements of the $R\,\infty^1$ and the values of a parameter is called a *parametric representation* of the $R\,\infty^1$.

If (x_1, y_1), (x_2, y_2) are two distinct values of a ratio parameter (x, y) of an $R\,\infty^1$, then all values of (x, y) are of the form

$$(\kappa_1 x_1 + \kappa_2 x_2, \kappa_1 y_1 + \kappa_2 y_2)$$

and the ratio variable (κ_1, κ_2) may be used as a ratio parameter of the $R\,\infty^1$ instead of (x, y). In fact the reasoning of 1·54, 1·55, 1·56 applies to ratio parameters, and shows that it is possible to choose a ratio parametric representation of an $R\,\infty^1$ in which any three assigned distinct elements correspond to the values $(0, 1)$, $(1, 0)$, $(1, 1)$ of the ratio parameter, or to choose a parametric representation in which the assigned elements correspond to the values 0, ∞, 1 of the parameter.

2·25. *Irrational systems*

Algebraic systems which are not rational are said to be *irrational* and are far more difficult to deal with. Actually any $A\,\infty^1$ has an integer associated with it which is unaltered by a birational transformation* and is called the *genus* of the $A\,\infty^1$. Rational ∞^1 systems are $A\,\infty^1$ for which the genus is zero; they form a small and comparatively uninteresting class among algebraic ∞^1 systems. The fascinating general theory of irrational ∞^1 systems is quite

* An algebraic transformation which gives a $(1, 1)$ correspondence is said to be birational.

beyond the scope of this book, and the general theory of irrational systems of freedom greater than one is at present the object of a concentrated attack by many geometers.

2·3. Conditions

2·31. *Conditions on ratio sets*

Consider all the ∞^n ratio sets $(x_1, x_2, \ldots, x_{n+1})$ of index $n+1$. A non-homogeneous polynomial equation in $x_1, x_2, \ldots, x_{n+1}$ has no geometrical significance. A homogeneous polynomial equation

$$\phi(x_1, x_2, \ldots, x_{n+1}) = 0,$$

is called a *condition* on the ratio sets, and in particular a homogeneous linear polynomial equation in $x_1, x_2, \ldots, x_{n+1}$ is called a *linear condition*. If $\phi_1 = 0, \phi_2 = 0, \ldots, \phi_r = 0$ are a set of conditions, it may happen that all the ratio sets $(x)_{n+1}$ which satisfy

$$\phi_1 = 0, \phi_2 = 0, \ldots, \phi_s = 0,$$

with $s < r$, also satisfy the remaining $r-s$ conditions; the r conditions are then said to be *dependent*, and the last $r-s$ conditions are said to be dependent on the first s conditions.

A linear condition

$$c_1 x_1 + c_2 x_2 + \ldots + c_{n+1} x_{n+1} = 0$$

is uniquely determined by the ratio set $(c_1, c_2, \ldots, c_{n+1})$, and a set of linear conditions are dependent IF the ratio sets of coefficients are dependent. The theory of linear conditions on $(x)_{n+1}$ is therefore the theory of linear equations in $(x)_{n+1}$, which has already been discussed. The sets $(x)_{n+1}$ which satisfy r independent linear conditions are the sets of an L_{n-r+1}, which is an $R\infty^{n-r}$.

2·32. *Conditions on the elements of an $R\infty^n$*

The elements of any $R\infty^n$ may be represented by the ratio sets $(x)_{n+1}$ and a condition on the elements of the $R\infty^n$ is a condition on the ratio sets $(x)_{n+1}$. Thus the elements of an $R\infty^n$ which satisfy r independent linear conditions form an $R\infty^{n-r}$. For example, the planes $[X, Y, Z, T]$ of space form an $R\infty^3$, and a linear condition

$$x_1 X + y_1 Y + z_1 Z + t_1 T = 0$$

is a CONDITION for the plane $[X, Y, Z, T]$ to contain the point P_1. Thus the planes which satisfy one linear condition are the ∞^2

planes of a sheaf. Also the planes which satisfy two independent linear conditions are the ∞^1 planes of a pencil, and ONE plane satisfies three independent linear conditions. As another example consider the homogeneous equation

$$\phi(x,y) \equiv a_n x^n + a_{n-1} x^{n-1} y + \ldots + a_0 y^n = 0.$$

The CONDITION for this equation to have a given ratio number (x_1, y_1) as a root is

$$a_n x_1^n + a_{n-1} x_1^{n-1} y_1 + \ldots + a_0 y_1^n = 0,$$

which is a linear condition on the ratio set $(a_n, a_{n-1}, \ldots, a_0)$. Also CONDITIONS for $\phi(x,y) = 0$ to have a root of multiplicity r are obtained by eliminating x, y from the equations

$$\frac{\partial^{r-1}}{\partial x^{r-1}} \phi(x,y) = \frac{\partial^{r-1}}{\partial x^{r-2}\, \partial y} \phi(x,y) = \ldots = \frac{\partial^{r-1}}{\partial y^{r-1}} \phi(x,y) = 0,$$

and are therefore algebraic conditions on the ratio set

$$(a_n, a_{n-1}, \ldots, a_0).*$$

It was shown in 1·78 that the lines of space may be identified with the ratio sets $\{l, m, n, l', m', n'\}$ which satisfy the condition

$$ll' + mm' + nn' = 0,$$

so the lines of space form an $A\,\infty^4$.

2·33. $A\,\infty^{n-1}$ *of an* $R\,\infty^n$

The ratio sets $(x)_{n+1}$ which satisfy a homogeneous polynomial equation

$$\phi(x_1, x_2, \ldots, x_{n+1}) = 0$$

of any degree m form an $A\,\infty^{n-1}$. For each solution $(y_1, y_2, \ldots, y_{n+1})$ determines the ratio set $(y_1, y_2, \ldots, y_n)$ of index n, and conversely, if a ratio set $(y_1, y_2, \ldots, y_n)$ is given, there are in general† m solutions $(x_1, x_2, \ldots, x_{n+1})$ for which the ratio set $(x_1, x_2, \ldots, x_n)$ is $(y_0, y_1, \ldots, y_n)$. Thus it is possible to establish an $(m, 1)$ correspondence between the solutions and the ratio sets of index n, which shows that the solutions form an $A\,\infty^{n-1}$.‡

If $m = 1$ the correspondence is $(1, 1)$ and the $A\,\infty^{n-1}$ is an $R\,\infty^{n-1}$. If $m > 1$, the $A\,\infty^{n-1}$ is not necessarily rational, but it is

* Note that the notation of partial differential coefficients is only used for convenience, and that the process of deducing a partial differential coefficient of a polynomial from the polynomial is an algebraic process.

† See 2·35, p. 50. ‡ See 2·22, p. 46.

actually known that if $m = 2$ the $A\,\infty^{n-1}$ is rational. If $m = 3$ it is known that the $A\,\infty^{n-1}$ is in general irrational if $n = 2$ and rational if $n = 3$.

2·34. *Difficulties*

In dealing with a set of non-linear conditions on a rational system and with conditions on irrational systems certain difficulties arise which are beyond the scope of this book. As a trivial example of one type of difficulty consider the $R\,\infty^3$ of points (x, y, z, t) of space, and the two conditions $xy = 0$ and $xz = 0$. The points which satisfy both these conditions form two distinct systems of different freedoms, namely the ∞^1 points of the line $y = z = 0$ and the ∞^2 points of the plane $x = 0$. Another type of difficulty will be met in 9·4, when poristic properties are discussed. It is not true that the elements of an $A\,\infty^n$ which satisfy r independent conditions always form an $A\,\infty^{n-r}$, though this is true 'in general'.

2·35. *'In general'*

The expression 'in general' which has been used in 2·33 and 2·34 has a precise meaning in geometry. The geometrical constructs which arise in G_a are determined by the sets of coefficients in certain algebraic equations. When we say that a statement about certain constructs is true 'in general', we do not mean that it is always true but that it is true unless the sets of coefficients involved satisfy certain conditions. We do not mean that the statement is true if the coefficients are suitably chosen, but that it is true if the coefficients are arbitrarily chosen.

For example, two lines $\{l_1, m_1, n_1, l_1', m_1', n_1'\}$ and $\{l_2, m_2, n_2, l_2', m_2', n_2'\}$ in space do not meet unless

$$l_1 l_2' + l_1' l_2 + m_1 m_2' + m_1' m_2 + n_1 n_2' + n_1' n_2 = 0,$$

as was shown in 1·78. We may therefore say that 'in general' two lines in space do not meet. Similarly a pencil of lines in space is determined by the point vertex of the pencil and the plane which contains the pencil, so two pencils of lines have a common line IF the vertex of each pencil lies in the plane of the other pencil. We may therefore say that in general two pencils of lines in space have no common line.

The word 'general' is also used by itself in a similar way. A general member of a system of constructs is one whose sets of coefficients are arbitrarily chosen and do not satisfy any condition. For instance a general line of a plane does not contain a particular given point of the plane, and a general line of space does not meet a particular given line of space.

2·4. Geometrical constructs in a plane

2·41. *Systems of points and lines*

In a plane there are ∞^2 points and ∞^2 lines, and the first constructs to be considered are those formed by algebraic systems of points and lines. An $A\,\infty^0$ of points or lines is a finite number of points or lines. An $A\,\infty^1$ of points is called a *curve* and consists of the points whose co-ordinates x, y, z satisfy a homogeneous polynomial equation $\phi(x,y,z) = 0$. The dual of a curve is an $A\,\infty^1$ of lines, which is called a *scroll* and consists of the lines whose coordinates X, Y, Z, satisfy a homogeneous polynomial equation $\Phi(X, Y, Z) = 0$. A curve is said to be *rational* IF its points form an $R\,\infty^1$, and a scroll is said to be *rational* IF its lines form an $R\,\infty^1$. Equations in point co-ordinates and line co-ordinates are called *point equations* and *line equations* respectively, so $\phi(x,y,z) = 0$ is the point equation of a curve and $\Phi(X, Y, Z) = 0$ is the line equation of a scroll.

An $A\,\infty^1$ of points in any space is called a curve, and curves which lie in a plane are called *plane curves* to distinguish them from curves which do not lie in a plane and are consequently called *twisted curves*. An $A\,\infty^1$ of lines in any space is called a scroll, but unfortunately the term 'plane scroll' cannot be used for a scroll of lines in a plane, for it is used to denote an $A\,\infty^1$ of planes in any space.

2·42. *Plane curves*

A curve γ whose point equation $\phi(x,y,z) = 0$ is of degree n, is said to be of *order* n. If $\phi(x,y,z)$ is the product of homogeneous polynomial factors $\phi_1(x,y,z)$, $\phi_2(x,y,z)$, ..., $\phi_r(x,y,z)$ the curve γ consists of the r curves whose point equations are $\phi_1(x,y,z) = 0$ $\phi_2(x,y,z) = 0$, ..., $\phi_r(x,y,z) = 0$, and is said to be degenerate or reducible. In general γ is irreducible. If $n = 1$, γ is a line. If $n = 2$ γ is said to be a *conic locus* or a *point conic*. If $n = 3, 4, 5, \ldots, \gamma$ is said to be cubic, quartic, quintic, ...

Let p be a line $[X, Y, Z]$, and let P_1, P_2 be two distinct points of p. The general point $\lambda P_1 + \mu P_2$ of p lies on γ IF

$$\phi(\lambda x_1 + \mu x_2, \lambda y_1 + \mu y_2, \lambda z_1 + \mu z_2) = 0. \qquad (\cdot 421)$$

This is a homogeneous polynomial equation of degree n in the ratio variable (λ, μ). It may happen that the equation is identically satisfied for all values of (λ, μ), when it can be shown that

$$Xx + Yy + Zz$$

is a factor of γ and p is a part of γ. If the equation is not identically satisfied, it has exactly n roots, with the convention that an r-ple root is to be counted r times. Each root (λ_i, μ_i) corresponds to a point Q_i of p which lies on γ, and if (λ_i, μ_i) is an r-ple root p is said to have r intersections with γ at Q_i. Thus any line which does not form a part of γ has exactly n intersections with γ. Also any line which has more than n intersections with γ is a part of γ.

2·43. *Scrolls in a plane*

Dually a scroll Γ whose line equation $\Phi(X, Y, Z) = 0$ is of degree N is said to be of *class* N. The scroll is said to be reducible or degenerate IF $\Phi(X, Y, Z)$ is the product of two or more homogeneous polynomial factors. If $N = 1$, Γ is a pencil of lines. If $N = 2$, Γ is said to be a *conic scroll* or a *line conic*. If $N = 3, 4, 5, \ldots$, Γ is said to be cubic, quartic, quintic, ...

Let P be a point (x, y, z), and let p_1, p_2 be two distinct lines through P. The general line $\lambda p_1 + \mu p_2$ through P belongs to Γ IF

$$\Phi(\lambda X_1 + \mu X_2, \lambda Y_1 + \mu Y_2, \lambda Z_1 + \mu Z_2) = 0.$$

This is a homogeneous polynomial of degree N in the ratio variable (λ, μ). If it is identically satisfied for all (λ, μ), $xX + yY + zZ$ is a factor of $\Phi(X, Y, Z)$ and the pencil of lines with vertex P is a part of Γ. If the equation is not identically satisfied, it has exactly N roots, and each root corresponds to ONE line through P which belongs to Γ. With the convention that a line through P which corresponds to an r-ple root is to be counted r times among the lines of Γ which pass through P, it follows that any pencil of lines which does not form a part of Γ has exactly N lines in common with Γ. Also a pencil of lines which contains more than N lines of Γ is a part of Γ.

2·44. *Double points and cusps of a plane curve**

The theory of plane curves of order greater than two, commonly known as 'higher plane curves', will not be discussed in this volume, but a few further remarks made at this stage without proof may give rise to a better appreciation of the theory of conics.

Consider an irreducible curve γ of order n with point equation $\phi(x, y, z) = 0$, which may be written in the form

$$az^n + (b_1x + b_2y)z^{n-1} + (c_1x^2 + 2c_2xy + c_3y^2)z^{n-2} + \ldots = 0. \quad (\cdot 441)$$

The point (0, 0, 1) lies on γ IF $a = 0$. We want to consider a point P of γ and we will suppose that the co-ordinate system is so chosen that P is the point (0, 0, 1), when γ will be given by ·441 with $a = 0$.

The lines $[\lambda, \mu, 0]$ through P form a pencil, with (λ, μ) as a ratio parameter. The line $[\lambda, \mu, 0]$ has point equation $\lambda x + \mu y = 0$ and parametric equations $x:y:z = \mu t: -\lambda t: 1$ in terms of a parameter t. The intersections of the line $[\lambda, \mu, 0]$ with γ are determined by the following equation in t:

$$(b_1\mu - b_2\lambda)t + (c_1\mu^2 - 2c_2\mu\lambda + c_3\lambda^2)t^2 + \ldots = 0. \quad (\cdot 442)$$

If $b_1\mu - b_2\lambda \neq 0$, the point P, given by $t = 0$, counts as one of the n intersections of the line $[\lambda, \mu, 0]$ with γ† and P is said to be a simple intersection. If $b_1\mu - b_2\lambda = 0$ and $c_1\mu^2 - 2c_2\mu\lambda + c_3\lambda^2 \neq 0$, P counts as two of the n intersections and the line $[\lambda, \mu, 0]$ is said to have two-point contact with γ at P, or to have two intersections with γ at P. Similarly the line $[\lambda, \mu, 0]$ is said to have m intersections with γ at P if $t = 0$ counts m times among the roots of (·442).

If b_1 and b_2 are not both zero, the point P is said to be a *simple point* of γ. In this case the general line through P has ONE intersection with γ at P, but ONE line through P, $b_1x + b_2y = 0$, has at least two intersections with γ at P and is called the *tangent line*, or more simply the *tangent* to γ at P. In general this tangent line will have exactly two intersections with γ at P, but, if

$$c_1b_2^2 - 2c_2b_2b_1 + c_3b_1^2 = 0,$$

the value (b_1, b_2) of (λ, μ) makes two of the coefficients of (·442) vanish. In this case three intersections of the tangent line with γ are at P. The tangent is then called an *inflexional tangent* and P is said to be an *inflexion* of the curve γ.

* Cf. R(1), pp. 141–147.

† See 2·42, p. 52.

If $b_1 = b_2 = 0$, all lines through P have at least two intersections with γ at P, and P is then said to be a *singular point* of γ. It can be shown that γ has a singular point at (x_1, y_1, z_1) IF the equations

$$\frac{\partial\phi(x,y,z)}{\partial x} = \frac{\partial\phi(x,y,z)}{\partial y} = \frac{\partial\phi(x,y,z)}{\partial z} = 0 \qquad (\cdot 443)$$

are satisfied by (x_1, y_1, z_1). In general the three equations ($\cdot$443) will have no common solution, so a general curve of order n has no singular point.

Suppose that $b_1 = b_2 = 0$ but that c_1, c_2, c_3 are not all zero. Suppose also that the equation

$$c_1\mu^2 - 2c_2\mu\lambda + c_3\lambda^2 = 0 \qquad (\cdot 444)$$

has distinct roots. The point P is then said to be a double point of γ. A general line $[\lambda, \mu, 0]$ through P has TWO intersections with γ at P but each of two particular lines, given by ($\cdot$444), has at least three intersections with γ at P. The equation of this pair of lines is

$$c_1x^2 + 2c_2xy + c_3y^2 = 0. \qquad (\cdot 445)$$

They are called the tangents at P. Note that a general line through P is not considered to be a tangent to γ although two of its intersections with γ coincide. One of the tangents may possibly have a fourth intersection with γ at P, in which case it is said to be an inflexional tangent. The situation may be described by saying that two separate parts or branches of γ cross each other at P, the two tangents being the tangents to the separate branches.

If the two roots of ($\cdot$444) coincide, there is only one tangent line to γ at P. This situation can arise in two ways. The curve γ may have two separate branches through P, which happen to have the same tangent line. Alternatively there may be only one branch of the curve through P, in which case P is said to be a *cusp*. If γ is a rational curve,* whose points are determined by the values of a parameter t, the distinction between the two cases is that in the former case P corresponds to two distinct values of t, while in the latter case P corresponds to only one value of t. An example of a curve with a cusp is the cubic curve

$$x^3 = y^2z,$$

which is determined by the parametric equations

$$x : y : z = t^2 : t^3 : 1$$

* See 2·23, p. 46.

and which has a cusp at (0, 0, 1) given by $t = 0$, the tangent being $y = 0$. An example of the other case is the quartic curve

$$x^3(x-y)-y^2z(x-z) = 0,$$

which is determined by the parametric equations

$$x:y:z = \lambda(1+\lambda)(1+\lambda^2):(1+\lambda^2)^2:\lambda(1+\lambda)^2.$$

This curve has a double point at (0, 0, 1), with only one tangent $y = 0$, but there are two separate branches of the curve at (0, 0, 1) given by the roots of $\lambda^2+1 = 0$.

We have been assuming that $a = b_1 = b_2 = 0$, but that c_1, c_2, c_3 are not all zero. Suppose now that $a = b_1 = b_2 = 0$ and $c_1 = c_2 = c_3 = 0$, but that the next set of coefficients in (·441) are not all zero, P is then said to be a *triple* point of γ. A general line through P has three intersections with γ at P, but there are three tangents at P, in general distinct, each of which has at least four intersections with γ at P. Similarly, if the highest power of z that occurs in (·441) is z^{n-r}, P is said to be an r-ple point, and there are in general r distinct tangents at P, each of which has at least $r+1$ intersections with γ at P, the general line through P having r intersections at P.

In the case of a conic locus the equation (·441) reduces to

$$az^2+(b_1x+b_2y)z+(c_1x^2+2c_2xy+c_3y^2) = 0.$$

The point (0, 0, 1) is a double point IF $a = b_1 = b_2 = 0$, but in this case the equation reduces to

$$c_1x^2+2c_2xy+c_3y_2 = 0,$$

which is a reducible locus consisting either of two distinct lines through (0, 0, 1) or of one line through (0, 0, 1) counted twice. In fact a conic locus has a double point P IF it consists of two lines through P. A conic locus has more than one double point IF it consists of a line counted twice, in which case every point of the line is a double point.

Similarly a curve of order n with an n-ple point at P reduces to n lines through P. In fact an irreducible curve of order n cannot have an n-ple point.

2·45. *Double lines and stationary lines of a scroll.*

Dually we consider an irreducible scroll Γ of class N given by a line equation $\Phi(X, Y, Z) = 0$, which may be written in the form

$$AZ^N+(B_1X+B_2Y)Z^{N-1}+(C_1X^2+2C_2XY+C_3Y^2)Z^{N-2}+\ldots = 0. \qquad (\cdot 451)$$

Consider a line p of Γ and suppose that the co-ordinate system is so chosen that p is $[0, 0, 1]$, when A will be zero. The points $(\lambda, \mu, 0)$ of p have line equations $\lambda X + \mu Y = 0$. Parametric equations determining the pencil of lines through the point $(\lambda, \mu, 0)$ of p are

$$X : Y : Z = \mu t : -\lambda t : 1,$$

where t is the parameter. Thus the lines of Γ which contain the point $(\lambda, \mu, 0)$ of p are determined by the following equation in t.

$$(B_1\mu - B_2\lambda)\,t + (C_1\mu^2 - 2C_2\mu\lambda + C_3\lambda^2)\,t^2 + \ldots = 0. \qquad (\cdot 452)$$

In general B_1 and B_2 are not both zero. In this case p is said to be a simple line of Γ. The line p counts as ONE of the lines of Γ that pass through a general point of p, but there is ONE point of p, namely $B_1X + B_2Y = 0$, such that p counts as two of the lines of Γ through the point. This point is called the *contact** of the line p.

When $B_1 = B_2 = 0$ but not all of C_1, C_2, C_3 are zero the line p is said to be a *double line* of Γ. The line p counts twice among the lines of Γ through a general point of p, but there are in general two distinct points of p such that p counts three times among the lines of Γ through each of them. These two points are called the contacts of p. There is also the dual case to that of a cusp, and in this case p is said to be a *stationary line* of Γ.

A conic scroll has a double line p IF it consists of two points on p. A conic scroll has more than one double line IF it consists of a point counted twice, when every line through the point is a double line.

2·46. *Tangent scroll and contact curve*

The tangents of an irreducible curve γ of order n form a scroll, which is called the *tangent scroll* of γ. If the point equation of γ is $\phi(x, y, z) = 0$, the general point $(\lambda Y + \mu Z, -\lambda X, -\mu Z)$ of a line $[X, Y, Z]$ for which $X \neq 0$ lies on γ IF

$$\phi(\lambda Y + \mu Z, -\lambda X, -\mu X) = 0.$$

In general this equation in (λ, μ) has n distinct roots† but the CONDITION for the equation to have a double root is one condition on $[X, Y, Z]$. In general the tangents of γ are the only lines that satisfy this condition. When γ has a singular point a general line through the singular point satisfies the condition, but is not a tangent. The class of the tangent scroll is called the class of γ, so

* Cf. $R(1)$, p. 65. † See 2·66, p. 65.

the class of an irreducible plane curve of order n is the number of its tangents that pass through a general point of the plane. In general this number is $n(n-1)$, but the class is reduced if the curve has double points or cusps.

Dually through a general point there pass N distinct lines of an irreducible scroll Γ of class N, but there is a curve formed by the contacts of Γ, which are points such that two of the N lines of Γ through them coincide. This curve is called the *contact curve* of Γ. The order of the contact curve is called the order of Γ. For a general scroll Γ the order is $N(N-1)$, but the order is reduced if Γ has double lines or stationary lines.

A general curve is the contact curve of its tangent scroll, so a general point of a curve γ may be called the contact, or point of contact, of the tangent to γ at the point. Dually a general scroll is the tangent scroll of its contact curve. In particular the tangent scroll of a general conic locus is a conic scroll, and dually the contact curve of a conic scroll is a conic locus. Thus the construct formed by the points and tangents of a general conic locus is also the construct formed by the lines and contacts of a conic scroll, and is therefore self-dual; this construct is called a *conic*. The tangent scroll of a general cubic curve is of order six, so the construct formed by the points and tangents of a general plane cubic curve is not self-dual. The self-duality of a conic is perhaps its most remarkable property.

2·47. *Freedom of plane curves*

The point equations of the conic loci in a plane are

$$ax^2+by^2+cz^2+2fyz+2gzx+2hxy = 0,$$

so the conic loci may be represented by the ratio sets (a, b, c, f, g, h), and consequently form a rational system of freedom five. Incidentally the conic loci of a plane may be represented by the points of a space of five dimensions. Similarly it is easily shown that a general homogeneous polynomial of degree n in x, y, z has

$$\tfrac{1}{2}(n+1)(n+2)$$

coefficients, so the curves of order n form a rational system of freedom $\frac{1}{2}n(n+3)$. Dually the scrolls of class N in a plane form a rational system of freedom $\frac{1}{2}N(N+3)$, and in particular the conic scrolls form an $R\infty^5$.

2·48. *Limiting arguments*

It is possible to regard the tangent to a plane curve γ at a point P as the limit of the line joining P to another point Q of γ as Q tends to P, and to regard the contact of a line p of a scroll Γ as the limit of the intersection of p with another line q of Γ as q tends to p in Γ.

If γ is a rational curve, its points may be represented by the values of a parameter λ and the lines through P may be represented by the values of a parameter μ, so this notion of limits is not difficult to justify. But the notion of one point of an irrational curve tending to another requires more consideration than can be given in this book.

As a matter of taste an algebraic geometer will prefer to use purely algebraic methods as far as possible, but it is legitimate to use limiting arguments and other non-algebraic methods to establish isolated algebraic theorems. It is essential, however, that the theorems used in establishing a geometrical construction shall be algebraic, so that the result of the construction may be known to be capable of algebraic expression. Suppose, for example, that some geometrical construction gives rise to a plane curve. Since we know that each step of the construction is capable of algebraic expression, the curve must be determined by a set of polynomial equations in the point co-ordinates x, y, z and certain other variables. The other variables may be eliminated, and we are able to say that the curve must be given by a single homogeneous polynomial equation $\phi(x, y, z) = 0$.

2·5. Algebraic correspondences

2·51. *Correspondence determined by a polynomial equation in* (x, y), (x', y')

Consider a polynomial $\phi(x, y;\ x', y')$, which is homogeneous of degree m in x, y and homogeneous of degree n in x', y', and may be written in the form

$$\phi(x, y;\ x', y') \equiv \sum_{i,j} a_{ij} x^i y^{m-i} x'^j y'^{n-j}, \qquad (\cdot 511)$$

where $i = 0, 1, 2, \ldots, m$ and $j = 0, 1, 2, \ldots, n$, and $m > 0$, $n > 0$. If the ratio variable (x', y') is given an assigned value (x'_0, y'_0), the polynomial equation $\phi(x, y;\ x'_0, y'_0)$ in (x, y) has m roots, with the convention of 2·14 about multiple roots, unless all the coefficients

$\sum_j a_{ij} x_0'^j y_0'^{n-j}$ vanish, which happens IF $(y_0'x' - x_0'y')$ is a factor of $\phi(x, y;\ x', y')$. Thus

The equation $\phi(x, y;\ x', y') = 0$ determines a correspondence between the ratio variables (x, y) and (x', y') in which a general value of (x', y') corresponds to a set of m values of (x, y), while a general value of (x, y) corresponds to a set of n values of (x', y'). This is called an (m, n) correspondence between (x, y) and (x', y') and will be denoted by $T(m, n)$.

A polynomial $\phi(x, y;\ x', y')$ will only be said to be a polynomial in the ratio variables (x, y) and (x', y') if it is homogeneous in x, y and also homogeneous in x' and y'.

2·52. *Reducible correspondences*

The polynomial $\phi(x, y;\ x', y')$ and the correspondence $T(m, n)$ are said to be *reducible* IF $\phi(x, y;\ x', y')$ is the product of two or more polynomial factors. Suppose that $T(m, n)$ is reducible and that

$$\phi(x, y;\ x', y') \equiv \phi_1(x, y;\ x', y')\,\phi_2(x, y;\ x', y') \ldots \phi_r(x, y;\ x', y'),$$

where $\phi_1, \phi_2, \ldots, \phi_r$ are irreducible polynomials in the ratio variables (x, y), (x', y'), of degree $m_1, m_2, \ldots, m_r$ in (x, y) and $n_1, n_2, \ldots, n_r$ in (x', y'). Then $m_1 + m_2 + \ldots + m_r = m$, and $n_1 + n_2 + \ldots + n_r = n$. Suppose that none of the numbers $m_1, m_2, \ldots, m_r$ and $n_1, n_2, \ldots, n_r$ is zero.

Each equation $\phi_i(x, y;\ x', y') = 0$ determines an irreducible correspondence $T_i(m_i, n_i)$ between (x, y) and (x', y'). The values of (x, y) which correspond to a particular value (x_0', y_0') of (x', y') in $T(m, n)$ are the solutions of the equations

$$\phi_1(x, y;\ x_0', y_0') = 0, \quad \phi_2(x, y;\ x_0', y_0') = 0, \quad \ldots, \quad \phi_r(x, y;\ x_0', y_0') = 0,$$

which give the values of (x, y) corresponding to (x_0', y_0') in $T_1, T_2, \ldots, T_r$. The values of (x', y') which correspond to (x_0, y_0) in $T(m, n)$ are similarly obtained as the sum of the sets of values of (x', y') which correspond to (x_0, y_0) in $T_1, T_2, \ldots, T_r$. The reducible correspondence $T(m, n)$ is therefore said to be the *sum* of these r correspondences·

2·53. *Improper correspondences*

The correspondence $T(m, n)$ given by $\phi(x, y;\ x', y') = 0$ is said to be *improper* IF $\phi(x, y;\ x', y')$ has as a factor a homogeneous polynomial in (x, y) or a homogeneous polynomial in (x', y'). Thus the correspondence of 2·52 is improper if any one of the numbers $m_1, m_2, \ldots, m_r, n_1, n_2, \ldots, n_r$ is zero. A reducible correspondence is in general proper, but an improper correspondence is always reducible, with the trivial exceptions of a $T(0,1)$ and a $T(1,0)$.

Consider an improper correspondence given by an equation of the form

$$\phi(x,y;\ x',y') \equiv f(x,y)\,\psi(x,y;\ x',y') = 0,$$

and suppose that $\psi(x,y;\ x',y')$ has no factor which is a polynomial in (x,y). Then every value of (x,y) other than the roots of $f(x,y) = 0$ corresponds to n values of (x',y'), but each root of $f(x,y) = 0$ corresponds to all values of (x',y'). Similarly, if a polynomial $g(x',y')$ is a factor of $\phi(x,y;\ x',y')$, each value of (x',y') which satisfies $g(x',y') = 0$ corresponds to all values of (x,y).

It was observed in 2·51 that a value (x_0', y_0') of (x',y') corresponds to m values of (x,y) unless $y_0'x' - x_0'y'$ is a factor of $\phi(x,y;\ x',y')$, when the correspondence is improper. Thus

In a proper (m,n) correspondence between (x,y) and (x',y') every value of (x',y') corresponds to m values of (x,y) and every value of (x,y) corresponds to n values of (x',y').

2·54. *Bilinear equations*

In particular a bilinear equation

$$B(x,y;\ x',y') \equiv axx' + bxy' + cx'y + dyy' = 0 \qquad (\cdot 541)$$

gives a $T(1,1)$, which is reducible IF $B(x,y;\ x',y')$ is of the form

$$B(x,y;\ x',y') \equiv (qx - py)(q'x' - p'y'), \qquad (\cdot 542)$$

which happens IF $\qquad bc = ad.$

But in this case the $T(1,1)$ is also improper; the value (p,q) of (x,y) corresponds to all values of (x',y') and the value (p',q') of (x',y') corresponds to all values of (x,y).

Thus the $T(1,1)$ given by (·541) is proper IF $bc \neq ad$, when the bilinear equation (·541) is also said to be proper. But if any one particular value of (x,y) or of (x',y') corresponds to two distinct values of the other ratio variable, then the $T(1,1)$ is improper and is given by an equation of the form (·542).

2·55. *Single variables*

The values of the ratio variables (x,y) and (x',y') may be represented by the values of the variables λ, λ' given by $\lambda = x/y$,

$\lambda' = x'/y'$. Provided neither y nor y' is a factor of $\phi(x, y;\, x', y')$ the $T(m, n)$ of 2·51 is equally well determined by the equation

$$\phi(\lambda, \lambda') \equiv \sum_{i,j} a_{ij} \lambda^i \lambda'^j = 0 \quad (i = 1, 2, \ldots, m;\, j = 1, 2, \ldots, n), \qquad (\cdot 551)$$

which is obtained from (·511) by substituting $\lambda = x/y$, $\lambda' = x'/y'$. A general value of λ' corresponds to m finite values of λ, but particular values of λ' may correspond to $\lambda = \infty$ in accordance with the convention explained in 2·13 and 2·14. The values of λ which correspond to $\lambda' = \infty$ are the roots of the equation

$$\sum_i a_{in} \lambda^i = 0 \quad (i = 1, 2, \ldots, m),$$

with the convention that, if this equation is of degree $m-h$, $\lambda = \infty$ is to be counted h times. Similarly the values of λ' which correspond to $\lambda = \infty$ are the roots of the equation

$$\sum_j a_{mj} \lambda'^j = 0 \quad (j = 1, 2, \ldots, n).$$

If y is a factor of $\phi(x, y;\, x', y')$, then $a_{mj} = 0$ for all j, so the degree of the polynomial $\phi(\lambda, \lambda')$ in λ is less than m, and the equation (·551) does not determine the $T(m, n)$. In fact the $T(m, n)$ is improper and the value $(1, 0)$ of (x, y) corresponds to all values of (x', y'); thus $\lambda = \infty$ corresponds to all values of λ', and this cannot be expressed by an equation of the form (·551). Similarly if y' is a factor of $\phi(x, y;\, x', y')$, the $T(m, n)$ is not determined by (·551). For example, the $T(1, 1)$ given by (·541) is equally well determined by

$$B(\lambda, \lambda') \equiv a\lambda\lambda' + b\lambda + c\lambda' + d = 0, \qquad (\cdot 552)$$

provided that neither $a = b = 0$ nor $a = c = 0$. Clearly the improper $T(1, 1)$ given by $cx'y + dyy' = 0$, $bxy' + dyy' = 0$, and $dyy' = 0$ are not determined by the equations

$$c\lambda' + d = 0, \quad b\lambda + d = 0, \quad \text{or} \quad d = 0.$$

These exceptional cases do not cause any serious difficulty because it is always possible to replace (x, y), (x', y') by other ratio variables (ξ, η), $\xi', \eta')$ related to (x, y), (x', y') by non-singular linear transformations,* and to choose the coefficients of the transformation so that in the $T(m, n)$ the values $(1, 0)$ of (ξ, η) and of (ξ', η') do not correspond to all values of the other ratio variable. Then on representing (ξ, η) and (ξ', η') by $\mu = \xi/\eta$ and $\mu' = \xi'/\eta'$, a polynomial equation in μ and μ' is obtained which determines the $T(m, n)$.

* See 1·55, p. 17.

2·56. *Algebraic (m,n) correspondences*

A great deal of the theory of rational ∞^1 systems depends on the following theorem.

Any algebraic correspondence between two ratio variables (x,y) and (x',y'), in which a general value of (x',y') corresponds to m distinct values of (x,y) and a general value of (x,y) corresponds to n distinct values of (x',y'), is a $T(m,n)$ determined by a polynomial equation $\phi(x,y;\, x',y') = 0$ of degree m in (x,y) and of degree n in (x',y').

Such a correspondence, being algebraic, must be determined by a set of polynomial equations from which any variables other than (x,y) and (x',y') may be eliminated, and is consequently determined by a single polynomial equation $\phi(x,y;\, x',y') = 0$. It seems obvious that this equation must be of degree m in (x,y) and of degree n in (x',y'), but theorems which seem obvious may be difficult to prove. A proof is given in 2·6.

It should be noticed that it is not assumed that *every* value of (x,y) corresponds to n values of (x',y'); indeed some particular values of (x,y) may correspond to all values of (x',y'); but it is assumed that a *general* value of (x,y) corresponds to n *distinct* values of (x',y').

2·57. If the ratio variables (x,y), (x',y') are represented by the variables $\lambda = x/y$ and $\mu = x'/y'$, and the exceptional cases mentioned in 2·55 are borne in mind, the theorem of 2·56 takes the following form.

Any algebraic correspondence between two variables λ and λ', in which a general value of λ' corresponds to m distinct finite values of λ and a general value of λ corresponds to n distinct finite values of λ', is a $T(m,n)$ determined by a polynomial equation $\phi(\lambda,\lambda') = 0$ of degree m in λ and n in λ'.

2·58. *Algebraic $(1,1)$ correspondences*

A particular case of 2·56 is the following theorem.

Any algebraic correspondence between two ratio variables (x,y) and (x',y'), in which a general value of (x',y') corresponds to ONE *value of (x,y) and a general value of (x,y) corresponds to* ONE *value of (x',y') is a $T(1,1)$ determined by a bilinear equation*

$$B(x,y;\, x',y') \equiv axx' + bxy' + cx'y + dyy' = 0.$$

The $T(1,1)$ is only improper if there is a value of (x,y) which corresponds to all values of (x',y'). Thus if it is known that *every* value of (x',y') corresponds to ONE value of (x,y) or that *every* value of (x,y) corresponds to ONE value of (x',y'), then the $T(1,1)$ is proper, and $bc \neq ad$.

Similarly, from 2·57,

Any algebraic correspondence between two variables λ and λ', in which a general value of λ corresponds to ONE *finite value of λ', and a general value of λ' corresponds to* ONE *finite value of λ, is determined by a bilinear equation*

$$B(\lambda,\lambda') \equiv a\lambda\lambda' + b\lambda + c\lambda' + d = 0.$$

2·6. Proof of correspondence theorem. Order of a curve

2·61. In the proof of 2·56 we are concerned with polynomial equations, rather than with the polynomials themselves, so two polynomials will be regarded as identical when one is obtained by multiplying the other by a constant. We shall assume the following theorem.

If $\phi(x,y;\ x',y')$ and $\psi(x,y;\ x',y')$ are two distinct polynomials in the ratio variables (x,y) and (x',y'), the equations $\phi(x,y;\ x',y') = 0$ and $\psi(x,y;\ x',y') = 0$ have only a finite number of common solutions unless the polynomials $\phi(x,y;\ x',y')$ and $\psi(x,y;\ x',y')$ have a common polynomial factor.

This is a known theorem of algebra, and though it seems obvious it is difficult to prove.† Stated in terms of the variables $\lambda = x/y$ and $\lambda' = x'/y'$ the theorem takes the following form.*

If $\phi(\lambda,\lambda')$ and $\psi(\lambda,\lambda')$ are two distinct polynomials in the variables λ and λ', the equations $\phi(\lambda,\lambda') = 0$ and $\psi(\lambda,\lambda') = 0$ have only a finite number of common solutions unless the polynomials $\phi(\lambda,\lambda')$ and $\psi(\lambda,\lambda')$ have a common polynomial factor.

2·62. If $\phi(x,y;\ x',y')$ is irreducible, then the polynomial $\frac{\partial}{\partial x}\phi(x,y;\ x',y')$ is of lower degree in (x,y) and can have no polynomial factor in common with $\phi(x,y;\ x',y')$, so the equations

$$\phi(x,y;\ x',y') = 0 \quad \text{and} \quad \frac{\partial}{\partial x}\phi(x,y;\ x',y') = 0$$

can only have a finite number of common solutions. Thus

* A proof of the theorem stated in this form will be found in M. Bôcher, *Introduction to Higher Algebra* (The Macmillan Company, New York, 1936, p. 210).

If $\phi(x, y;\ x', y')$ is irreducible, the equation $\phi(x, y;\ x', y') = 0$, regarded as an equation in (x, y), only has a multiple root for a finite number of values of (x', y').

2·63. If $\phi(x, y; x', y')$ is the product of any number of distinct irreducible polynomial factors

$$\phi_1(x, y;\ x', y'),\ \phi_2(x, y;\ x', y'), \ldots, \phi_r(x, y;\ x', y'),$$

then the equation

$$\phi(x, y;\ x', y') \equiv \phi_1(x, y;\ x', y')\,\phi_2(x, y;\ x', y') \ldots \phi_r(x, y;\ x', y') = 0,$$

regarded as an equation in (x, y), only has multiple roots for a finite number of values of (x', y'). For a multiple root can only occur as a multiple root of one of the equations $\phi_i(x, y;\ x', y') = 0$ or as a common root of two of these equations. Thus

The equation $\phi(x, y;\ x', y') = 0$, regarded as an equation in (x, y), has a multiple root for an infinite number of values of (x', y') only if $\phi(x, y;\ x', y')$ has a repeated polynomial factor, when the equation has a multiple root for every value of (x', y'). Thus, if $\phi(x, y;\ x', y')$ has no repeated factor, then for a general value of (x', y') the equation

$$\phi(x, y;\ x', y') = 0$$

in (x, y) has distinct roots.

2·64. The theorem of 2·56 can now be proved. The correspondence, being algebraic, is determined by a polynomial equation

$$\psi(x, y;\ x', y') = 0.$$

Suppose that this equation is of degree h in (x, y) and of degree k in (x', y'). If $\psi(x, y;\ x', y')$ has no repeated factors, it follows from 2·63 that a general value of (x', y') corresponds to h distinct values of (x, y), so $h = m$. Similarly $k = n$.

If $\psi(x, y;\ x', y')$ contains one or more repeated factors it may be expressed in the form

$$\psi(x, y;\ x', y') = \psi_1^{p_1}\,\psi_2^{p_2} \ldots \psi_r^{p_r},$$

where $\psi_1, \psi_2, \ldots, \psi_r$ denote r distinct irreducible polynomials $\psi_i(x, y;\ x', y')$ and $p_i \geqslant 1$, $(i = 1, 2, \ldots, r)$. The correspondence is then equally well determined by the equation

$$\phi(x, y;\ x', y') \equiv \psi_1 \psi_2 \ldots \psi_r = 0,$$

and since $\phi(x, y;\ x', y')$ has no repeated factors it follows as above that $\phi(x, y;\ x', y')$ is of degree m in (x, y) and of degree n in (x', y'). Thus the theorem is established.

2·65. *Note on the assumptions in* 2·56

In the theorem of 2·56 it is assumed that a *general* value of (x', y') corresponds to *m distinct* values of (x, y). The word 'distinct' is essential, but it is not assumed that *every* value of (x', y') corresponds to m values or even to a finite number of values of (x, y), so improper correspondences in which certain values of (x', y') correspond to all values of (x, y) are not excluded.

2·66. *Criterion for a curve of order n*

It is perhaps tempting to say that an algebraic plane curve which is met by a general line in n distinct points is a curve of order n. But the reducible cubic curve $x^2y = 0$ is met by a general line in only two distinct points, so the possibility of repeated factors in the point equation of a curve must be removed if the theorem is to be true.

A plane curve γ, whose point equation $\phi(x, y, z) = 0$ is of degree n, is said to be a *simple* curve of order n IF the polynomial

$$\phi(x, y, z) \equiv \sum a_{ij} x^i y^j z^{n-i-j} \quad (i, j = 0, 1, 2, \ldots, n),$$

has no repeated factor. In particular an irreducible curve is simple. Suppose then that γ is simple and that the triangle of reference is so chosen that the vertex X does not lie on γ and consequently $a_{n0} \neq 0$.

The points (x, y, z) for which $z \neq 0$ may be represented by the non-homogeneous co-ordinates λ, μ given by

$$x = \lambda z, \quad y = \mu z,*$$

and this substitution for x, y in $\phi(x, y, z) = 0$ gives

$$z^n \phi(\lambda, \mu) = 0,$$

where
$$\phi(\lambda, \mu) \equiv \sum a_{ij} \lambda^i \mu^j.$$

So the points of γ other than those on $z = 0$ are given by

$$\phi(\lambda, \mu) = 0.$$

Also $\phi(\lambda, \mu)$ is of degree n in λ, since $a_{n0} \neq 0$, and has no repeated factors, since any repeated factor of $\phi(\lambda, \mu)$ would correspond to a repeated factor of $\phi(x, y, z)$. It follows from 2·63 that for a general value μ_0 of μ the equation

$$\phi(\lambda, \mu_0) = 0$$

has n distinct roots in λ. But the roots of this equation give the intersections of the line $y = \mu_0 z$ with γ, so

* See 1·62, p. 19.

If a simple curve γ of order n does not contain a point A, then a general line through A meets γ in n distinct points.

Conversely:

If a simple curve γ, which does not contain A, meets a general line through A in n distinct points, then γ is of order n.

For γ, being algebraic and simple, is given by some homogeneous polynomial equation in x, y, z with no repeated factors, and if this equation is of degree n', the above argument shows that a general line through A meets γ in n' distinct points, so $n' = n$. It also follows that:

A simple curve of order n is met by a general line in n distinct points, and conversely a simple curve which is met by a general line in n distinct points is of order n.

2·67. *A false criterion*

If A is a general point of an irreducible curve γ of order n, a general line through A has $n-1$ distinct intersections with γ other than A, while ONE line through A, the tangent at A, has two intersections with γ at A and consequently has only $n-2$ other intersections with γ.* In particular, as will be shown later, if A is a general point of a general conic locus s, ONE line through A has no further intersection with s, while each other line through A has ONE intersection with s distinct from A. It is not true, however, that a curve which has this property is necessarily a conic locus. For consider the cubic curve $x^3 = y^2 z$, with A as $(0, 0, 1)$. The line $y = 0$ has no intersection with the curve other than A, but for all $\lambda \neq 0$ the line $y = \lambda x$ meets the curve in ONE point distinct from A, namely $(\lambda^2, \lambda^3, 1)$. This is the cubic curve with a cusp which was mentioned in 2·44.

2·7. Cross-ratio. Apolarity

2·71. *Definitions*

If (x_1, y_1), (x_2, y_2), (x_3, y_3), (x_4, y_4) are four values of a ratio variable (x, y), the ratio number

$$\{(x_1y_2 - x_2y_1)(x_3y_4 - x_4y_3),\ (x_1y_4 - x_4y_1)(x_3y_2 - x_2y_3)\} \quad (\cdot 711)$$

* See 2·44, 2·46, pp. 53, 56.

is called the *cross** of the ordered set (x_1, y_1), (x_2, y_2), (x_3, y_3), (x_4, y_4). Since $(0, 0)$ is not a ratio number, the cross does not exist IF

$$(x_1y_2 - x_2y_1)(x_3y_4 - x_4y_3) = (x_1y_4 - x_4y_1)(x_3y_2 - x_2y_3) = 0,$$

i.e. IF three of the ratio numbers (x_i, y_i) are identical.

Just as a single variable is used to represent a ratio variable, so the number

$$\frac{(x_1y_2 - x_2y_1)(x_3y_4 - x_4y_3)}{(x_1y_4 - x_4y_1)(x_3y_2 - x_2y_3)} \qquad (\cdot 712)$$

is used to represent the cross, and is called the *cross-ratio* of the ordered set of ratio numbers (x_i, y_i).

If the variable $\lambda = x/y$ is used to represent the ratio variable (x, y), and if $\lambda_1, \lambda_2, \lambda_3, \lambda_4$ are the values of λ corresponding to (x_i, y_i) $(i = 1, 2, 3, 4)$, then provided no one of $\lambda_1, \lambda_2, \lambda_3, \lambda_4$ is ∞ the cross is

$$\{(\lambda_1 - \lambda_2)(\lambda_3 - \lambda_4), (\lambda_1 - \lambda_4)(\lambda_3 - \lambda_2)\} \qquad (\cdot 713)$$

and the cross-ratio is

$$\frac{(\lambda_1 - \lambda_2)(\lambda_3 - \lambda_4)}{(\lambda_1 - \lambda_4)(\lambda_3 - \lambda_2)}. \qquad (\cdot 714)$$

The use of ∞ as a value of λ has already been explained. If $\lambda_i = \infty$, (x_i, y_i) is $(1, 0)$, and the values of the cross and cross-ratio must be obtained by returning to the definitions (·711) and (·712).

The cross-ratio (·712) is also said to be the cross-ratio of the ordered set $\lambda_1, \lambda_2, \lambda_3, \lambda_4$ of values of λ, and is denoted by the symbol $(\lambda_1, \lambda_2, \lambda_3, \lambda_4)$ or by $(\lambda_1\lambda_2\lambda_3\lambda_4)$.†

2·72. *Cross-ratio of four numbers*

Since any four numbers a, b, c, d may be regarded as values of a variable λ, the cross-ratio (a, b, c, d) is defined by (·714) to be

$$\frac{(a-b)(c-d)}{(a-d)(c-b)},$$

provided no three of a, b, c, d are equal. If a, b, c are distinct, it is legitimate to say that

$$(a, b, c, a) = \infty,$$

because the cross of $(a, 1)$, $(b, 1)$, $(c, 1)$, $(a, 1)$ is $(1, 0)$. Also

$$(\infty, a, b, c) = \frac{b-c}{b-a},$$

* In $R(1)$ this is called the homogeneous cross-ratio.

† Some writers use the symbol $(\lambda_1, \lambda_3; \lambda_2, \lambda_4)$.

because the cross of $(1,0), (a,1), (b,1), (c,1)$ is $(b-c, b-a)$. Similarly

$$(a,\infty,b,c) = \frac{b-c}{a-c},$$

$$(a,b,\infty,c) = \frac{a-b}{a-c},$$

and
$$(a,b,c,\infty) = \frac{a-b}{c-b}.$$

2·73. *Permutations of four numbers*

Different cross-ratios are obtained by arranging four given numbers a, b, c, d in the twenty-four possible ways. But, if two of the numbers are interchanged and the remaining two are also interchanged, the cross-ratio is unaltered. So each arrangement gives the same cross-ratio as three other arrangements, and there are at most six distinct cross-ratios. Also the cross-ratio given by any particular arrangement is equal to the cross-ratio given by some arrangement in which a is the first number.

If $(a,b,c,d) = \rho$ it is easily shown that

$$(a,c,b,d) = 1-\rho \quad \text{and} \quad (a,d,c,b) = 1/\rho,$$

whence it follows that

$$(a,d,b,c) = \frac{1}{1-\rho},$$

$$(a,c,d,b) = 1-\frac{1}{\rho} = \frac{\rho-1}{\rho},$$

and
$$(a,b,d,c) = \frac{\rho}{\rho-1}.$$

Thus the cross-ratios given by all the twenty-four arrangements have been obtained in terms of ρ.

2·74. *Three particular cases*

In general the six cross-ratios

$$\rho,\ 1-\rho,\ \frac{1}{\rho}, \frac{1}{1-\rho}, \frac{\rho-1}{\rho}, \frac{\rho}{\rho-1}$$

are distinct, and the only cases in which two or more of them are equal are the following.

(i)
$$(a,b,c,d) = (a,d,c,b) = 1,$$
$$(a,c,b,d) = (a,c,d,b) = 0,$$
$$(a,d,b,c) = (a,b,d,c) = \infty.$$

In this case $(a-c)(b-d) = 0$, so either $a = c$ or $b = d$.

(ii) $$(a,b,c,d) = (a,d,c,b) = -1,$$
$$(a,c,b,d) = (a,c,d,b) = 2,$$
$$(a,d,b,c) = (a,b,d,c) = \tfrac{1}{2}.$$

In this case the four numbers a, b, c, d are said to be *harmonic.*

(iii) $$(a,b,c,d) = (a,c,d,b) = (a,d,b,c) = -\omega,$$
$$(a,d,c,b) = (a,b,d,c) = (a,c,b,d) = -\omega^2,$$

where $\omega^3 = 1$ and $\omega \neq 1$. In this case the four numbers a, b, c, d are said to be *equianharmonic.*

Four ratio numbers (x_1, y_1), (x_2, y_2), (x_3, y_3), (x_4, y_4) are said to be harmonic or equianharmonic IF the numbers x_1/y_1, x_2/y_2, x_3/y_3, x_4/y_4 are harmonic or equianharmonic.

2·75. *Apolarity of pairs of ratio numbers*

Any two ratio numbers may be regarded as the roots of a quadratic equation, for (x_1, y_1), (x_2, y_2) are the roots of

$$y_1 y_2 x^2 - (y_1 x_2 + y_2 x_1)\,xy + x_1 x_2 y^2 = 0.$$

The two unordered pairs of values of the ratio variable which are given by the equations

$$ax^2 + 2bxy + cy^2 = 0, \quad a'x^2 + 2b'xy + c'y^2 = 0 \qquad (\cdot 751)$$

are said to be *apolar* IF

$$ac' + a'c = 2bb'. \qquad (\cdot 752)$$

Also the two pairs of values of the variable $\lambda = x/y$ which are given by the equations

$$a\lambda^2 + 2b\lambda + c = 0, \quad a'\lambda^2 + 2b'\lambda + c' = 0 \qquad (\cdot 753)$$

are said to be apolar IF ($\cdot$752) is satisfied. When two pairs of values of (x, y) or of λ are apolar, each pair is said to be *harmonically separated* by the other, or to be apolar to the other.

Bars will be used to denote unordered pairs, as was stated in 1·27. The pairs $|\,(x_1, y_1), (x_2, y_2)\,|$, $|\,(x_1', y_1'), (x_2', y_2')\,|$ are apolar IF

$$2(y_1 y_2 x_1' x_2' + y_1' y_2' x_1 x_2) = (y_1 x_2 + y_2 x_1)(y_1' x_2' + y_2' x_1'),$$

i.e. IF

$$(x_1 y_1' - x_1' y_1)(x_2 y_2' - x_2' y_2) + (x_1 y_2' - x_2' y_1)(x_2 y_1' - x_1' y_2) = 0. \quad (\cdot 754)$$

But this equation states that the cross-ratio of the ratio numbers (x_1, y_1), (x_1', y_1'), (x_2, y_2), (x_2', y_2') is -1 provided that it exists, i.e.

provided that no three of the ratio numbers are identical. We deduce a theorem which is more conveniently stated in terms of numbers than of ratio numbers, namely

If $\lambda_1, \lambda_2, \lambda_1', \lambda_2'$ *are four numbers such that* $(\lambda_1, \lambda_1', \lambda_2, \lambda_2') = -1$, *then the pairs* $|\lambda_1, \lambda_2|$, $|\lambda_1', \lambda_2'|$ *are apolar.*

It should be noticed that the pairs $|\mu, \mu|$, $|\mu, \mu'|$ are apolar, though the cross-ratio (μ, μ, μ, μ') does not exist. Also the equation (·752) is unaltered by interchanging the equations (·751) or (·753), so the order of the pairs, as well as the order of the ratio numbers or numbers in each pair, is immaterial. It follows that if $(a, b, c, d) = -1$, and if a', b', c', d' is some rearrangement of a, b, c, d in which the unordered pairs $|a', c'|$, $|b', d'|$ are either $|a, c|$, $|b, d|$ or $|b, d|$, $|a, c|$, then $(a', b', c', d') = -1$, which agrees with 2·74 (ii).

2·76. *Harmonic conjugates*

The ratio number (x_0', y_0') is said to be *conjugate to* (x_0, y_0) with respect to the pair $|(x_1, y_1), (x_2, y_2)|$ IF the pairs

$$|(x_0, y_0), (x_0', y_0')|, \ |(x_1, y_1), (x_2, y_2)|$$

are apolar. Thus (x_0, y_0) is conjugate to (x_0', y_0') IF (x_0', y_0') is conjugate to (x_0, y_0), and (x_0, y_0), (x_0', y_0') are then said to be *conjugate* with respect to or relative to the pair $|(x_1, y_1), (x_2, y_2)|$. If $(x_1, y_1), (x_2, y_2)$ are the roots of

$$ax^2 + 2bxy + cy^2 = 0, \tag{·761}$$

(x_0, y_0) and (x_0', y_0') are the roots of

$$y_0 y_0' x^2 - (y_0 x_0' + y_0' x_0) xy + x_0 x_0' y^2 = 0,$$

and are therefore conjugate relative to $|(x_1, y_1), (x_2, y_2)|$ IF

$$a x_0 x_0' + b(x_0 y_0' + x_0' y_0) + c y_0 y_0' = 0.$$

This equation is of course symmetrical in (x_0, y_0) and (x_0', y_0') and may be written

$$(a x_0 + b y_0) x_0' + (b x_0 + c y_0) y_0' = 0. \tag{·762}$$

When (x_0, y_0) is given, the equation determines ONE ratio number (x_0', y_0'), unless

$$a x_0 + b y_0 = b x_0 + c y_0 = 0,$$

which can only happen if the roots of (·761) coincide in (x_0, y_0). Thus

ONE *ratio number is conjugate to* (x_0, y_0) *relative to the roots of*

$$f(x, y) \equiv ax^2 + 2bxy + cy^2 = 0,$$

unless these roots coincide in (x_0, y_0). *This ratio number is called the harmonic conjugate of* (x_0, y_0) *relative to the roots of* $f(x, y) = 0$, *and is given by*

$$(ax_0 + by_0)\,x + (bx_0 + cy_0)\,y = 0,$$

or by

$$\left(x_0\frac{\partial}{\partial x} + y_0\frac{\partial}{\partial y}\right)f(x, y) = 0.$$

If the roots of $f(x, y) = 0$ *coincide in* (x_0, y_0), *then every ratio number is conjugate to* (x_0, y_0) *relative to the roots, and the harmonic conjugate of* (x_0, y_0) *is said to be indeterminate.*

If (x, y) is represented by $\lambda = x/y$, the harmonic conjugate of λ_0 relative to the roots of

$$f(\lambda) \equiv a\lambda^2 + 2b\lambda + c = 0$$

is given by

$$(a\lambda_0 + b)\,\lambda + (b\lambda_0 + c) = 0,$$

and is indeterminate if the roots of $f(\lambda) = 0$ coincide in λ_0.

2·8. Algebraic (1, 1) correspondences

2·81. *Forward and reverse correspondence*

Consider a $T(1, 1)$ between (x, y) and (x', y') given by

$$B(x, y;\, x', y') \equiv axx' + bxy' + cx'y + dyy' = 0, \qquad (\cdot 811)$$

and suppose for the moment that it is proper, i.e. that $bc \neq ad$.* The $T(1, 1)$ determines two operations.

(i) Each value (x_1, y_1) of (x, y) determines ONE value (x_1', y_1') of (x', y'). This operation we agree to call the *forward correspondence*, and we denote it by T. The operation is expressed symbolically by writing

$$(x_1, y_1) \rightarrow (x_1', y_1') \quad \text{or} \quad (x_1', y_1') = T(x_1, y_1).$$

(ii) Each value (x_1', y_1') of (x', y') determines ONE value (x_1, y_1) of (x, y). This operation is called the *reverse correspondence*, and is denoted by T^{-1}. It is expressed symbolically by writing

$$(x_1', y_1') \leftarrow (x_1, y_1) \quad \text{or} \quad (x_1, y_1) = T^{-1}(x_1', y_1').$$

Thus

$$T^{-1}\,T(x_1, y_1) = T^{-1}(x_1', y_1') = (x_1, y_1)$$

and

$$T\,T^{-1}(x_1', y_1') = T(x_1, y_1) = (x_1', y_1').$$

2·82. *Invariance of cross-ratio and apolarity*

Let $(x_i, y_i) \rightarrow (x_i', y_i')$ $(i = 1, 2, 3, 4)$, in a proper $T(1, 1)$ given by (·811). Let λ_i, λ_i' be the values of $\lambda = x/y$, $\lambda' = x'/y'$ corresponding to (x_i, y_i) and (x_i', y_i'). The $T(1, 1)$ is also given by the transformation

$$x = cx' + dy', \quad y = -ax' - by', \qquad (\cdot 821)$$

* See 2·54, p. 60.

which is non-singular since $bc \neq ad$. On substituting for x_i, y_i, x_j, y_j from (·821) it follows that

$$(x_i y_j - x_j y_i) = (ad - bc)(x'_i y'_j - x'_j y'_i).$$

It then follows from the definitions of 2·71, since $ad - bc \neq 0$, that if the cross-ratio $(\lambda_1, \lambda_2, \lambda_3, \lambda_4)$ exists (i.e. if no three of $\lambda_1, \lambda_2, \lambda_3, \lambda_4$ are equal), then $(\lambda'_1, \lambda'_2, \lambda'_3, \lambda'_4)$ also exists and

$$(\lambda_1, \lambda_2, \lambda_3, \lambda_4) = (\lambda'_1, \lambda'_2, \lambda'_3, \lambda_4).$$

Thus:

If $\lambda_i \to \lambda'_i$ in a proper $T(1, 1)$ between λ and λ', and if $(\lambda_1, \lambda_2, \lambda_3, \lambda_4)$ exists, then $(\lambda'_1, \lambda'_2, \lambda'_3, \lambda'_4)$ exists and

$$(\lambda_1, \lambda_2, \lambda_3, \lambda_4) = (\lambda'_1, \lambda'_2, \lambda'_3, \lambda'_4).$$

The reader will have no difficulty in stating the equivalent theorem for ratio variables.

It is easily verified also that, if the unordered pairs of ratio numbers $|(x_1, y_1), (x_2, y_2)|$, $|(x_3, y_3), (x_4, y_4)|$ are apolar, then the corresponding unordered pairs $|(x'_1, y'_1), (x'_2, y'_2)|$, $|(x'_3, y'_3), (x'_4, y'_4)|$ are also apolar.

2·83. *Improper* $T(1, 1)$

The $T(1, 1)$ given by (·811) is improper IF $bc = ad$, when (·811) may be written in the form

$$B(x, y;\ x', y') \equiv (qx - py)(q'x' - p'y') = 0. \qquad (\cdot 831)$$

Thus in an improper $T(1, 1)$ there is ONE value (p, q) of (x, y), called the *singular value*, which corresponds to all values of (x', y'); there is also ONE singular value (p', q') of (x', y') which corresponds to all values of (x, y). Any value of (x, y) other than the singular value corresponds to ONE value of (x', y'), namely the singular value of (x', y'); any value of (x', y') other than the singular value corresponds to ONE value of (x, y), namely the singular value of (x, y). It follows that:

If in a $T(1, 1)$ between (x, y) and (x', y') three distinct values of (x, y) correspond to three distinct values of (x', y'), then the $T(1, 1)$ is proper.

The theorem of 2·82 is not true for an improper $T(1, 1)$, for the cross-ratios $(\lambda_1, \lambda_2, \lambda_3, \lambda_4)$ and $(\lambda'_1, \lambda'_2, \lambda'_3, \lambda'_4)$ cannot both exist. Also an improper $T(1, 1)$ is not determined by the equations (·821), for the value $(d, -c)$ of (x', y'), which is identical with $(b, -a)$ since $bc = ad$, would give $x = y = 0$.*

2·84. *Determination of a $T(1,1)$*

The $T(1,1)$ given by (·811) is determined by the ratio set (a,b,c,d). If (x_i, y_i), (x_i', y_i'), $(i = 1,2,3)$, are given values of (x,y), (x',y'), then $(x_i, y_i) \rightarrow (x_i', y_i')$ in the $T(1,1)$ IF the ratio set (a,b,c,d) satisfies the three equations

$$ax_i x_i' + bx_i y_i' + cx_i' y_i + dy_i y_i' = 0. \qquad (\cdot 841)$$

These equations certainly have a solution, and if they are independent they have ONE solution.

Suppose first that (x_1, y_1), (x_2, y_2), (x_3, y_3) are distinct and that (x_1', y_1'), (x_2', y_2'), (x_3', y_3') are distinct, and let $T_0(1,1)$ be a $T(1,1)$ given by a solution of (·841). Then by 2·83 $T_0(1,1)$ is proper, so by 2·82 any corresponding values of (x,y) and (x',y') satisfy the equation

$$(\lambda_1, \lambda_2, \lambda_3, \lambda) = (\lambda_1', \lambda_2', \lambda_3', \lambda'). \qquad (\cdot 842)$$

But this equation is equivalent to

$$(x_1 y_2 - x_2 y_1)(x_3 y - x y_3)(x_1' y' - x' y_1')(x_3' y_2' - x_2' y_3')$$
$$= (x_1' y_2' - x_2' y_1')(x_3' y' - x' y_3')(x, y - x y_1)(x_3 y_2 - x_2 y_3),$$

which is bilinear in (x,y) and (x',y') and is therefore the bilinear equation which determines $T_0(1,1)$. Thus:

There is ONE *$T(1,1)$ between (x,y) and (x',y') in which three distinct given values (x_1, y_1), (x_2, y_2), (x_3, y_3) of (x,y) correspond to three distinct given values (x_1', y_1'), (x_2', y_2'), (x_3', y_3') of (x',y'). This $T(1,1)$ is proper, and is given by the equation*

$$(\lambda_1, \lambda_2, \lambda_3, \lambda) = (\lambda_1', \lambda_2', \lambda_3', \lambda'),$$

where $\lambda = x/y$, $\lambda' = x'/y'$.

2·85. *Particular cases*

The cases in which the given values of (x,y) and (x',y') are not distinct are of three types. For convenience of writing the variable λ, λ' will be used to represent (x,y), (x',y').

(i) $$\lambda_1 = \lambda_2, \quad \lambda_1' = \lambda_2'.$$

* The equations (·831) could be replaced by

$$x : y = (cx' + dy') : -(ax' + by'),$$

and $x : y = 0 : 0$ might be interpreted as meaning that the ratio number (x, y) is indeterminate. If it were also agreed that a non-existent cross-ratio is to be regarded as an indeterminate number, which might be regarded as 'equal' to any number, then it might be said that the theorem of 2·82 applies also to an improper $T(1, 1)$. But this would be a misuse of the notion of equality.

This case is trivial, as two of the equations (·841) are identical. In fact only two pairs of corresponding values of λ and λ' are given.

(ii) $$\lambda_1 = \lambda_2, \quad \lambda_1' \neq \lambda_2', \quad \lambda_3 \neq \lambda_1.$$

Any $T(1,1)$ satisfying the conditions must be improper since λ_1 corresponds to the distinct values λ_1', λ_2' of λ'. Also λ_1 is the singular value of λ. Since $\lambda_3 \neq \lambda_1$, λ_3' is the singular value of λ'. Thus there is ONE $T(1,1)$, namely the improper $T(1,1)$ given by

$$(\lambda - \lambda_1)(\lambda' - \lambda_3') = 0 \quad \text{or} \quad (y_1 x - x_1 y)(y_3' x' - x_3' y') = 0.$$

(iii) $$\lambda_1 = \lambda_2 = \lambda_3 \quad \text{and} \quad \lambda_1', \lambda_2', \lambda_3'$$

are not all equal.

Any $T(1,1)$ satisfying the conditions is improper, and λ_1 is the singular value of λ. There are ∞^1 $T(1,1)$, namely those given by

$$(y_1 x - x_1 y)(y_0' x' - x_0' y') = 0,$$

where (x_0', y_0') is any value of (x', y').

2·86. An immediate consequence of the theorem of 2·84 is the following:

The CONDITION *that there should exist a $T(1,1)$ between (x,y) and (x',y') in which four distinct given values (x_1,y_1), (x_2,y_2), (x_3,y_3), (x_4,y_4) of (x,y) correspond to four distinct given values (x_1',y_1'), (x_2',y_2'), (x_3',y_3'), (x_4',y_4') of (x',y') is*

$$(\lambda_1, \lambda_2, \lambda_3, \lambda_4) = (\lambda_1', \lambda_2', \lambda_3', \lambda_4'),$$

where $$\lambda_i = x_i/y_i, \quad \lambda_i' = x_i'/y_i'.$$

CHAPTER III

RATIONAL SYSTEMS OF FREEDOM ONE

3·1. Representation and cross-ratio

3·11. *Notation*

A rational ∞^1 system was defined in 2·23, and the notion of ratio parametric and parametric representations of an $R\infty^1$ was explained in 2·24. In this chapter E will denote an $R\infty^1$, e will denote a general element of E, and $e_1, e_2, \ldots$ will denote particular elements of E. All the elements e of E are said to form a range, which will be denoted by $\{e\}$. An ordered set of elements $e_1, e_2, \ldots, e_r$ is said to form a subrange, which will be denoted by $\{e_1, e_2, \ldots, e_r\}$. It was remarked in 2·23 that the points of a line and the lines of a pencil form rational ∞^1 systems, and a reader to whom the idea of an $R\infty^1$ is new may find the following theory more easy to understand if he regards E as a line p, e as a general point P of p, and $e_1, e_2, \ldots$ as particular points $P_1, P_2, \ldots$ of p.

3·12. *Different representations*

By the definition of an $R\infty^1$ the elements e of E may be represented by the values of a ratio parameter (x, y), or by the values of the parameter $\lambda = x/y$, and it was shown in 2·24 that this representation is not unique. But if (x, y) and (x', y') are any two ratio parameters of E, *every* value of (x, y) determines ONE element of E, which determines ONE value of (x', y'), and conversely *every* value of (x', y') determines ONE value of (x, y). Thus, by 2·58

The values of any two ratio parameters (x, y), (x', y') of an $R\infty^1$ which correspond to the same element of the $R\infty^1$ are connected by a proper bilinear equation

$$B(x, y;\ x', y') \equiv axx' + bxy' + cx'y + dyy' = 0, \qquad (\cdot 121)$$

in which $bc \neq ad$.

The values of any two parameters λ, λ' of an $R\infty^1$ which correspond to the same element of the $R\infty^1$ are connected by a proper bilinear equation

$$B(\lambda, \lambda') \equiv a\lambda\lambda' + b\lambda + c\lambda' + d = 0, \qquad (\cdot 122)$$

in which $bc \neq ad$.

Since $bc \neq ad$, the equations $B(x, y; x', y') = 0$ and $B(\lambda, \lambda') = 0$ are equivalent to

$$x = cx' + dy', \quad y = -ax' - by' \qquad (\cdot 123)$$

and

$$\lambda = -\frac{c\lambda' + d}{a\lambda' + b}, \qquad (\cdot 124)$$

so the equations 1·551 and 1·561 give all transformations of coordinates in space of one dimension.

3·13. *Cross-ratio on an* $R\infty^1$

Let (x, y), (x', y') be two ratio parameters of E, and let e_1, e_2, e_3, e_4 be any four elements of E no three of which are identical. Then since the $T(1, 1)$ between (x, y) and (x', y') is proper, it follows from 2·82 that the cross-ratio of the four values of (x, y) which correspond to e_1, e_2, e_3, e_4 is equal to the cross-ratio of the four values of (x', y') which correspond to e_1, e_2, e_3, e_4. In fact, if $\lambda_1, \lambda_2, \lambda_3, \lambda_4$ and $\lambda_1', \lambda_2', \lambda_3', \lambda_4'$ are the values of the parameters $\lambda = x/y$ and $\lambda' = x'/y'$ which correspond to e_1, e_2, e_3, e_4, then

$$(\lambda_1, \lambda_2, \lambda_3, \lambda_4) = (\lambda_1', \lambda_2', \lambda_3', \lambda_4').$$

It follows that, if in any particular ratio parametric or parametric representation of an $R\infty^1$ the cross-ratio of the four values of the ratio parameter or parameter which correspond to four elements e_1, e_2, e_3, e_4 is ρ, then in all possible ratio parametric or parametric representations the cross-ratio of the four corresponding values of the ratio parameter or parameter is also ρ. This number ρ is therefore determined by the ordered set of elements e_1, e_2, e_3, e_4 of the $R\infty^1$, and it is called the *cross-ratio of* e_1, e_2, e_3, e_4 *on* (*or in*) *the* $R\infty^1$, and is denoted by (e_1, e_2, e_3, e_4). Thus

The cross-ratio on (*or in*) *an* $R\infty^1$ *of four elements of the system is the cross-ratio of the four corresponding values of any ratio parameter or parameter of the* $R\infty^1$.

Four elements of an $R\infty^1$ are said to be harmonic or equianharmonic when the four corresponding values of any parameter of the $R\infty^1$ are harmonic or equianharmonic.*

3·14. *Apolarity on an* $R\infty^1$

Consider two unordered pairs of elements of E, $|e_1, e_3|$ and $|e_2, e_4|$. Let the corresponding pairs of values of a ratio parameter (x, y) be the roots of

$$a_1x^2 + 2b_1xy + c_1y^2 = 0, \quad a_2x^2 + 2b_2xy + c_2y^2 = 0. \qquad (\cdot 141)$$

* See 2·74, p. 69

These pairs of values of (x, y) are apolar IF

$$a_1 c_2 + a_2 c_1 = 2b_1 b_2.\text{*} \qquad (\cdot 142)$$

Any other ratio parameter (x', y') of E is connected with (x, y) by equations of the form

$$x = px' + qy', \quad y = rx' + sy', \qquad (\cdot 143)$$

where $ps \neq qr$.† So, on substitution in ·141, the pairs of corresponding values of (x', y') are the roots of

$$(a_i p^2 + 2b_i pr + c_i r^2) x'^2 + 2\{a_i pq + b_i(ps + qr) + c_i rs\} x'y' + (a_i q^2 + 2b_i qs + c_i s^2) y'^2 = 0,$$

where $i = 1, 2$. It is easily seen that the CONDITION for these pairs of values of (x', y') to be apolar reduces to

$$(ps - qr)^2 (a_1 c_2 + a_2 c_1 - 2b_1 b_2) = 0. \qquad (\cdot 144)$$

Since $ps \neq qr$ it follows that:

If the pairs of values of one particular ratio parameter of E which correspond to $|\, e_1, e_3 \,|$, $|\, e_2, e_4 \,|$ *are apolar, then the pairs of values of any other ratio parameter of E which correspond to* $|\, e_1, e_3 \,|$ *and* $|\, e_2, e_4 \,|$ *are also apolar.*

When this happens the pairs $|\, e_1, e_3 \,|$, $|\, e_2, e_4 \,|$ are said to be *apolar on E*, and each pair is said to be harmonically separated by the other, or apolar to the other. It follows from 2·75 that, if $(e_1, e_2, e_3, e_4) = -1$, then $|\, e_1, e_3 \,|$ and $|\, e_2, e_4 \,|$ are apolar. Conjugate elements of E relative to a pair of elements of E and the harmonic conjugate of an element of E relative to a pair of elements of E are defined as in 2·76 by using a ratio parameter (x, y), and it follows from the theorem just established that these definitions do not depend on the choice of a ratio parameter.

3·15. *Choice of representation*

If (x, y) is a ratio parameter of E, any other ratio parameter (x', y') may be obtained from (x, y) by some choice of the coefficients a, b, c, d of (·121). Any three distinct elements of E correspond to distinct values of (x, y), so by 2·84

If e_1, e_2, e_3 *are three given distinct elements of an* $R\infty^1$, *there is* ONE *ratio parametric representation of the* $R\infty^1$ *in which* e_1, e_2, e_3 *correspond to any three given distinct ratio numbers.*

* See 2·75, p. 69.

† See equation (·123), p. 76.

The same theorem holds for parameters, and the ONE parametric representation of E in which e_1, e_2, e_3 correspond to given distinct values $\lambda_1, \lambda_2, \lambda_3$ of λ is determined by the equation

$$(\lambda_1, \lambda_2, \lambda_3, \lambda) = (e_1, e_2, e_3, e),$$

where e is a variable element of E. In particular, taking $\lambda_1 = \infty$, $\lambda_2 = 1$, $\lambda_3 = 0$, the cross-ratio

$$\lambda = (e_1, e_2, e_3, e)$$

is a parameter of E whose values ∞, 1, 0 correspond to e_1, e_2, e_3.

3·16. *Note on cross-ratio*

In 3·13 the cross-ratio of four things is only defined when the four things are regarded as elements of a definite $R\infty^1$. For instance, if A, B, C, D are four points in a plane, and if γ is a rational curve* through the points, then the four points have a cross-ratio on γ, which is the cross-ratio of the four values of any parameter of γ corresponding to A, B, C, D. If γ' is another rational curve through the points, then the points also have a cross-ratio on γ', and these two cross-ratios will not in general be equal. In fact the cross-ratio of four things depends on the choice of an $R\infty^1$ which contains them.

If A, B, C, D are four points of a line p, one rational curve through the points is the line p itself, and the points have a cross-ratio on p. But it is actually possible to construct rational quartic curves through A, B, C, D on which the points have different cross-ratios. So even in this simple case the symbol (A, B, C, D) has no meaning unless we say what rational curve through the points is being considered.

3·17. *Conventions in a plane*

In this volume we shall only be concerned with curves of order one and two, and no confusion will be caused if we agree to the following conventions.

(i) If A, B, C, D are four points of a line p, then (A, B, C, D) denotes the cross-ratio of the points on p.

(ii) If a, b, c, d are four coplanar lines through a point P, then (a, b, c, d) denotes the cross-ratio of the lines in the pencil of lines in the plane with vertex P.

* See 2·41, p. 51.

(iii) If P, A, B, C, D are five general points in a plane π,

$$P(A, B, C, D)$$

denotes the cross-ratio of the lines PA, PB, PC, PD in the pencil of lines of π with vertex P.

(iv) If p, a, b, c, d are five general lines of a plane, $p(a, b, c, d)$ denotes the cross-ratio of the points pa, pb, pc, pd on p.

3·18. *Conventions in space*

Similarly we make the following conventions in space.

(i) If α, β, γ, δ are four general planes and p a general line, $p(\alpha, \beta, \gamma, \delta)$ denotes the cross-ratio of the points $p\alpha$, $p\beta$, $p\gamma$, $p\delta$ on p.

(ii) If α, β, γ, δ are four planes through a point P, and π is a plane through P, $\pi(\alpha, \beta, \gamma, \delta)$ denotes the cross-ratio of the lines $\pi\alpha$, $\pi\beta$, $\pi\gamma$, $\pi\delta$ in the pencil of lines of π with vertex P.

(iii) If α, β, γ, δ are four planes through a line p, and π is a general plane, $(\alpha, \beta, \gamma, \delta)$ denotes the cross-ratio of the planes in the pencil of planes with vertex p, while $\pi(\alpha, \beta, \gamma, \delta)$ denotes the cross-ratio of the lines $\pi\alpha$, $\pi\beta$, $\pi\gamma$, $\pi\delta$ in the pencil of lines of π with vertex at the point $p\pi$.

(iv) If a, b, c, d are four lines meeting a line p, then $p(a, b, c, d)$ might mean either the cross-ratio of the planes $[pa]$, $[pb]$, $[pc]$, $[pd]$ in the pencil of planes with vertex p, or the cross-ratio of the points (pa), (pb), (pc), (pd) on the line p. In general these cross-ratios are different, so the first will be denoted by $p[a, b, c, d]$ and the second by $p(a, b, c, d)$.

(v) If four lines a, b, c, d meet a line p, if P is a general point of p, and if π is a general plane through p, then $P(a, b, c, d)$ denotes the cross-ratio of the planes Pa, Pb, Pc, Pd in the pencil of planes with vertex p, while $\pi(a, b, c, d)$ denotes the cross-ratio of the points πa, πb, πc, πd on p.

(vi) If A, B, C, D are four general points and p is a general line, $p(A, B, C, D)$ denotes the cross-ratio of the planes pA, pB, pC, pD in the pencil of planes with vertex p.

3·2. Relation between two rational ∞^1 systems

3·21. *Related* $R\infty^1$

In dealing with rational ∞^1 systems it is convenient to use the word *relation* to mean 'proper algebraic (1, 1) correspondence'. Thus two $R\infty^1$, E and E' are said to be *related* IF there is a proper

$T(1, 1)$ between their elements. It is also said that E and E' are in proper $T(1, 1)$. Clearly two $R\infty^1$ which are related to the same $R\infty^1$ are related to each other. Also, if (x, y) is a ratio parameter of E and (x', y') is a ratio parameter of E', a relation between E and E' is equivalent to a relation between (x, y) and (x', y'). Thus, from 2·58

Any relation between two $R\infty^1$, E and E', is determined by a proper bilinean equation

$$B(x, y;\ x', y') \equiv axx' + bxy' + cx'y + dyy' = 0,$$

where (x, y) and (x', y') are any ratio parameters of E and E'. Alternatively any relation between E and E' is determined by a proper bilinean equation

$$B(\lambda, \lambda') \equiv a\lambda\lambda' + b\lambda + c\lambda' + d = 0,$$

where λ and λ' are any parameters of E and E'.

3·22. From 2·82:

If two $R\infty^1$, E and E', are related, the cross-ratio of any four elements of E, no three of which coincide, is equal to the cross-ratio of the four corresponding elements of E'.

3·23. From 2·84:

If e_1, e_2, e_3 are three distinct elements of E and e_1', e_2', e_3' are three distinct elements of E', there is ONE *relation between E and E' in which $e_1 \to e_1'$, $e_2 \to e_2'$, $e_3 \to e_3'$, and this relation is determined by the equation*

$$(e_1, e_2, e_3, e) = (e_1', e_2', e_3', e').$$

3·24. The elements of an $R\infty^1$ form a range. When two $R\infty^1$, E and E', are related, the corresponding elements e and e' are said to form related ranges, and this is expressed by writing

$$\{e\} \equiv \{e'\}.$$

Thus if $\{e\} \equiv \{e'\}$ and $\{e'\} \equiv \{e''\}$, then $\{e\} \equiv \{e''\}$. Also two subranges $e_1, e_2, \ldots, e_r$ and $e_1', e_2', \ldots, e_r'$ in E and E' are said to be related IF $e_i \to e_i'$, $i = 1, 2, \ldots, r$ is some relation between E and E'. A subrange formed by a set of lines or planes of a pencil of lines or planes will be called a *subpencil*.

Any $T(1, 1)$ between two $R\infty^1$ is called a *homography* or a *projectivity*, an improper $T(1, 1)$ being called an improper or degenerate homography or projectivity. Thus a relation is a proper homography or projectivity. Related ranges are sometimes said to be homographically related or projectively related, or to be homographic or projective, or even to be linearly related.

3·25. *A pencil of lines. Perspectivity*

As an example, consider a pencil of lines p with vertex V in a plane π, and let q be any line of π not passing through V. Each line p through V meets q in ONE point Q, while each point Q of q lies on ONE line p of the pencil. So there is a proper $(1, 1)$ correspondence between the lines p and the points Q. Also this correspondence can be expressed by algebraic equations, so the lines p are related to the points Q. Thus, by 3·22, if four lines p_1, p_2, p_3, p_4 of the pencil meet q in the four points Q_1, Q_2, Q_3, Q_4, then

$$q(p_1, p_2, p_3, p_4) = (Q_1, Q_2, Q_3, Q_4) = (p_1, p_2, p_3, p_4) \\ = V(Q_1, Q_2, Q_3, Q_4),$$

using the conventions of 3·17.

If q' is another line of π which does not contain V, and if the lines p of the pencil meet q' in the points Q', then the range $\{Q'\}$ is related to the pencil of lines p and therefore to the range $\{Q\}$, so

$$(Q_1, Q_2, Q_3, Q_4) = (Q'_1, Q'_2, Q'_3, Q'_4).$$

The point qq' is a common self-corresponding point of the related ranges $\{Q\}$, $\{Q'\}$, and this particular type of relation between two coplanar lines, which is determined by projection from V, is called a perspectivity.*

Conversely any relation between two coplanar lines q, q' in which the meet O of q, q' is a self-corresponding point, is a perspectivity. For any two distinct points Q_1, Q_2 of q other than O correspond to two distinct points Q'_1, Q'_2 of q' other than O. If V is the point $(Q_1 Q'_1, Q_2 Q'_2)$ projection from V gives a relation between q and q' in which $O \to O$, $Q_1 \to Q'_1$, $Q_2 \to Q'_2$. But by 3·23 there is ONE relation in which $O \to O$, $Q_1 \to Q'_1$, $Q_2 \to Q'_2$, so the given relation is the perspectivity obtained by projection from V. Thus:

If qq' is a self-corresponding point of related ranges on two distinct coplanar lines q and q', the joins of corresponding points of the ranges are concurrent.

* See 1·82, p. 38.

3·26. *General relation between two lines*

Any relation between two distinct lines in a plane π may be obtained by two successive projections.

For let q, q' be the lines meeting in O, and let A, A' be any two corresponding points distinct from O. Let V be any point of AA' distinct from A, A' and let q'' be any line of π through A' other than q'. The points Q of q project from V into the points Q'' of q'', and the range $\{Q''\}$ is related to $\{Q\}$ and therefore to $\{Q'\}$, where $\{Q'\}$ is the range of points of q' related to $\{Q\}$ by the given relation. But A' is a self-corresponding point of the related ranges $\{Q''\}$ and $\{Q'\}$, so by 3·25 the relation between $\{Q''\}$ and $\{Q'\}$ is obtained by projection

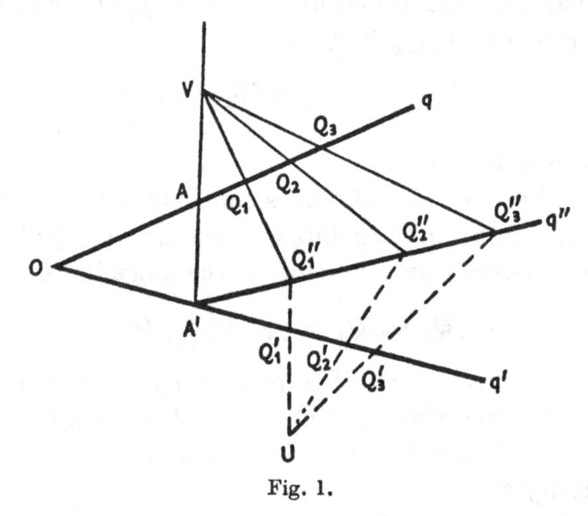

Fig. 1.

from some vertex U. Thus the given relation between q and q' is determined by first projecting a general point Q from V into a point Q'' of q'' and then projecting Q'' from U into a point Q' of q'. This theorem shows that it is natural to call a relation between two lines a projectivity (Fig. 1).

A neater method of obtaining a general relation between q and q' by two successive projections is as follows. If O is regarded as a point of q it corresponds to a point M' of q'. Similarly if O is regarded as a point of q' it corresponds to a point L of q. Let A, A' be any other pair of corresponding points. Then the given relation is obtained by first projecting the points Q of q into the points R of LM' from A' and then projecting the points R of LM' into the points Q' of q' from A, for this construction determines a relation between q and q' in which $A \to A'$, $O \to M'$, $L \to O$, and by 3·23 there is ONE relation with this property (Fig. 2).

3·27. *A pencil of planes*

As another example consider a pencil of planes π in space with a line vertex l. Let q be a line which does not meet l, and let α be a plane which does not contain l. Each plane π of the pencil meets q in ONE point Q and α in one line a. Conversely each point Q of q lies in ONE plane π of the pencil, and each line a of α through the point αl lies in ONE plane π of the pencil, so the pencil of planes π is related to

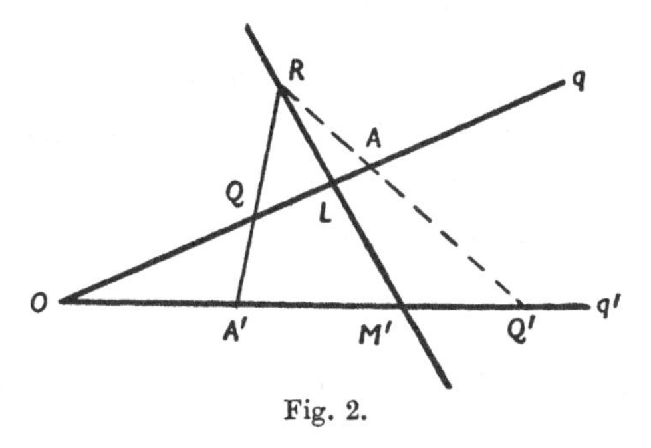

Fig. 2.

the points Q of q and to the pencil of lines a in α with vertex αl. Thus, if π_i, Q_i, a_i denote corresponding elements,

$$(\pi_1, \pi_2, \pi_3, \pi_4) = (Q_1, Q_2, Q_3, Q_4) = (a_1, a_2, a_3, a_4),$$

using the conventions of 3·18. Also

$$(\pi_1, \pi_2, \pi_3, \pi_4) = l(Q_1, Q_2, Q_3, Q_4) = l[a_1, a_2, a_3, a_4],$$
$$(Q_1, Q_2, Q_3, Q_4) = q(\pi_1, \pi_2, \pi_3, \pi_4),$$
$$(a_1, a_2, a_3, a_4) = \alpha(\pi_1, \pi_2, \pi_3, \pi_4).$$

3·3. Homography on an $R\infty^1$

3·31. *Definition*

The bilinear equation $B(x, y; x', y') = 0$ has already been interpreted in two ways. In 3·1 one $R\infty^1$ was considered, and (x, y), (x', y') were regarded as the values of two different ratio parameters of the $R\infty^1$ corresponding to the same element of the $R\infty^1$; the bilinear equation, when proper, gave a change of ratio parameter. In 3·2 (x, y), (x', y') were regarded as ratio parameters of two $R\infty^1$, E and E', and the bilinear equation, when proper, gave a relation between E and E'.

We now consider one ratio parameter (ξ, η) of one $R\infty^1$, E, and regard (x, y) and (x', y') as values of (ξ, η) corresponding to two elements e and e' of E. The equation

$$B(x, y;\, x', y') \equiv axx' + bxy' + cx'y + dyy' = 0$$

now determines a $T(1,1)$ between the elements e and e' of E, which will be called a *homography* on E. We shall usually suppose that the homography is proper, i.e. that $bc \neq ad$, unless the contrary is stated. If $\rho = \xi/\eta$ is the parameter of E representing the ratio parameter (ξ, η), and if $\lambda = x/y$, $\lambda' = x'/y'$ are the values of ρ representing the values (x, y), (x', y') of (ξ, η), then the homography is also determined by the equation

$$B(\lambda, \lambda') \equiv a\lambda\lambda' + b\lambda + c\lambda' + d = 0.$$

It follows from 2·82 and 3·13 that if four positions e_1, e_2, e_3, e_4 of e on E correspond to four positions e_1', e_2', e_3', e_4' of e', then

$$(e_1, e_2, e_3, e_4) = (e_1', e_2', e_3', e_4').$$

3·32. *Forward and reverse correspondence*

As in 2·81 the $T(1,1)$ between e and e', which is called a homography on E, determines two operations.

(i) Each position e_1 of e on E corresponds to ONE value (x_1, y_1) of (x, y), which determines ONE value (x_1', y_1') of (x', y'), which corresponds to one position e_1' of e'. This operation, denoted by T, is called the forward correspondence, and is expressed symbolically by writing $e_1 \rightarrow e_1'$, or $e_1' = Te_1$. The position e_1' of e' is given by the value $(bx_1 + dy_1,\ -ax_1 - cy_1)$ of (ξ, η) or by the value $-\dfrac{bx_1 + dy_1}{ax_1 + cy_1}$ of ρ.

(ii) Each position e_1' of e' corresponds to ONE value (x_1', y_1') of (x', y'), which determines ONE value (x_1, y_1) of (x, y), which corresponds to ONE position e_1 of e. This operation, denoted by T^{-1}, is called the reverse correspondence and is expressed symbolically by writing $e_1' \leftarrow e_1$, or $e_1 = T^{-1}e_1'$. The position e_1 of e is given by the value $(cx_1' + dy_1',\ -ax_1' - by_1')$ of (ξ, η) or by the value $-\dfrac{cx_1' + dy_1'}{ax_1' + by_1'}$ of ρ.

3·33. *General* $T(1,1)$ *on an* $R\infty^1$

From 2·58 any algebraic correspondence between elements e, e' of an $R\infty^1$, E, in which a general position of e determines ONE position of e', while a general position of e' determines ONE position of e, is a homography given by a bilinear equation $B(x, y;\ x', y') = 0$. Also, if e_1, e_2, e_3 are distinct elements of E and e_1', e_2', e_3' are also distinct elements of E, then by 2·84 there is ONE homography on E in which $e_1 \to e_1'$, $e_2 \to e_2'$, $e_3 \to e_3'$, this homography being determined by the equation

$$(e_1, e_2, e_3, e) = (e_1', e_2', e_3', e').$$

In an homography on E the corresponding elements e and e' are said to form homographic ranges, and we write

$$\{e\} \equiv \{e'\}.$$

It is also convenient to say that the subranges $e_1, e_2, \ldots, e_r$ and $e_1', e_2', \ldots, e_r'$ on E are homographic, IF $e_i \to e_i'$ $(i = 1, 2, \ldots, r)$, in some homography on E, and this is expressed by writing

$$\{e_1, e_2, \ldots, e_r\} \equiv \{e_1', e_2', \ldots, e_r'\}.$$

The homography is of course determined by three pairs of corresponding elements of the subranges, and the cross-ratio of any four elements of one subrange is equal to the cross-ratio of the four corresponding elements of the other subrange.

3·34. *Self-corresponding elements*

In the homography on E determined by $B(x, y;\ x', y') = 0$ or by $B(\lambda, \lambda') = 0$, the elements e which coincide with their corresponding elements e' are given by

$$ax^2 + (b+c)\,xy + dy^2 = 0, \qquad (\cdot 341)$$

or by

$$a\lambda^2 + (b+c)\,\lambda + d = 0. \qquad (\cdot 342)$$

These are called the *self-corresponding elements* of the homography on E.* Thus in general there are TWO distinct self-corresponding elements, but these coincide IF $(b+c)^2 = 4ad$. If the homography has more than two distinct self-corresponding elements, the equations (·341), (·342) must be identities, so $a = b+c = d = 0$, and every element of E is self-corresponding. This particular homography, which is given by $xy' - x'y = 0$ or by $\lambda = \lambda'$, is called *identity*.

3·35. *Distinct self-corresponding elements*

Consider a general homography on E with two distinct self-corresponding elements d_1, d_2. Choose a ratio parameter (ξ, η) of E

* They will not be called double elements of the homography, as this would lead to confusion in Chapter X, when a $T(2, 2)$ is considered.

so that d_1, d_2 correspond to the values $(0, 1)$, $(1, 0)$ of (ξ, η). (This is possible in infinitely many ways by 3·15.) Then the homography is determined by an equation of the form $B(x, y;\ x', y') = 0$, but the roots of (·341) are $(0, 1)$ and $(1, 0)$, so $a = d = 0$. Thus the homography is determined by an equation of the form

$$bxy' + cx'y = 0, \tag{·351}$$

or by

$$b\lambda + c\lambda' = 0. \tag{·352}$$

It follows that, if $e \to e'$ in the homography, then

$$(d_1, e, d_2, e') = (0, \lambda, \infty, \lambda') = \frac{\lambda}{\lambda'} = -\frac{c}{b}. \tag{·353}$$

Thus:

In any homography on an $R\infty^1$ which has TWO *distinct self-corresponding elements d_1, d_2, and in which e, e' are any pair of corresponding elements, the cross-ratio (d_1, e, d_2, e') is constant.*

In fact the homography is determined by an equation of the form

$$(\lambda_1, \lambda, \lambda_2, \lambda') = m, \tag{·354}$$

where m is a constant and λ_1, λ_2 are the values of the parameter λ which give the two self-corresponding elements. This constant m is sometimes called the *cross-ratio of the homography*, though it depends on the order in which the self-corresponding elements are taken. The bilinear equation (·354) in λ, λ' is proper unless $m = 0$ or $m = \infty$.

3·36. *Coincident self-corresponding elements*

If the self-corresponding elements coincide in d, we may choose a ratio parameter (ξ, η) so that d corresponds to the value $(0, 1)$ of (ξ, η). The homography is then given by $B(x, y;\ x', y') = 0$, but the roots of (·341) coincide in $(0, 1)$, so $b + c = d = 0$. The homography is therefore determined by an equation of the form

$$axx' + b(xy' - x'y) = 0, \tag{·361}$$

or by

$$a\lambda\lambda' + b(\lambda - \lambda') = 0. \tag{·362}$$

The bilinear equation (·361) is proper unless $b = 0$.

3·37. *Degenerate homography*

The homography given by $B(x, y;\ x', y') = 0$ is improper or degenerate IF $bc = ad$, when the bilinear equation is of the form

$$(q_1 x - p_1 y)(q_2 x' - p_2 y') = 0. \tag{·371}$$

The self-corresponding elements are the singular elements d_1, d_2 given by the values $(p_1, q_1), (p_2, q_2)$ of (ξ, η), and $(\cdot 371)$ expresses that either e is d_1 or e' is d_2. The degenerate homography is determined by the equation

$$(d_1, \lambda, d_2, \lambda') = 0,$$

but it cannot be said that e and e' form homographic ranges on E.

3·4. Involutory homography on an $R\infty^1$

3·41. *Involutory pairs of elements*

The forward and reverse correspondences T and T^{-1} associated with a general homography on E were defined in 3·32. Each element e of E determines ONE element e' such that $e \rightarrow e'$ in T; this element e' is denoted by Te and, if e is (x, y), then e' is

$$(bx + dy, \ -ax - cy).$$

Also each element e of E determines ONE element e'' such that $e \leftarrow e''$ in T, or $e \rightarrow e''$ in T^{-1}; this element e'' is denoted by $T^{-1}e$ and, if e is (x, y), then e'' is $(cx + dy, \ -ax - by)$. The two elements e' and e'' are in general distinct, but they coincide IF

$$(bx + dy)(ax + by) = (ax + cy)(cx + dy),$$

i.e. IF $$(b - c)\{ax^2 + (b + c)xy + dy^2\} = 0, \qquad (\cdot 411)$$

i.e. IF either $b = c$ or e is a self-corresponding element of the homography.*

Two elements e, e' of E are said to form an *involutory* pair of the homography IF $e \rightarrow e'$ in T and $e' \rightarrow e$ in T, i.e. IF e'' coincides with e'. It follows that, if $b \neq c$, the pair e, e' is involutory IF e is a self-corresponding element, in which case e' coincides with e.

3·42. *Involutory homography*

A homography on E is said to be *involutory* IF every pair of corresponding elements e, e' is involutory, i.e. IF $(\cdot 411)$ is identically satisfied for all (x, y), i.e. IF either $b = c$ or $a = b + c = d = 0$. But in the latter case the homography is identity so

The homography on E given by $B(x, y;\ x', y') = 0$ is involutory IF *either $b = c$, when the equation is symmetrical in (x, y) and (x', y'), or the homography is identity.*

* See 3·34, p. 85.

Thus an involutory homography on E other than identity is given by a bilinear equation of the form

$$pxx' + q(xy' + x'y) + ryy' = 0,$$

or by
$$p\lambda\lambda' + q(\lambda + \lambda') + r = 0.$$

In an involutory homography the forward and reverse correspondences T and T^{-1} are the same.

3·43. It follows from 3·41 that:

If any one pair of distinct elements of E is an involutory pair in a homography on E, then the homography is involutory.

Thus, if e_1, e_2, e_3 are distinct, e'_1, e'_2, e'_3 are distinct, and e_1, e'_1 are distinct, then the ONE homography on E in which $e_1 \to e'_1$, $e_2 \to e'_2$, $e_3 \to e'_3$ is given by

$$(e_1, e_2, e_3, e) = (e'_1, e'_2, e'_3, e'),$$

and is involutory IF $e'_1 \to e_1$, i.e. IF

$$(e_1, e_2, e_3, e'_1) = (e'_1, e'_2, e'_3, e_1).$$

3·44. It was shown in 3·35 that the equation of a general homography on E with two distinct self-corresponding elements may be reduced to the form

$$bxy' + cx'y = 0,$$

and this gives an involutory homography IF either $b = c$ or $b + c = 0$, the latter case giving identity. Thus:

A homography on E other than identity which has two distinct self-corresponding elements d_1, d_2 is involutory IF *the constant cross-ratio* $(d_1, e, d_2, e') = -1$.

Also the homography given by (·362) is involutory IF either $b = 0$ or $a = 0$, the latter case giving identity. Thus:

An involutory homography with only one self-corresponding element is given by the equation

$$xx' = 0,$$

where $(0, 1)$ *is the self-corresponding element.*

It also follows that:

If the pair of self-corresponding elements $|d_1, d_2|$ of a homography other than identity is apolar to any one pair $|e_1, e'_1|$ of corresponding elements, then the homography is involutory, and every pair of corresponding elements (e, e') is apolar to $|d_1, d_2|$.

3·45. *Pairs containing an element*

In 3·41 the element e belongs to two pairs of corresponding elements, namely e, e' and e'', e. Thus:

In a homography on E each element other than the self-corresponding elements belongs to two distinct pairs of corresponding elements, unless the homography is involutory.

3·46. *Example*

The following example illustrates the theory. Let p, q be two distinct lines through a point A; let H, K be two general points of the plane $[pq]$, and let HK meet p, q in B, C, as in Fig. 3. P is a variable point of p, PH meets q in Q, and QK meets p in P'. By 3·25 the ranges $\{P\}$ and $\{Q\}$ are related, and $\{Q\}$ and $\{P'\}$ are related,

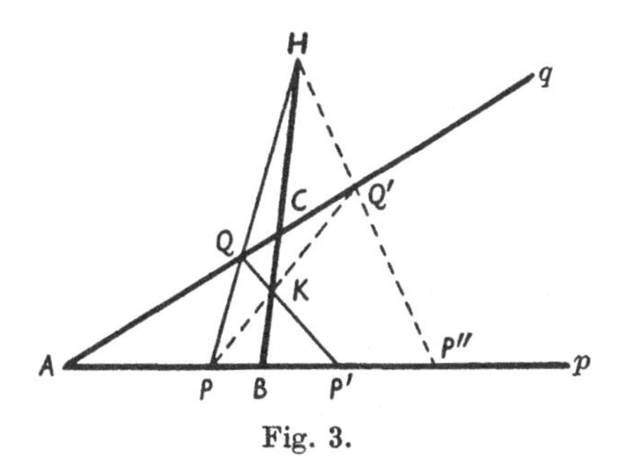

Fig. 3.

so P and P' form homographic ranges on p. Denote the $(1, 1)$ correspondence $P \to P'$ by T.

The unique point P'' such that $P'' \to P$ in T is determined as follows. If PK meets q in Q', then $Q'H$ meets p in P''. Thus the reverse correspondence T^{-1} is determined by a construction similar to that which gives T but with H and K interchanged.

The self-corresponding points of the homography on p are A and B, so it follows from 3·43 that, if P' and P'' coincide for any one position of P other than A and B, then the homography is involutory and P' coincides with P'' for all positions of P. Also

$$(A, P, B, P') = Q(A, P, B, P') = (C, H, B, K)^*$$

so by 3·44 the homography is involutory IF H and K harmonically separate B and C.

* See 3·17 and 3·25, pp. 78, 81.

The homography obtained by the above construction will have coincident self-corresponding points IF HK contains A; identity may be obtained by allowing H and K to coincide. Moreover any homography on p may be obtained by a similar construction. For suppose that, in a given homography on p, B is a self-corresponding point and $P_1 \to P'_1$, $P_2 \to P'_2$. Choose any two points H and K collinear with B, call the line $[(P_1H, P'_1K), (P_2H, P'_2K)]\, q$, and let q meet p in A. Then the construction of figure 3 gives a homography $P \to P'$ on p in which B is a self-corresponding point and $P_1 \to P'_1$, $P_2 \to P'_2$. This homography, having three pairs of corresponding points in common with the given homography, is the given homography, by 3·33.

3·5. Involutions on an $R\infty^1$

3·51. *Unordered pairs of values of* (x, y)

The pairs of corresponding elements in a homography on an $R\infty^1$ are essentially ordered pairs, because of the difference between the forward and reverse correspondences. We now consider unordered pairs of elements of an $R\infty^1$, and we first consider unordered pairs of values of a ratio variable (x, y).

A ratio set (l, m, n) determines a quadratic equation

$$lx^2 + mxy + ny^2 = 0$$

in (x, y) whose roots are an unordered pair of values of (x, y). Also any unordered pair $|(x_1, y_1), (x_2, y_2)|$ of values of (x, y) are the roots of the equation

$$y_1y_2x^2 - (x_1y_2 + x_2y_1)xy + x_1x_2y^2 = 0$$

and determine a ratio set (l, m, n) given by

$$l:m:n = y_1y_2 : -(x_1y_2 + x_2y_1) : x_1x_2.$$

Thus:

The unordered pairs of values of a ratio variable (x, y) *may be identified with the ratio sets* (l, m, n) *and therefore form an* $R\infty^2$.

3·52. *Involution in* (x, y)

An *involution* among the values of (x, y) is defined as follows:

An algebraic system of pairs of values of (x, y) *is called an involution in* (x, y) IF *it has the properties*: (1) *Each value of* (x, y) *belongs to* ONE *pair*. (2) *Some pair consists of two distinct values of* (x, y).

The following theorem is easily established.

The pairs of any involution in* (x, y) *are the roots of the* ∞^1 *equations*

$$lx^2+mxy+ny^2 = 0 \qquad (\cdot 521)$$

obtained by giving the ratio set (l, m, n) *all values which satisfy a linear equation*

$$al+bm+cn = 0. \qquad (\cdot 522)$$

For in any involution, by property (1), if (x_1, y_1) is any value of (x, y), ONE value (x_2, y_2) is algebraically determined such that $|(x_1, y_1), (x_2, y_2)|$ is a pair of the involution. Also, if (x_2, y_2) is given, (x_1, y_1) is uniquely determined. Thus there is a homography between (x_1, y_1) and (x_2, y_2), which is involutory by property (1).† This homography is not identity, by property (2), and is therefore‡ given by an equation of the form

$$px_1x_2+q(x_1y_2+x_2y_1)+ry_1y_2 = 0.$$

But this equation is equivalent to the statement that the coefficients l, m, n in the equation

$$lx^2+mxy+ny^2 = 0,$$

whose roots are (x_1, y_1) and (x_2, y_2), satisfy

$$pn-qm+rl = 0,$$

so, on taking $a:b:c = r:-q:p$, the theorem is established. It is clear conversely that, for any (a, b, c), the ∞^1 pairs of roots of the equations

$$lx^2+mxy+ny^2 = 0,$$

for which

$$al+bm+cn = 0$$

do actually form an involution as defined above.

3·53. The argument of 3·52 shows that:

The pairs of any involution in (x, y) *are corresponding values in an involutory homography in* (x, y) *which is given by an equation of the form*

$$px_1x_2+q(x_1y_2+x_2y_1)+ry_1y_2 = 0.$$

Conversely the pairs of corresponding values of (x, y) *in any involutory homography in* (x, y) *other than identity are the pairs of an involution.*

3·54. *Involution on an* $R\infty^1$

An involution on an $R\infty^1$, E, *is defined to be an algebraic system of pairs of elements of* E *such that:* (1) *Each element of* E *belongs to* ONE *pair.* (2) *Some pair consists of distinct elements.*

* The pairs of an involution are sometimes called mates.

† See 3·45, p. 89. ‡ See 3·42, p. 88.

If (x, y) is a ratio parameter of E an involution on E corresponds to an involution in (x, y), and is therefore determined by equations of the form (·521), (·522). It also follows from the definition of an involution in (x, y) that, if two ratio variables (x, y) and (x', y') are related, then an involution in (x, y) corresponds to an involution in (x', y'). Thus

If two $R\infty^1$ are related, an involution on one of them corresponds to an involution on the other.

For example the pairs of planes of an involution in a pencil of planes meet a general line in pairs of points of an involution on the line. An involution in a pencil of planes, or in a pencil of lines is called an *involution pencil.*

3·55. *Determination of an involution*

Let (x, y) be a ratio parameter of an $R\infty^1$, E. Any involution I on E is given by

$$lx^2 + mxy + ny^2 = 0, \qquad (\cdot 551)$$

$$al + bm + cn = 0, \qquad (\cdot 552)$$

and the pairs of elements of I are corresponding elements in the involutory homography on E given by

$$cx_1x_2 - b(x_1y_2 + x_2y_1) + ay_1y_2 = 0. \qquad (\cdot 553)$$

If the parameter $\lambda = x/y$ is used to represent E, these equations are replaced by

$$l\lambda^2 + m\lambda + n = 0, \qquad (\cdot 554)$$

$$al + bm + cn = 0, \qquad (\cdot 555)$$

and

$$c\lambda_1\lambda_2 - b(\lambda_1 + \lambda_2) + a = 0. \qquad (\cdot 556)$$

Any two distinct pairs of elements of E belong to ONE *involution on E.*

For if the pairs are the roots of

$$l_1x^2 + m_1xy + n_1y^2 = 0,$$

$$l_2x^2 + m_2xy + n_2y^2 = 0,$$

then

$$l_1 : m_1 : n_1 \neq l_2 : m_2 : n_2,$$

so the equations

$$al_1 + bm_1 + cn_1 = 0,$$

$$al_2 + bm_2 + cn_2 = 0,$$

are satisfied by ONE ratio set (a, b, c). In fact the equation (·552), which determines the involution, is

$$\begin{vmatrix} l & m & n \\ l_1 & m_1 & n_1 \\ l_2 & m_2 & n_2 \end{vmatrix} = 0,$$

while (·553) is

$$\begin{vmatrix} y_1 y_2 & -(x_1 y_2 + x_2 y_1) & x_1 x_2 \\ l_1 & m_1 & n_1 \\ l_2 & m_2 & n_2 \end{vmatrix} = 0.$$

It also follows that:

Two involutions on E which have two distinct pairs of elements in common are identical.

3·56. *The three pairs of elements of E which are given by*

$$l_i x^2 + m_i xy + n_i y^2 = 0 \quad (i = 1, 2, 3),$$

belong to an involution on E IF

$$\begin{vmatrix} l_1 & m_1 & n_1 \\ l_2 & m_2 & n_2 \\ l_3 & m_3 & n_3 \end{vmatrix} = 0.$$

For the three pairs belong to an involution IF there is a ratio set (a, b, c) which satisfies the three equations

$$al_i + bm_i + cn_i = 0 \quad (i = 1, 2, 3).$$

Also, if (a_1, b_1, c_1), (a_2, b_2, c_2) are two distinct ratio sets, there is ONE ratio set (l, m, n) which satisfies the equations

$$a_1 l + b_1 m + c_1 n = 0,$$
$$a_2 l + b_2 m + c_2 n = 0,$$

so

Two distinct involutions on E have ONE *common pair.*

3·57. *Double elements*

If e_0 is an element of E, it would seem natural that the pair $|e_0, e_0|$, which consists of e_0 counted twice, should be called a double element. We shall say, less precisely, that the element e_0 is a *double element* of an involution I on E IF the pair $|e_0, e_0|$ is a pair of I. It follows from (·553) that the double elements of I are given by

$$cx^2 - 2bxy + ay^2 = 0; \quad (·571)$$

also, by 2·75, the equation

$$al + bm + cn = 0$$

is a CONDITION for the pairs of elements given by (·571) and by

$$lx^2 + mxy + ny^2 = 0$$

to be apolar. Thus:

In a general involution I on E there are two distinct double elements d_1, d_2, and the pairs of I are just those pairs of elements of E which are apolar to $|d_1, d_2|$.

3·58. *Involution with a fixed element*

The double elements of I coincide in the element d_0 given by (x_0, y_0) IF (·571) is

$$y_0^2 x^2 - 2x_0 y_0 xy + x_0^2 y^2 = 0,$$

i.e. IF

$$a : b : c = x_0^2 : x_0 y_0 : y_0^2,$$

and the equation (·553) is then

$$(y_0 x_1 - x_0 y_1)(y_0 x_2 - x_0 y_2) = 0,$$

which expresses that either (x_1, y_1) or (x_2, y_2) is (x_0, y_0). So the pairs of I consist of d_0 and another element of E; they are still just those pairs of elements of E which are apolar to the pair $|d_0, d_0|$ of double elements of I. The involution I given by (·551), (·552) is of this type IF $b^2 = ac$, i.e. IF the associated involutory homography is improper, and it is said to be an *improper involution*, an *involution with a fixed element*, or a *parabolic involution.*

It was shown in 3·55 that an involution on E is uniquely determined by any two distinct pairs, and in particular by two distinct double elements. But an involution is also uniquely determined by its double elements even when they are coincident, for when the roots of (·571) are given the ratio set (a, b, c) is uniquely determined. It should be noticed that, whereas a homography on E can have coincident self-corresponding elements without being improper, an involutory homography other than identity which has coincident double elements is necessarily improper.

Even in cases of coincidence it follows from the above that, if $|e_1, e_2|$ is a pair of the involution on E which has e_3, e_4 as double elements, then $|e_1, e_2|$ and $|e_3, e_4|$ are apolar, and $|e_3, e_4|$ is therefore a pair of the involution on E which has e_1, e_2 as double elements. Also, since any two distinct pairs of elements of E belong to ONE involution, which has a pair of double elements,

ONE *pair of elements of E is apolar to any two given distinct pairs of E.*

3·6. Pairs of elements of an $R\infty^1$

3·61. *An involution is rational*

If (l_1, m_1, n_1), (l_2, m_2, n_2) are two independent solutions of the equation

$$al+bm+cn = 0,$$

all the solutions of the equation form the linear ∞^1 system

$$(\kappa_1 l_1+\kappa_2 l_2, \kappa_1 m_1+\kappa_2 m_2, \kappa_1 n_1+\kappa_2 n_2),$$

and the solutions may be identified with the values of the ratio variable (κ_1, κ_2).* Thus:

If two distinct pairs of any involution I on E are given by

$$l_1 x^2+m_1 xy+n_1 y^2 = 0, \qquad (\cdot 611)$$

$$l_2 x^2+m_2 xy+n_2 y^2 = 0, \qquad (\cdot 612)$$

then all pairs of I are given by equations of the form

$$\kappa_1(l_1 x^2+m_1 xy+n_1 y^2)+\kappa_2(l_2 x^2+m_2 xy+n_2 y^2) = 0. \qquad (\cdot 613)$$

The pairs of I are related to the values of the ratio variable (κ_1, κ_2) and therefore form an $R\infty^1$ with (κ_1, κ_2) as a ratio parameter.

3·62. *Double elements*

By 2·14 the equation $(\cdot 613)$ in (x, y) has (x_0, y_0) as a double root IF

$$\kappa_1(2l_1 x_0+m_1 y_0)+\kappa_2(2l_2 x_0+m_2 y_0) = 0,$$

$$\kappa_1(m_1 x_0+2n_1 y_0)+\kappa_2(m_2 x_0+2n_2 y_0) = 0. \qquad (\cdot 621)$$

Thus, eliminating κ_1, κ_2, the double elements of I are given by

$$(l_1 m_2-l_2 m_1)\,x^2+2(l_1 n_2-l_2 n_1)\,xy+(m_1 n_2-m_2 n_1)\,y^2 = 0. \qquad (\cdot 622)$$

Also, eliminating x_0, y_0 from $(\cdot 621)$, the values of (κ_1, κ_2) which correspond to the double elements of I are given by

$$\begin{vmatrix} 2(l_1\kappa_1+l_2\kappa_2), & m_1\kappa_1+m_2\kappa_2 \\ m_1\kappa_1+m_2\kappa_2, & 2(n_1\kappa_1+n_2\kappa_2) \end{vmatrix} = 0,$$

or by

$$(4l_1 n_1-m_1^2)\,\kappa_1^2+2(2l_1 n_2+2l_2 n_1-m_1 m_2)\,\kappa_1\kappa_2+(4l_2 n_2-m_2^2)\,\kappa_2^2 = 0. \qquad (\cdot 623)$$

The involution I is improper IF $(\cdot 622)$ has a double root, i.e. IF

$$(l_1 n_2-l_2 n_1)^2 = (l_1 m_2-l_2 m_1)\,(m_1 n_2-m_2 n_1), \qquad (\cdot 624)$$

so this equation is a CONDITION for the equations $(\cdot 611)$, $(\cdot 612)$ to have a common root.

* See 1·42 or 1·63, pp. 13, 20.

3·63. *Apolar pairs of an involution*

Since the pairs of any involution I on E may be given by the equations (·613), and form an $R\infty^1$ with (κ_1, κ_2) as a ratio parameter, the cross-ratio of four pairs of I in this $R\infty^1$ is defined by 3·13 and is equal to the cross-ratio of the four corresponding values of (κ_1, κ_2), or of any other ratio parameter of the $R\infty^1$. Let $|e_1, e_3|$, $|e_2, e_4|$ be the pairs of I which are given by (·611), (·612); these pairs correspond to the values $(1, 0)$, $(0, 1)$ of (κ_1, κ_2). Also the double elements of I correspond to the roots of (·623), which harmonically separate $(1, 0)$ and $(0, 1)$ IF the coefficient of $\kappa_1\kappa_2$ in (·623) vanishes, i.e. IF

$$2l_1 n_2 + 2l_2 n_1 = m_1 m_2. \qquad (\cdot 631)$$

But this is also a CONDITION for the pairs $|e_1, e_3|$, $|e_2, e_4|$ to be apolar on E. Thus:

Two pairs $|e_1, e_3|$, $|e_2, e_4|$ of an involution I on E are apolar on E IF *the corresponding values of any ratio parameter (ξ, η) of the $R\infty^1$ formed by the pairs of I harmonically separate the values of (ξ, η) which correspond to the double elements of I.**

3·64. *Related involutions*

If (ξ, η) is any ratio parameter of the $R\infty^1$ formed by the pairs of the involution I given by (·613), then (ξ, η) is connected with (κ_1, κ_2) by equations of the form

$$\kappa_1 = p\xi + q\eta, \quad \kappa_2 = r\xi + s\eta,$$

with $ps \neq qr$, so (·613) may be written in the form

$$\xi(l_1' x^2 + m_1' xy + n_1' y^2) + \eta(l_2' x^2 + m_2' xy + n_2' y^2) = 0,$$

where $l_1' = pl_1 + rl_2, \quad m_1' = pm_1 + rm_2, \quad \ldots, \quad n_2' = qn_1 + sn_2.$

If there is a relation between the two $R\infty^1$ formed by the pairs of two involutions, the involutions are said to be related; any ratio parameter (ξ, η) of either $R\infty^1$ may be used as a ratio parameter of the other, corresponding pairs of the involutions being given by the same value of (ξ, η). Thus the pairs of two related involutions I and I' on the same $R\infty^1$, E, may be given by equations of the form

$$\xi(l_1 x^2 + m_1 xy + n_1 y^2) + \eta(l_2 x^2 + m_2 xy + n_2 y^2) = 0, \qquad (\cdot 641)$$

$$\xi(l_1' x^2 + m_1' xy + n_1' y^2) + \eta(l_2' x^2 + m_2' xy + n_2' y^2) = 0, \qquad (\cdot 642)$$

corresponding pairs of I and I' being given by the same value of

* Another proof of this theorem will be given in 4·65.

(ξ, η). The CONDITION for the pairs given by (·641), (·642) to be apolar on E is a quadratic equation in (ξ, η) so

In any relation between two involutions I and I' on E, two (possibly coincident) pairs of I are apolar on E to the corresponding pairs of I'.

3·65. *Representation of pairs of elements of an $R\infty^1$*

With a change of notation, 3·51 shows that the unordered pairs of values of a ratio variable (ξ, η) given by the equations

$$X\xi^2 + Y\xi\eta + Z\eta^2 = 0 \qquad (\cdot 651)$$

may be represented by the ratio sets (X, Y, Z). This representation will be denoted by R, (ξ, η) will be regarded as a ratio parameter of an $R\infty^1$, E, and X, Y, Z will be regarded as line co-ordinates of lines p with point equations

$$Xx + Yy + Zz = 0$$

in a plane π in which x, y, z are point co-ordinates.

The pair of elements of E given by (·651) contains a particular element e_1 IF

$$X\xi_1^2 + Y\xi_1\eta_1 + Z\eta_1^2 = 0,$$

i.e. IF the line p of π contains the point $(\xi_1^2, \xi_1\eta_1, \eta_1^2)$, which will be called P_1. Thus each element e of E determines ONE point P of π, given by

$$x : y : z = \xi^2 : \xi\eta : \eta^2,$$

and, as e varies in E, P describes the conic locus γ whose point equation is

$$y^2 = zx.$$

Conversely each point of γ arises from ONE element of E, so there is a proper (1, 1) correspondence between the elements of E and the points of γ, which will be denoted by T.*

The intersections of a line p of π with γ are obtained by substituting $x : y : z = \xi^2 : \xi\eta : \eta^2$ in the equation

$$Xx + Yy + Zz = 0,$$

so they are given by

$$X\xi^2 + Y\xi\eta + Z\eta^2 = 0.$$

But this equation also gives the pair of elements of E which is represented in R by the line p. So in T each pair of elements of E corresponds to the intersections of γ with the line which corresponds to the pair in R. Also any involution on E is obtained by making X, Y, Z satisfy a linear equation

$$aX + bY + cZ = 0,$$

* The curve γ is rational because its points may be identified with the values of the ratio variable (ξ, η).

which is equivalent to making the line p of π pass through the point (a, b, c). Thus:

The representation R of pairs of elements of E by lines of a plane π establishes a $(1, 1)$ correspondence T between the elements of E and the points of a conic locus γ in π. In T each pair of elements of E corresponds to the pair of intersections of γ with the corresponding line in R, and the pairs of elements of an involution on E correspond in T to the pairs of intersections of γ with the lines of a pencil.

3·66. *Method of transformation*

It follows from 3·65 that the theory of pairs of elements of any $R\infty^1$ may be reduced to the simple theory of the intersections of a conic locus with the lines of its plane, which will be discussed in the next chapter. This notion is typical of the trend of modern geometry. In this chapter theorems applicable to any $R\infty^1$, E, have been proved by applying elementary algebra to ratio parameters of E. But if E' is any other $R\infty^1$, and if a relation is established between E and E', any theorem about E which is essentially a theorem about the values of a ratio parameter of E is equivalent to a theorem about E'. It may well happen that a theorem about E of this nature may be more easily proved by choosing E' suitably and using geometrical arguments to establish the equivalent theorem about E' than by direct application of algebra. Thus 3·65 suggests that a theorem about pairs of elements of E may be most easily proved by taking E' to be a conic locus. Similarly theorems about sets of three elements of E may often be most easily proved by taking E' to be a rational cubic curve in space.

The same notion applies to irrational ∞^1 systems and to systems of freedom greater than one. A geometrical problem is not necessarily solved in the form in which it arises, but transformations are used to reduce the problem to an equivalent one which is either known or more easy to solve. For example, in the theory of irrational curves it is shown that any curve may be transformed into one of a set of standard curves, called canonical curves, so that the theory of these standard curves is applicable to all curves. The property of rational ∞^1 systems which makes their theory so easy is the fact that any two $R\infty^1$ may be related to each other.

CHAPTER IV

THE CONIC

4·1. Conic locus and conic scroll

4·11. *Proper conic locus, line-pair, repeated line*

In 2·42 a conic locus or a point conic was defined to be an $A\,\infty^1$ of points in a plane given by a quadratic point equation, which may be written

$$s \equiv ax^2+by^2+cz^2+2fyz+2gzx+2hxy = 0. \qquad (\cdot 111)$$

The quadratic form s may be the product of two linear factors, when the conic locus consists of the points of two lines p_1, p_2; the conic locus is then a *line-pair* and is denoted by $\| p_1, p_2 \|$; the point $p_1 p_2$ is called the *vertex* of the line-pair. In particular s may be the square of a linear form, when the conic consists of the points of one line p_1 counted twice; the conic locus is thus the line-pair $\| p_1, p_1 \|$ and is called a *repeated line*; every point of p_1 is a vertex of the repeated line. A line-pair which is composed of distinct lines is called a *proper line-pair*, to distinguish it from a *repeated line*. When s is not the product of two linear factors, the conic locus is said to be a *proper conic locus*.

The quadratic form s will sometimes be denoted by

$$(abcfgh \between xyz)^2$$

and the conic locus $s = 0$ will be denoted by s. Until Chapter IX we shall use the notation

$$s_{12} \equiv s_{21} \equiv (ax_1+hy_1+gz_1)\,x_2+(hx_1+by_1+fz_1)\,y_2+(gx_1+fy_1+cz_1)\,z_2$$
$$\equiv (ax_2+hy_2+gz_2)\,x_1+(hx_2+by_2+fz_2)\,y_1+(gx_2+fy_2+cz_2)\,z_1, \qquad (\cdot 112)$$

$$s_1 \equiv (ax_1+hy_1+gz_1)\,x+(hx_1+by_1+fz_1)\,y+(gx_1+fy_1+cz_1)\,z. \qquad (\cdot 113)$$

If p is any line, and P_1, P_2 are two distinct points of p, the general point of p is

$$(\lambda x_1+\mu x_2, \lambda y_1+\mu y_2, \lambda z_1+\mu z_2),$$

and lies on s IF

$$\lambda^2 s_{11}+2\lambda\mu s_{12}+\mu^2 s_{22} = 0. \qquad (\cdot 114)$$

Since (λ, μ) is a ratio parameter of p, the intersections of p with s correspond to the roots of (·114), regarded as a quadratic equation in (λ, μ).

If P_1 is a *double point* of s, i.e. if s is a line-pair with vertex P_1 or a repeated line through P_1,* and if P_2 is a *general* point of the plane, the equation (·114) has two roots coinciding in $(1, 0)$;† in this case $s_{11} = s_{12} = 0$. Since $s_{12} = 0$ for a general P_2 it follows from (·112) that the equations

$$\left.\begin{aligned} ax_1 + hy_1 + gz_1 &= 0, \\ hx_1 + by_1 + fz_1 &= 0, \\ gx_1 + fy_1 + cz_1 &= 0, \end{aligned}\right\} \tag{·115}$$

are separately true, and these also imply $s_{11} = 0$, which is the CONDITION for P_1 to lie on s. From (·115) it follows that the determinant

$$\delta \equiv \begin{vmatrix} a & h & g \\ h & b & f \\ g & f & c \end{vmatrix} \tag{·116}$$

vanishes. Conversely, if $\delta = 0$, the equations (·115) can be solved to give a point P_1 which is such that $s_{11} = s_{12} = 0$ for *general* P_2 and which is therefore, by (·114), a double point of s. If δ is of rank two the solution of (·115) is unique; if it is of rank one the equations are identical and there is a line of points P_1 each of which is a double point of s. Thus:

The curve $s = 0$ is a proper conic locus, a proper line-pair, or a repeated line according as the determinant δ is of rank three, two, or one. If δ is of rank three, the equations

$$\left.\begin{aligned} ax + hy + gz &= 0, \\ hx + by + fz &= 0, \\ gx + fy + cz &= 0, \end{aligned}\right\} \tag{·117}$$

are independent. If δ is of rank two they have ONE *solution which is the vertex of the line pair $s = 0$. If δ is of rank one all three equations give the same line, and $s = 0$ is this line repeated.*

4·12. *Chords and tangents of a proper conic locus*

If s is a proper conic locus the equation (·114) cannot be an identity, as this would imply that p is part of s; therefore p has two intersections with s, which will be denoted by (p, s). In general these intersections are distinct, but if (·114) has a double root the intersections (p, s) coincide in a point P; the line p is then said to be a tangent of s.

* See 2·44, p. 55.

† If the line P_1P_2 forms part of s the equation (·114) is an identity.

If P_1 is any point on a proper conic locus s, ONE *line through P_1 is a tangent of s; this line is called the tangent to s at P_1, and its point equation is $s_1 = 0$.*

For let p be any line through P_1 and let P_2 be any point of p distinct from P_1. Then $s_{11} = 0$, since P_1 lies on s, and $(1, 0)$ is one of the roots of (·114). Thus p is a tangent of s IF $(1, 0)$ is a double root of (·114), i.e. IF $s_{12} = 0$, i.e. IF P_2 lies on $s_1 = 0$. But $s_1 = 0$ is the point equation of a line through P_1, since $s_{11} = 0$, so P_2 lies on $s_1 = 0$ IF p is $s_1 = 0$. In fact each point P_1 of s determines ONE line p_1 whose intersections with s coincide in P_1. Thus:

If P_1, P_2 are two points, distinct or coincident, which lie on a proper conic locus s, there is ONE *line whose two intersections with s are P_1 and P_2; this line is called the chord P_1P_2, and is the join of P_1 and P_2 when these points are distinct; if P_2 is identical with P_1, the chord P_1P_2 or P_1P_1 is the tangent to s at P_1.*

4·13. *A proper conic locus is rational*

Let A be any fixed point of a proper conic locus s. By 4·12 each line p through A determines ONE point P of s such that the intersections (p, s) are A and P; for a general position of p this point P is distinct from A, but when p is the tangent to s at A the point P coincides with A. Also each point P of s determines one line p through A such that the intersections (p, s) are A and P; when P is A the line p is the tangent at A. A correspondence has thus been established between the points of s and the pencil of lines with vertex A in which each point of s corresponds to ONE line of the pencil and each line of the pencil corresponds to ONE point of s. Since the lines of a pencil may be identified with the values of a ratio variable it follows that the points of s form an $R\infty^1$; in other words a proper conic locus is a rational curve.*

4·14. *By a suitable choice of a system of co-ordinates the points of any proper conic locus may be determined by the equations*

$$x : y : z = \xi^2 : \xi\eta : \eta^2, \tag{·141}$$

where (ξ, η) is a ratio variable, or by the equations

$$x : y : z = \lambda^2 : \lambda : 1, \tag{·142}$$

where λ is a variable.

* See 1·63 (ix), 2·23, 2·41, pp. 20, 46, 51.

For let X, Z, U be three distinct points of a proper conic s and let the tangents to s at X and Z meet in Y. Since s is proper the lines YZ, ZX, XY have no intersections with s other than X and Z, so U does not lie on any of these lines. It is therefore possible to choose a system of point co-ordinates x, y, z in which XYZ is the triangle of reference and U is the unit point.* Suppose that with this system of co-ordinates the point equation of s is

$$s \equiv (abcfgh \between xyz)^2 = 0.$$

Then by 4·12 the tangents to s at X and Z are

$$ax+hy+gz = 0, \quad gx+fy+cz = 0,$$

and since these are known to be $z = 0$, $x = 0$, it follows that

$$a = h = f = c = 0.$$

Since U is $(1, 1, 1)$ and lies on s

$$b+2g = 0,$$

so the equation of s is $\quad y^2 - zx = 0,$

which is equivalent to $\quad x:y = y:z.$

Thus for each value of (ξ, η) or of λ the point (x, y, z) determined by (·141) or (·142) lies on s and conversely each point of s is so determined by ONE value of (ξ, η) or of λ. This is another proof of the theorem of 4·13, that a proper conic locus is a rational curve.

4·15. *Proper conic scroll, point-pair, repeated point*

The dual of a conic locus, called a conic scroll or a line conic, was defined in 2·43 to be an $A\,\infty^1$ of lines in a plane given by a quadratic line equation, which may be written

$$S \equiv AX^2 + BY^2 + CZ^2 + 2FYZ + 2GZX + 2HXY = 0. \quad (\cdot 151)$$

If S is the product of two linear factors the conic scroll consists of the lines of two pencils with vertices P_1, P_2; the conic scroll is then called a *point-pair* and is denoted by $\| P_1, P_2 \|$; the line P_1P_2 is called the *axis* of the point-pair. In particular if S is the square of a linear form the conic scroll consists of the lines of one pencil counted twice; if P_1 is the vertex of this pencil, the conic scroll is the point-pair $\| P_1, P_1 \|$ and is called a *repeated point*; every line through P_1 is an axis of the repeated point. A point-pair which is composed of distinct pencils of lines is called a *proper point-pair*. When S is not the product of two linear factors, the conic scroll $S = 0$ is said to be a *proper conic scroll.*

* See 1·64, p. 21.

The form S of (·151) is denoted by $(ABCFGH\between XYZ)^2$, and the conic scroll $S = 0$ is denoted by S. The symbols S_1 and S_{12} have similar interpretations to those of s_1 and s_{12} but with capital instead of small letters.

If P is any point, and p_1, p_2 are two distinct lines through P, the general line through P is

$$[\lambda X_1 + \mu X_2, \quad \lambda Y_1 + \mu Y_2, \quad \lambda Z_1 + \mu Z_2],$$

and belongs to S IF

$$\lambda^2 S_{11} + 2\lambda\mu S_{12} + \mu^2 S_{22} = 0. \qquad (\cdot 152)$$

Since (λ, μ) is a ratio parameter of the pencil of lines with vertex P, the lines of S through P correspond to the roots of (·152) in (λ, μ).

If p_1 is a *double line* of S, i.e. if S is a point-pair with axis p_1 or a repeated point on p_1,* and if p_2 is a *general* line of the plane, the equation (·152) has two roots coinciding in $(1, 0)$; in this case $S_{11} = S_{12} = 0$. Since $S_{12} = 0$ for general p_2 it follows from the definition of S_{12} that the equations

$$\left.\begin{aligned} AX_1 + HY_1 + GZ_1 &= 0, \\ HX_1 + BY_1 + FZ_1 &= 0, \\ GX_1 + FY_1 + CZ_1 &= 0, \end{aligned}\right\} \qquad (\cdot 153)$$

are separately true, and these also imply $S_{11} = 0$, which is the CONDITION for p_1 to belong to S. From (·153) it follows that the determinant

$$\Delta \equiv \begin{vmatrix} A & H & G \\ H & B & F \\ G & F & C \end{vmatrix} \qquad (\cdot 154)$$

vanishes. Conversely, if $\Delta = 0$, the equations (·153) can be solved to give a line p_1 which is such that $S_{11} = S_{12} = 0$ for *general* p_2 and which is therefore, by (·152), a double line of S. If Δ is of rank two the solution of (·153) is unique; if it is of rank one the equations are identical and there is a pencil of lines p_1 each of which is a double line of S. Thus

The scroll $S = 0$ is a proper conic scroll, a proper point-pair, or a repeated point according as the determinant Δ is of rank three, two or one. If Δ is of rank three, the equations

$$\left.\begin{aligned} AX + HY + GZ &= 0, \\ HX + BY + FZ &= 0, \\ GX + FY + CZ &= 0, \end{aligned}\right\} \qquad (\cdot 155)$$

* See 2·45, p. 56.

are independent. If Δ is of rank two they have ONE *solution which is the axis of the point-pair $S = 0$. If Δ is of rank one all three equations give the same point, and $S = 0$ is this point repeated.*

The above theorem follows at once from that of 4·11 by the principle of duality,* but the dual argument has been given in full for the reason explained in 1·67.

4·16. *Contacts of a proper conic scroll*

If S is a proper conic scroll the equation (·152) cannot be an identity, as this would imply that all lines of the pencil with vertex P belong to S; this pencil therefore has two lines in common with S, which will be denoted by $[P, S]$. In general these lines are distinct, but if (·152) has a double root the lines $[P, S]$ coincide in a line p; the point P is then said to be a contact of S.

By an argument dual to that of 4·12 we establish the following results.

If p_1 is any line of a proper conic scroll S, ONE *point of p_1 is a contact of S; this point is called the contact of p_1 with S and its equation is $S_1 = 0$.*

If p_1, p_2 are two lines, distinct or coincident, which belong to a proper conic scroll S, there is ONE *point P_{12} such that the lines $[P_{12}, S]$ are p_1 and p_2; this point P_{12} is the meet p_1p_2 of p_1 and p_2 when these lines are distinct; if p_2 is identical with p_1, the point P_{12} or P_{11} is the contact of p_1 with S_1, and may be denoted by p_1p_1.*†

4·17. *A proper conic scroll is rational*

This follows at once by dualizing the argument of 4·13. If a is any fixed line of a proper conic scroll S each point P on a determines ONE line p of S such that the lines $[P, S]$ are a and p, p coinciding with a when P is the contact of a with S; conversely each line p of S determines ONE point P of a such that the lines $[P, S]$ are a and p, P being the contact of a when p and a coincide. There is thus established a $(1, 1)$ correspondence between the lines p of S and the points P of a; since the points of a may be identified with the values of a ratio variable it follows that the lines of S form an $R\infty^1$.

* See 1·66, p. 24.

† Unfortunately there is no established name for the point P_{12}, which is the dual of the chord P_1P_2 of a proper conic locus s.

4·2. Proper conic locus

4·21. *Cross-ratio on a proper conic locus*

Since the points of a proper conic locus s form an $R\infty^1$, as was shown in 4·13 and 4·14, the cross-ratio on s of four points P_1, P_2, P_3, P_4 of s is defined by 3·13, and is equal to the cross-ratio $(\lambda_1, \lambda_2, \lambda_3, \lambda_4)$ of the corresponding values of any parameter λ of the $R\infty^1$. As long as only one proper conic locus s is being considered the cross-ratio of four points P_1, P_2, P_3, P_4 on s may be denoted by (P_1, P_2, P_3, P_4) without ambiguity.* A parameter of the $R\infty^1$ formed by the points of s will be called a parameter of s.

4·22. *If A and B are two fixed points of a proper conic locus s, and if P is a variable point of s, the chords AP, BP form related pencils with vertices A, B. The line AB when regarded as a line of the first or second pencil corresponds to the tangent to s at B or A respectively.*

For, if $\{P\}$ denotes the range formed by the points P of s, then it follows from 4·13 that

$$\{P\} \equiv A\{P\}, \qquad (\cdot 221)$$

where $A\{P\}$ denotes the pencil formed by the chords AP of s. Also in this relation the point A of $\{P\}$ corresponds to the chord AA of s, which is the tangent to s at A. Similarly

$$\{P\} \equiv B\{P\}, \qquad (\cdot 222)$$

so

$$A\{P\} \equiv \{P\} \equiv B\{P\}. \qquad (\cdot 223)$$

In these relations the chord AB of $A\{P\}$ corresponds to the point B of $\{P\}$ which corresponds to the chord BB of $B\{P\}$, which is the tangent to s at B. By 3·22 it follows from (·223) that:

If P_1, P_2, P_3, P_4 are any four points of s, then

$$A(P_1, P_2, P_3, P_4) = (P_1, P_2, P_3, P_4) = B(P_1, P_2, P_3, P_4).$$

If P_1 is A, then $A(P_1, P_2, P_3, P_4)$ is to be interpreted as the cross-ratio of the subpencil $\{a, AP_2, AP_3, AP_4\}$, where a is the tangent to s at A.

4·23. *The point equations of the corresponding lines of any two related pencils of lines may be written in the form*

$$\lambda\alpha + \alpha' = 0, \quad \lambda\beta + \beta' = 0, \qquad (\cdot 231)$$

corresponding lines being given by the same value of λ.

For, if A and B are the vertices of the pencils, and if any two distinct lines through A and any two distinct lines through B have

* See 3·16, p. 78.

point equations $\alpha = 0$, $\alpha' = 0$ and $\beta_1 = 0$, $\beta_1' = 0$, then the general lines of the pencils are

$$\lambda\alpha + \alpha' = 0 \quad \text{and} \quad \mu\beta_1 + \beta_1' = 0.$$

Also λ and μ are parameters of the pencils, so the relation is determined by a proper bilinear equation

$$a\lambda\mu + b\lambda + c\mu + d = 0,$$

which is equivalent to $$\mu = -\frac{b\lambda + d}{a\lambda + c}.$$

The line $\mu\beta_1 + \beta_1' = 0$ which corresponds to $\lambda\alpha + \alpha' = 0$ is therefore

$$\lambda(b\beta_1 - a\beta_1') + (d\beta_1 - c\beta_1') = 0,$$

so on taking $$\beta \equiv b\beta_1 - a\beta_1', \quad \beta' \equiv d\beta_1 - c\beta_1',$$

the theorem is established. Note that the lines $\beta = 0$, $\beta' = 0$ are distinct because $bc \neq ad$.

It is essential to the above argument that the correspondence between the two pencils is a relation and not an improper (1, 1) correspondence. For, if $bc = ad$, the lines $\beta = 0$, $\beta' = 0$ are identical, so the equation of a general line through B cannot be written in the form $\lambda\beta + \beta' = 0$.

4·24. *When AB is not a self-corresponding line of two related pencils with distinct vertices A and B, the locus of meets of corresponding lines is a proper conic locus s through A and B.*

For by 4·23 the equations of corresponding lines may be written in the form

$$\lambda\alpha + \alpha' = 0, \quad \lambda\beta + \beta' = 0,$$

and on eliminating λ we see that the locus of meets is the conic locus s given by

$$\begin{vmatrix} \alpha & \alpha' \\ \beta & \beta' \end{vmatrix} = 0$$

or $$\alpha\beta' - \alpha'\beta = 0.$$

One pair of corresponding lines is AB and a line b_0 through B, which is distinct from AB; the meet of this pair is B, so B lies on s. Also any line through B other than b_0 corresponds to a line through A distinct from AB; so any line through B other than b_0 meets s in ONE point other than B, while b_0 has no further intersection with s. If s were a line pair, no point of s could have this property, so s is proper and b_0 is the tangent to s at B. Similarly A lies on s and the tangent to s at A is the line through A corresponding to AB regarded as a line through B. This generation of a conic locus by related pencils is called a *projective generation*.

If instead of a relation between the pencils there is an improper (1, 1) correspondence, one line a_0 through A corresponds to all lines through B, while one line b_0 through B corresponds to all lines through A. The locus of meets of corresponding lines is then the line-pair $\| a_0, b_0 \|$.

4·25. *When AB is a self-corresponding line of two related pencils with distinct vertices A and B, the locus of meets of corresponding lin es other than AB is a line.*

For the line $\alpha = 0$ of (·231) may be any line through A, and $\beta = 0$ is then the corresponding line through B. Choosing $\alpha = 0$ to be AB, the equations of corresponding lines are of the form

$$\lambda\alpha + \alpha' = 0, \quad \lambda\alpha + \alpha'' = 0.$$

The locus of meets of corresponding lines is then

$$\alpha(\alpha' - \alpha'') = 0,$$

i.e. AB and the line $\alpha' - \alpha'' = 0$.

Alternatively, if p is the line joining the meets P_1, P_2 of two pairs of corresponding lines of the pencils, and if P is a variable point of p, then the lines AP, BP form related pencils in which $AP_1 \to BP_1$, $AP_2 \to BP_2$, $AB \to BA$. So by 3·23 this relation is the given relation, and all pairs of corresponding lines in the given relation meet on p.

4·26. *Three points P_1, P_2, P_3 are collinear* IF

$$A(B, P_1, P_2, P_3) = B(A, P_1, P_2, P_3),$$

where A and B are any two points such that AB does not contain P_1, P_2 or P_3.

This follows immediately from 4·25, and an equivalent statement is

If a_1, a_2, a_3 are three lines through A and b_1, b_2, b_3 are three lines through B, all these lines being distinct from AB, then the points a_1b_1, a_2b_2, a_3b_3 are collinear IF $(c, a_1, a_2, a_3) = (c, b_1, b_2, b_3)$, *where c denotes AB.*

4·27. *Five general points lie on* ONE *conic locus*

For, if $A_1, A_2, \ldots, A_5$ are the points, the lines joining A_1 and A_2 to the remaining three points determine two related pencils with vertices A_1 and A_2. By 4·24 the locus of meets of corresponding lines of these pencils is a conic locus through the five points, and by 4·22 any conic locus through the five points may be so obtained.

If three of the points, A_1, A_2, A_3, lie on a line, there is still ONE conic locus consisting of this line and A_4A_5. But if four of the points lie on a line, there are ∞^1 conic loci consisting of this line and any line through the fifth point.

If A_1, A_2, A_3, A_4 are four general points, and a_1 is a general line through A_1, ONE *conic locus through the four points touches a_1 at A_1.*

For there is ONE relation between the pencils of lines with vertices A_1 and A_2 in which the lines a_1, A_1A_3, A_1A_4 correspond to A_2A_1, A_2A_3, A_2A_4. The meets of corresponding lines of the related pencils form a conic locus through the four points and touching a_1 at A_1, and any such conic locus may be so obtained. In particular, if a_1 is the line A_1A_2, there is still ONE conic locus, the line-pair $\| A_1A_2, A_3A_4 \|$. Similarly

If A_1, A_2, A_3 are independent points, there is ONE *conic locus through A_1, A_2, A_3 with prescribed tangents at A_1 and A_2.*

4·28. *If A_1, A_2, A_3, A_4 are four general points, and P, Q are two other points such that*

$$P(A_1, A_2, A_3, A_4) = Q(A_1, A_2, A_3, A_4),$$

then the six points lie on a conic locus.

For the lines joining P, Q to A_1, A_2, A_3 determine a relation between the pencils of lines with vertices P, Q, and the locus of meets of corresponding lines is the conic locus through P, Q, A_1, A_2, A_3. But since $P(A_1, A_2, A_3, A_4) = Q(A_1, A_2, A_3, A_4)$, the line PA_4 corresponds to QA_4, so A_4 lies on this conic locus.

4·29. *If A_1, A_2, A_3, A_4 are four general points, the locus of a point P such that the cross-ratio $P(A_1, A_2, A_3, A_4)$ has a constant value is a conic locus through A_1, A_2, A_3, A_4.*

For if P_0 is one such point, any point P such that

$$P(A_1, A_2, A_3, A_4) = P_0(A_1, A_2, A_3, A_4)$$

lies on the conic locus through P_0, A_1, A_2, A_3, A_4, by 4·28.

4·3. Proper conic scroll

4·31. *Cross-ratio on a proper conic scroll*

The dual theorems to those of 4·2 follow without further proof by the principle of duality, and we also know that proofs of these dual theorems may be obtained by replacing each step in the reasoning of 4·2 by its dual. But a consideration of the dual reasoning will lead to a greater familiarity with the dual attitudes to plane geometry which were described in 1·66, and which should in practice be borne in mind simultaneously.

The cross-ratio on a proper conic scroll S of four lines p_1, p_2, p_3, p_4 of S is defined by 3·13, is denoted by (p_1, p_2, p_3, p_4), and is equal to the cross-ratio $(\lambda_1, \lambda_2, \lambda_3, \lambda_4)$ of the corresponding values of any parameter λ of S.

4·32. *If a and b are two fixed lines of a proper conic scroll S, and if p is a variable line of S, the points ap, bp form related ranges on a and b. The point ab when regarded as a point of the range on a or b corresponds to the contact with S of b or a respectively.**

For, if $\{p\}$ denotes the range formed by the lines p of S_1 it follows from 4·17 that

$$\{p\} \equiv a\{p\},$$

where $a\{p\}$ denotes the range on a formed by the points ap. The line a of $\{p\}$ corresponds to the point aa, which is the contact of S with a. Similarly

$$\{p\} \equiv b\{p\},$$

so

$$a\{p\} \equiv \{p\} \equiv b\{p\}.$$

In these relations, the point ab of $a\{p\}$ corresponds to the line b of $\{p\}$, which corresponds to the point bb of $b\{p\}$, which is the contact of b with S.

It follows that:

If p_1, p_2, p_3, p_4 are any four lines of S, then

$$a(p_1, p_2, p_3, p_4) = (p_1, p_2, p_3, p_4) = b(p_1, p_2, p_3, p_4).$$

If p_1 is a, then $a(p_1, p_2, p_3, p_4)$ is to be interpreted as the cross-ratio of the subrange $\{A, ap_2, ap_3, ap_4\}$, where A is the contact of S with a.

* In the above statement the points aa and bb are to be interpreted as the contacts of S with a and b, in accordance with 4·16.

4·33. *The line equations of corresponding points of any two related ranges on two lines may be written in the form*

$$\lambda A + A' = 0, \quad \lambda B + B' = 0, \qquad (\cdot 331)$$

corresponding points being given by the same value of λ.

For if a and b are the two lines, if any two distinct points of a are

$$A \equiv a_1 X + a_2 Y + a_3 Z = 0, \quad A' \equiv a_1' X + a_2' Y + a_3' Z = 0,$$

and if any two distinct points of b are

$$B_1 \equiv b_{11} X + b_{12} Y + b_{13} Z = 0, \quad B_1' \equiv b_{11}' X + b_{12}' Y + b_{13}' Z = 0,$$

then the general points of a and b are

$$\lambda A + A' = 0, \quad \mu B_1 + B_1' = 0.$$

The relation is determined by a proper bilinear equation

$$p\lambda\mu + q\lambda + r\mu + s = 0,$$

in the parameters λ, μ and by taking

$$B \equiv qB_1 - pB_1', \quad B' \equiv sB_1 - rB_1',$$

the theorem is established. The points $B = 0$, $B' = 0$ of b are distinct because $qr \neq ps$.

If instead of a relation there is an improper (1, 1) correspondence between a and b, then $qr = ps$ and the points $B = 0$, $B' = 0$ are identical.

4·34. *When ab is not a self-corresponding point of related ranges on two distinct lines a and b, the joins of corresponding points form a proper conic scroll S containing a and b.*

For the line equations of corresponding points may be written in the form

$$\lambda A + A' = 0, \quad \lambda B + B' = 0,$$

and the scroll formed by the joins of corresponding points is therefore the conic scroll S given by

$$AB' - A'B = 0.$$

One pair of corresponding points is ab and a point B_0 of b distinct from ab; the join of this pair is b, so b belongs to S. Also any point of b other than B_0 lies on ONE line of S other than b, while B_0 lies on no line of S other than b. If S were a point pair, no line of S could have this property, so S is proper and B_0 is the contact of S with b. Similarly a belongs to S and the contact of S with a is the point of a which corresponds to ab regarded as a point of b. This generation of a conic scroll by related ranges on two lines is called a projective generation.

If instead of a relation there is an improper (1, 1) correspondence between a and b, the joins of corresponding lines form a point-pair $\| A_0, B_0 \|$, where A_0 and B_0 are the singular points on a and b.

4·35. *When ab is a self-corresponding point of related ranges on two distinct lines, the joins of corresponding points other than ab form a pencil of lines.*

For when ab is chosen as the point $A = 0$ of (·331), the line equations of corresponding points are of the form

$$\lambda A + A' = 0, \quad \lambda A + A'' = 0.$$

The scroll formed by the joins of corresponding points is then

$$A(A' - A'') = 0,$$

i.e. the pencil of lines with vertex ab and the pencil with vertex at the point $A' - A'' = 0$.

An alternative proof, dual to that of 4·25, has already been given in 3·25.

4·36. *Three lines p_1, p_2, p_3 are concurrent* IF

$$a(b, p_1, p_2, p_3) = b(a, p_1, p_2, p_3),$$

where a and b are any two lines such that ab does not lie on p_1, p_2 or p_3.

This follows immediately from 3·25 and an equivalent statement is

If A_1, A_2, A_3 are three points of a and B_1, B_2, B_3 are three points of b, all these points being distinct from ab, then the lines A_1B_1, A_2B_2, A_3B_3 are concurrent IF $(C, A_1, A_2, A_3) = (C, B_1, B_2, B_3)$, *where C denotes ab.*

4·37. *Five general lines belong to* ONE *conic scroll*

For if $a_1, a_2, \ldots, a_5$ are the lines, the meets of a_1 and a_2 with the remaining three lines determine a relation between a_1 and a_2. By 4·34 the joins of corresponding points form a conic scroll containing the five lines, and by 4·32 any such conic scroll may be so obtained. If a_1, a_2, a_3 meet in a point there is still ONE conic scroll, consisting of the lines through this point and the lines through a_4a_5. But if a_1, a_2, a_3, a_4 meet in a point there are ∞^1 conic scrolls consisting of the lines through this point and the lines through any point of a_5.

If a_1, a_2, a_3, a_4 are four general lines, and A_1 is a general point of a_1, ONE *conic scroll containing the four lines has A_1 as its contact with a_1.*

For there is ONE relation between a_1 and a_2 in which A_1, a_1a_3, a_1a_4 correspond to a_2a_1, a_2a_3, a_2a_4. The joins of corresponding lines form a conic scroll containing the four lines and having A_1 as its contact with a_1, and any such conic scroll may be so obtained. In particular, if A_1 is a_1a_2, there is still ONE conic scroll, the point-pair $\| A_1A_2, A_3A_4 \|$. Similarly

If a_1, a_2, a_3 are independent lines, ONE *conic scroll contains a_1, a_2, a_3 and has prescribed contacts with a_1 and a_2.*

4·38. *If a_1, a_2, a_3, a_4 are four general lines, and p, q are two other lines such that*

$$p(a_1, a_2, a_3, a_4) = q(a_1, a_2, a_3, a_4),$$

then the six lines belong to a conic scroll.

For the meets of p, q with a_1, a_2, a_3 determine a relation between p, q, and the joins of corresponding points form a conic scroll containing p, q, a_1, a_2, a_3. But since $p(a_1, a_2, a_3, a_4) = q(a_1, a_2\ a_3, a_4)$, the point pa_4 corresponds to qa_4, so a_4 belongs to this conic scroll.

4·39. *If a_1, a_2, a_3, a_4 are four general lines, the lines p such that the cross-ratio $p(a_1, a_2, a_3, a_4)$ has a constant value form a conic scroll containing a_1, a_2, a_3, a_4.*

For if p_0 is one such line, any line p such that

$$p(a_1, a_2, a_3, a_4) = p_0(a_1, a_2, a_3, a_4)$$

belongs to the conic scroll which contains p_0, a_1, a_2, a_3, a_4, by 4·38.

4·4. Parametric equations

4·41. *General quadratic parametric equations*

Consider the equations

$$x : y : z = (a_1\lambda^2 + b_1\lambda + c_1) : (a_2\lambda^2 + b_2\lambda + c_2) : (a_3\lambda^2 + b_3\lambda + c_3), \tag{·411}$$

and suppose first that the determinant

$$m = \begin{vmatrix} a_1 & b_1 & c_1 \\ a_2 & b_2 & c_2 \\ a_3 & b_3 & c_3 \end{vmatrix}$$

does not vanish, so that the equations may be solved for $\lambda^2:\lambda:1$, giving

$$\lambda^2:\lambda:1 = (A_1x+A_2y+A_3z):(B_1x+B_2y+B_3z):(C_1x+C_2y+C_3z), \quad (\cdot 412)$$

where $A_1, A_2, \ldots, C_3$ are the cofactors of $a_1, a_2, \ldots, c_3$ in m. Then each value of λ determines ONE point (x, y, z), and the points (x, y, z) obtained by allowing λ to take all values are the points of the conic locus s given by

$$(B_1x+B_2y+B_3z)^2 = (A_1x+A_2y+A_3z)(C_1x+C_2y+C_3z). \quad (\cdot 413)$$

Also each point of s corresponds to ONE value of λ, by ($\cdot$412), so a parameter of s.

Since

$$\begin{vmatrix} a_1 & b_1 & c_1 \\ a_2 & b_2 & c_2 \\ a_3 & b_3 & c_3 \end{vmatrix}\begin{vmatrix} A_1 & A_2 & A_3 \\ B_1 & B_2 & B_3 \\ C_1 & C_2 & C_3 \end{vmatrix} = \begin{vmatrix} m & 0 & 0 \\ 0 & m & 0 \\ 0 & 0 & m \end{vmatrix},$$

it follows that

$$\begin{vmatrix} A_1 & A_2 & A_3 \\ B_1 & B_2 & B_3 \\ C_1 & C_2 & C_3 \end{vmatrix} = m^2 \neq 0,$$

so the transformation

$$\left.\begin{aligned} x' &= A_1x+A_2y+A_3z, \\ y' &= B_1x+B_2y+B_3z, \\ z' &= C_1x+C_2y+C_3z, \end{aligned}\right\} \quad (\cdot 414)$$

is non-singular and therefore gives a transformation of co-ordinates.* With x', y', z' as co-ordinates s is given by

$$x':y':z' = \lambda^2:\lambda:1,$$

and is therefore the proper conic locus

$$y'^2 = z'x'.$$

Thus:

If $m \neq 0$ the equations ($\cdot$411) *are parametric equations of a proper conic locus.*†

4·42. *Particular cases*

Suppose that the determinant m is of rank two and that the polynomials

$$a_1\lambda^2+b_1\lambda+c_1, \quad a_2\lambda^2+b_2\lambda+c_2, \quad a_3\lambda^2+b_3\lambda+c_3 \quad (\cdot 421)$$

* See 1·65, p. 22.

† Parametric equations of a conic locus, or of any rational curve, are equations which express the co-ordinates of the points of the curve as polynomials in a parameter of the curve.

have no common factor.* Then there are numbers κ_1, κ_2, κ_3 not all zero such that

$$\kappa_1 a_1 + \kappa_2 a_2 + \kappa_3 a_3 = 0,$$
$$\kappa_1 b_1 + \kappa_2 b_2 + \kappa_3 b_3 = 0,$$
$$\kappa_1 c_1 + \kappa_2 c_2 + \kappa_3 c_3 = 0,$$

so all the points (x, y, z) given by (·411) lie on the line

$$\kappa_1 x + \kappa_2 y + \kappa_3 z = 0.$$

A general point of this line corresponds to two values of λ.

If m is of rank two, and the polynomials (·421) have a common factor $\lambda - \lambda_0$, the value λ_0 of λ does not determine a point (x, y, z), but all other values of λ determine points of a certain line. If m is of rank one, there are two values of λ for which all three polynomials (·421) vanish; neither of these determines a point, while all other values of λ determine the same point.

4·43. *If P_1, P_2, P_3 are three distinct points of a proper conic locus s, it is possible to choose a co-ordinate system so that s is given by*

$$x : y : z = \lambda^2 : \lambda : 1$$

and P_1, P_2, P_3 correspond to distinct assigned values λ_1, λ_2, λ_3 of λ.

For by 3·15 it is possible to choose a parameter λ of s so that P_1, P_2, P_3 correspond to the values λ_1, λ_2, λ_3 of λ. Also by 4·22 and 4·23 the points of s are given by equations of the form

$$\mu\alpha + \alpha' = 0, \quad \mu\beta + \beta' = 0, \tag{·431}$$

where μ is a parameter of s. But by 3·12 μ is related to λ by an equation of the form

$$\mu = -\frac{b\lambda + d}{a\lambda + c},$$

so, on substitution in (·431), the points of s are given by

$$\lambda(b\alpha - a\alpha') + (d\alpha - c\alpha') = 0, \quad \lambda(b\beta - a\beta') + (d\beta - c\beta') = 0. \tag{·432}$$

Parametric equations of s in terms of λ are obtained by solving these two equations for $x : y : z$. Since the equations are linear in x, y, z and in λ, the solution is of the form (·411), and the determinant

* Note that, with the usual attitude to ∞, this means that a_1, a_2, a_3 are not all zero; for if they were all zero the corresponding forms in the ratio variable (ξ, η) given by $\xi = \lambda\eta$ would have η as a common factor.

m is not zero because s is proper. It follows from 4·41 that a co-ordinate system x', y', z' can be chosen so that s is given by

$$x':y':z' = \lambda^2:\lambda:1,$$

and this proves the theorem. In 4·14 the parameter λ was actually chosen so that the points X, Z, U were given by the values ∞, 0, 1 of λ.

It follows from 4·42 that if $m = 0$ the equations (·441) will in general determine a repeated line, but that they can never determine a proper line-pair. The reduction to the form

$$x':y':z' = \lambda^2:\lambda:1$$

breaks down when $m = 0$ because the transformation (·414) does not then give a transformation of co-ordinates.

4·44. *If p_1, p_2, p_3 are three distinct lines of a proper conic scroll S, it is possible to choose a co-ordinate system so that S is given by*

$$X:Y:Z = \lambda^2:\lambda:1$$

and p_1, p_2, p_3 correspond to distinct assigned values $\lambda_1, \lambda_2, \lambda_3$ of λ.

This is the dual of 4·43, and the reader should go through the dual proof.

4·45. *Tangents of $x:y:z = \lambda^2:\lambda:1$*

Consider the proper conic locus s given by $x:y:z = \lambda^2:\lambda:1$. The point $(\lambda_i^2, \lambda_i, 1)$ of s given by $\lambda = \lambda_i$ will be called the point λ_i. The intersections of the line $[X_1, Y_1, Z_1]$ with s are given by the roots of

$$X_1\lambda^2 + Y_1\lambda + Z_1 = 0.$$

If these are to be λ_1 and λ_2, $[X_1, Y_1, Z_1]$ must be $[1, -(\lambda_1+\lambda_2), \lambda_1\lambda_2]$. Thus:

The chord $\lambda_1\lambda_2$ is

$$x - (\lambda_1+\lambda_2)\,y + \lambda_1\lambda_2 z = 0,$$

and the tangent at λ is

$$x - 2\lambda y + \lambda^2 z = 0.$$

4·46. *Contacts of $X:Y:Z = \lambda^2:\lambda:1$*

Dually consider the proper conic scroll S given by

$$X:Y:Z = \lambda^2:\lambda:1.$$

The lines of S through the point (x_1, y_1, z_1) are given by the roots of

$$x_1\lambda^2 + y_1\lambda + z_1 = 0,$$

so

The meet of the lines λ_1, λ_2 of S is the point

$$X - (\lambda_1 + \lambda_2)\,Y + \lambda_1\lambda_2\,Z = 0$$

and the contact of the line λ with S is the point

$$X - 2\lambda Y + \lambda^2 Z = 0.$$

4·47. *Tangents of $x:y:z = f(\lambda):g(\lambda):h(\lambda)$*

If the quadratic polynomials (·421) are denoted by $f(\lambda)$, $g(\lambda)$, $h(\lambda)$, and if $m \neq 0$, so that the curve given by $x:y:z = f(\lambda):g(\lambda):h(\lambda)$ is a proper conic locus s with λ as a parameter, then if $\lambda_1 \neq \lambda_2$ the chord $\lambda_1\lambda_2$ of s is

$$\begin{vmatrix} x & y & z \\ f(\lambda_1) & g(\lambda_1) & h(\lambda_1) \\ f(\lambda_2) & g(\lambda_2) & h(\lambda_2) \end{vmatrix} = 0, \qquad (\cdot 471)$$

since this is the equation of a line which is satisfied by the distinct points λ_1 and λ_2. Also the line $[X, Y, Z]$ is the tangent to s at λ_1 IF the equation

$$Xf(\lambda) + Yg(\lambda) + Zh(\lambda) = 0 \qquad (\cdot 472)$$

has λ_1 as a double root, i.e. IF the equations (·472) and

$$X\frac{d}{d\lambda}f(\lambda) + Y\frac{d}{d\lambda}g(\lambda) + Z\frac{d}{d\lambda}h(\lambda) = 0 \qquad (\cdot 473)$$

are both satisfied by $\lambda = \lambda_1$. But the point (x, y, z) lies on the line $[X, Y, Z]$ IF

$$Xx + Yy + Zz = 0, \qquad (\cdot 474)$$

so, by elimination of X, Y, Z from (·472), (·473), (·474), the point equation of the tangent to s at λ is

$$\begin{vmatrix} x & y & z \\ f(\lambda) & g(\lambda) & h(\lambda) \\ \frac{d}{d\lambda}f(\lambda) & \frac{d}{d\lambda}g(\lambda) & \frac{d}{d\lambda}h(\lambda) \end{vmatrix} = 0. \qquad (\cdot 475)$$

It will be found that the coefficients of x, y, z in (·475) are quadratic in λ, as the cubic terms vanish.

Alternatively, if a proper conic locus s is given by ratio parametric equations

$$x:y:z = f(\xi, \eta):g(\xi, \eta):h(\xi, \eta),$$

where $f(\xi, \eta)$, $g(\xi, \eta)$, $h(\xi, \eta)$ are quadratic forms in the ratio parameter (ξ, η), then the equation

$$Xf(\xi, \eta) + Yg(\xi, \eta) + Zh(\xi, \eta) = 0$$

has (ξ_1, η_1) as a double root IF

$$X\frac{\partial}{\partial\xi}f(\xi, \eta) + Y\frac{\partial}{\partial\xi}g(\xi, \eta) + Z\frac{\partial}{\partial\xi}h(\xi, \eta) = 0$$

and
$$X\frac{\partial}{\partial\eta}f(\xi, \eta) + Y\frac{\partial}{\partial\eta}g(\xi, \eta) + Z\frac{\partial}{\partial\eta}h(\xi, \eta) = 0$$

are both satisfied by (ξ_1, η_1). It follows that the tangent at (ξ, η) is

$$\begin{vmatrix} x & y & z \\ \frac{\partial}{\partial\xi}f(\xi, \eta) & \frac{\partial}{\partial\xi}g(\xi, \eta) & \frac{\partial}{\partial\xi}h(\xi, \eta) \\ \frac{\partial}{\partial\eta}f(\xi, \eta) & \frac{\partial}{\partial\eta}g(\xi, \eta) & \frac{\partial}{\partial\eta}h(\xi, \eta) \end{vmatrix} = 0. \qquad (\cdot 476)$$

4·48. *Note on parametric equations*

If $f(\lambda)$, $g(\lambda)$, $h(\lambda)$ are polynomials of any degree, the equations

$$x : y : z = f(\lambda) : g(\lambda) : h(\lambda) \qquad (\cdot 481)$$

will in general define a curve γ. Clearly each value of λ determines ONE point of γ, but it is not necessarily true that each point of the curve corresponds to ONE value of λ. If a general point of γ corresponds to more than one value of λ, then λ is not a parameter of γ, and the equations (·481) are not parametric equations. For example, the equations $x : y : z = \lambda^4 : \lambda^2 : 1$ determine the conic locus $y^2 = zx$ but are not parametric equations since a general point of the conic locus corresponds to two values of λ. It is in fact true that a curve γ given by equations of the form (·481) is rational, so that a parameter μ can be found which is in (1, 1) correspondence with the points of γ.* In the example a parameter μ is defined by $\mu = \lambda^2$.

4·5. Involutions. Proper conics

4·51. *The lines of a pencil meet a proper conic locus s in pairs of points of an involution on s.*

This follows directly from the definition of 3·54, since the condition (1) and (2) are satisfied. The involution is said to be *cut out* by

* See, for example, E. Bertini, *Geometria Proiettiva degli Iperspazi* (1923), pp. 340–343.

the pencil of lines. If the vertex U of the pencil lies on s, the involution has U as a fixed point.

If s is $x:y:z = \lambda^2:\lambda:1$, and U is (a, b, c), the pairs of the involution are given by the roots of

$$l\lambda^2 + m\lambda + n = 0, \tag{·511}$$

where

$$al + bm + cn = 0, \tag{·512}$$

for the line whose intersections with s are the roots of (·511) is

$$lx + my + nz = 0, \tag{·513}$$

which contains U IF l, m, n satisfy (·512). Also the pairs $|\lambda_1, \lambda_2|$ of the involution are given by

$$c\lambda_1\lambda_2 - b(\lambda_1 + \lambda_2) + a = 0,\text{*} \tag{·514}$$

which is a CONDITION for the chord $\lambda_1\lambda_2$, whose equation is

$$x - (\lambda_1 + \lambda_2)y + \lambda_1\lambda_2 z = 0,\dagger \tag{·515}$$

to contain U.

4·52. *The joins of pairs of points of any involution I on a proper conic locus s are the lines of a pencil, whose vertex U is called the centre of the involution.*

For by 3·55 an involution on s is determined by two pairs of points, so, if the joins of two pairs of I meet in U, the involution cut out by lines through U is the same as I. Alternatively I is given by equations of the form (·511), (·512), and (·512) is the CONDITION for the line (·513) to contain the point (a, b, c).

4·53. *The pairs of lines of a proper conic scroll S which pass through the points of a given line u form an involution in S.*

This is the dual of 4·51. If S is $X:Y:Z = \lambda^2:\lambda:1$, and u is $[A, B, C]$, the pairs of lines of the involution are given by the roots of

$$L\lambda^2 + M\lambda + N = 0, \tag{·531}$$

where

$$AL + BM + CN = 0; \tag{·532}$$

for the point such that the lines of S through it are given by (·531) is

$$LX + MY + NZ = 0, \tag{·533}$$

which lies on u IF L, M, N satisfy (·532). Also the pairs $\|\lambda_1, \lambda_2\|$ of the involution are given by

$$C\lambda_1\lambda_2 - B(\lambda_1 + \lambda_2) + A = 0,$$

* Cp. equations (3·554), (3·555), (3·556), p. 92.

† See 4·45, p. 115.

which is a CONDITION for the meet of the lines λ_1, λ_2 of S, whose equation is

$$X-(\lambda_1+\lambda_2)\,Y+\lambda_1\lambda_2 Z = 0,*$$

to lie on u.

4·54. *The meets of pairs of lines of any involution in a proper conic scroll lie on a line, which is called the axis of the involution.*

This is the dual of 4·52.

4·55. *The tangents of a proper conic locus s form a proper conic scroll, whose contacts are the points of s.*

In other words the tangent scroll of a proper conic locus s is a proper conic scroll, whose contact curve is s.† For the equations of s may be taken to be

$$x:y:z = \lambda^2:\lambda:1$$

and the tangent to s at the point λ is given by

$$X:Y:Z = 1:-2\lambda:\lambda^2,$$

so the tangents of s form the proper conic scroll

$$Y^2-4ZX = 0.$$

Also the point (x_1, y_1, z_1) is a contact of this scroll IF the two tangents through (x_1, y_1, z_1) coincide, i.e. IF the equation

$$x_1-2y_1\lambda+z_1\lambda^2 = 0$$

has equal roots in λ, i.e. IF

$$y_1^2 = z_1x_1,$$

i.e. IF (x_1, y_1, z_1) lies on s.

4·56. *The contacts of a proper conic scroll S form a conic locus whose tangents are the lines of S.*

This is the dual of 4·55. If S is given by $X:Y:Z = \mu^2:\mu:1$, the contacts of S are the points $(1, -2\mu, \mu^2)$ which form the proper conic locus $y^2 = 4zx$, whose tangents are the lines $[\mu^2, \mu, 1]$ of S.

4·57. *The points and tangents of a proper conic locus form a self-dual configuration.*

For by 4·55 they are also the contacts and lines of a proper conic scroll, which is the dual configuration. This self-dual configuration is called a *proper conic*, and a proper conic may be determined either by its points or by its tangents.

* See 4·46, p. 116. † See 2·46, p. 56.

The tangents of a proper line-pair $\| p_1, p_2 \|$, i.e. the lines whose two intersections coincide, are the lines through the vertex $p_1 p_2$; so the points and tangents of a proper-line pair do not form a self-dual configuration. Dually the contacts of a proper point-pair $\| P_1, P_2 \|$, i.e. the points P such that the two lines PP_1, PP_2 of the point pair through P coincide, are the points of the axis $P_1 P_2$. Every line is a tangent of a repeated line $\| p, p \|$, and every point is a contact of a repeated point $\| P, P \|$.

Line-pairs and point-pairs are also called conics, but a line-pair is essentially a degenerate conic locus, whereas a point-pair is essentially a degenerate conic scroll.

4·58. *Relation between points and tangents*

If S is the tangent scroll of a proper conic locus s, then each point P is the point of contact of ONE tangent p, and each tangent p is the tangent at ONE point of s. Thus the range of points $\{P\}$ on s is related to the range of lines $\{p\}$ in S.

4·6. Pole and polar

4·61. *Conjugate points relative to a conic locus*

Any line l meets a conic locus s in two points L, L' which may be distinct or coincident. Two points P_1, P_2 on l are said to be *conjugate* relative to s IF the pairs $|P_1, P_2|$ and $|L, L'|$ are apolar on l.* When P_1 and P_2 are conjugate relative to s each point is said to be conjugate to the other relative to s, for each point is conjugate to the other on l relative to $|L, L'|$.† A coincident pair of points $|P_1, P_1|$ on l is apolar to $|L, L'|$ on l IF P_1 is either L or L'; so a point is conjugate to itself, or self-conjugate, relative to s IF it lies on s. Also, if L and L' are coincident, any point of l is conjugate to L relative to the pair $|L, L|$; so a point P of s is conjugate to all points of the tangent to s at P.

4·62. *Polar of a point relative to a proper conic locus*

Suppose first that s is a proper conic locus. If P_1 does not lie on s, lines l through P_1 meet s in pairs of points $|L, L'|$ of an involution on s whose double points H and K are the points of contact of the two tangents through P_1, as in Fig. 4. On each line l there is ONE point P_2 conjugate to P_1 relative to s, namely the harmonic conjugate

* See 3·14, p. 76. † See 2·76, p. 70.

of P_1 relative to L and L' on l. This point lies on HK; for, since the points of s are related to the lines joining them to H, the pairs of lines $|HL, HL'|$ form an involution pencil with HK and HP_1 as double lines, and so HP_1 and HK harmonically separate HL and HL'; consequently, if l is distinct from P_1H and P_1K, the line HP_2 is HK; if l is P_1H or P_1K, then P_2 is H or K.

If P_1 lies on s, one of the intersections with s of any line l through P_1 is P_1 itself, so the point P_2 is unique and coincides with P_1 unless l is the tangent at P_1, when all points of l are conjugate to P_1. Thus:

The locus of points which are conjugate to a given point P relative to a proper conic locus s is a line p, which is called the polar line of P relative to s. If P does not lie on s, p is the join of the points of contact of the two tangents of s through P, which is called the chord of contact of P.

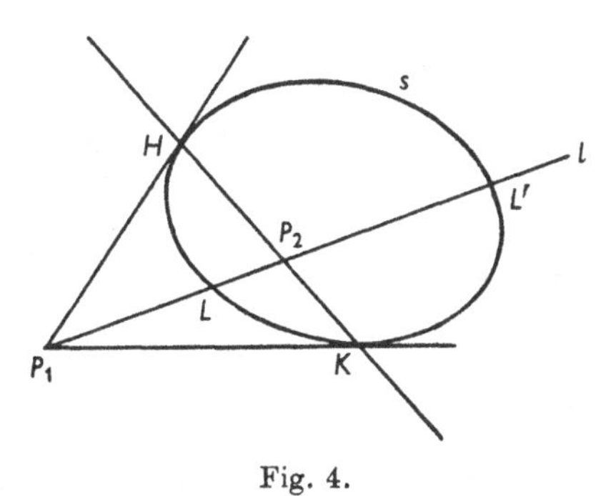

Fig. 4.

If P lies on s, p is the tangent to s at P. A point P is self-conjugate relative to s IF *P lies on s. If the polar line of P_1 contains P_2, then P_1 and P_2 are conjugate relative to s, so the polar line of P_2 contains P_1. Pairs of points of any line p which are conjugate relative to s form an involution on p whose double points are the intersections of p with s.*

4·63. *If s is a proper conic locus, any line p is the polar line of* ONE *point P, which is called the pole of p. The polars of the points of p form a related pencil of lines through P, and the poles of the lines through P form a related range of points on p.*

(i) For, if p meets s in distinct points H and K as in Fig. 5, then P is the meet of the tangents to s at H and K. Also the pairs of points $|Q, Q'|$ of p which are conjugate relative to s are the pairs of the involution I on p which has H and K as double points, and since

Q is conjugate to P and to Q' relative to s, the polar of Q relative to s is PQ'. As Q varies on p,

$$\{Q\} \equiv \{Q'\} \equiv P\{Q'\},$$

so the range of points Q on p is related to the pencil with vertex P formed by their polar lines PQ'. Also, if l is any line through P meeting p in Q, and if Q' is the mate of Q in I, then Q' being conjugate to P and Q relative to s, is the pole of l relative to s. Again as l varies,

$$\{l\} \equiv \{Q\} \equiv \{Q'\},$$

so the poles of lines through P form a related range of points on p. Incidentally Q', being the pole of l, is the meet of the tangents to s at the points (ls).

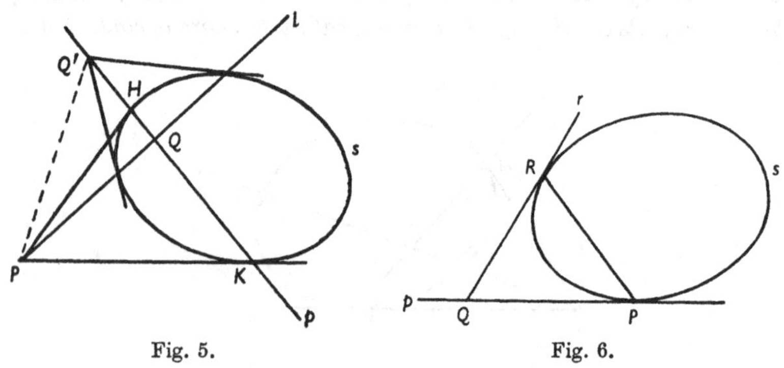

Fig. 5. Fig. 6.

(ii) If p is a tangent of s and P is its point of contact as in Fig. 6, let R be the point of contact of the other tangent r to s through a point Q of p. Then the polar of Q is PR, and as Q varies on p

$$\begin{aligned} \{Q\} &\equiv \{r\} \text{ in the tangent scroll of } s^* \\ &\equiv \{R\} \text{ on } s\dagger \\ &\equiv P\{R\}. \end{aligned}$$

So the polars of the points Q form a related pencil of lines through P, and, since PR is a general line through P, the poles of lines through P form a related range on p.

4·64. *Self-polar triangles*

A triangle is said to be self-polar relative to a conic locus s IF each pair of vertices are conjugate relative to s; if s is proper this will happen IF each side of the triangle is the polar of the opposite vertex.

* By 4·32, p. 109. † By 4·58, p. 120.

Thus, if p is the polar of a point P which does not lie on s, and if $|Q, R|$ is any pair of distinct conjugate points on p, then PQR is a self-polar triangle relative to s. In 4·63 (i) the triangles PQQ' are self-polar relative to s. There are ∞^2 points P and ∞^1 pairs of conjugate points on the polar of P, so a proper conic has ∞^3 self-polar triangles.

4·65. *Conjugate lines*

A line p_2 is said to be *conjugate* to p_1 relative to a proper conic locus s IF the pole P_1 of p_1 relative to s lies on p_2. When this happens the pole P_2 of p_2 relative to s is conjugate to P_1 and therefore lies on p_1, so p_1 is also conjugate to p_2. The two lines p_1 and p_2 are then said to be *conjugate.*

Two distinct lines are conjugate relative to s IF *they harmonically separate the two tangents of s through their point of intersection. A line is self-conjugate relative to s* IF *it is a tangent.*

For let two distinct lines p_1, p_2 meet in P, and suppose first that P does not lie on s. Let the polar line p of P relative to s meet p_1, p_2 and s in Q_1, Q_2 and H, K. Then, by the argument of 4·63 (i), the pole of p_1 relative to s lies on p and is the harmonic conjugate of Q_1 relative to $|H, K|$; so p_2 is conjugate to p_1 IF $|Q_1, Q_2|$ and $|H, K|$ are apolar on p, i.e. IF p_1 and p_2 harmonically separate PH, PK, which are the tangents to s through P.

If P is on s, p is the tangent at P, and the two distinct lines p_1, p_2 through P are conjugate relative to s IF one of them is p, i.e. IF the pairs $|p_1, p_2|$, $|p, p|$ are apolar in the pencil of lines through P. Finally a line p, with pole P, is conjugate to itself IF P lies on p, i.e. IF P is self-conjugate, i.e. IF P lies on s, i.e. IF p is a tangent.

It follows from the definition of conjugacy that

Two lines are conjugate relative to s IF *their poles are conjugate relative to s.*

Also

Two lines are conjugate relative to s IF *their intersections with s are apolar pairs of points on s.*

For, if L_1, L_1' and L_2, L_2' are the intersections of p_1 and p_2 with s, the pairs $|L_1, L_1'|$, $|L_2, L_2'|$ are apolar on s IF $|L_2, L_2'|$ is a pair of the involution on s which has L_1 and L_1' as double points, i.e. IF p_2 contains the pole of p_1.

These properties of conjugate lines provide an alternative proof of the theorem of 3·63, for, as was explained in 3·66, this theorem is established for all $R\infty^1$ when it is proved to be true for any particular $R\infty^1$. Consider then the $R\infty^1$, E, formed by the points, P, of a proper conic s. Any involution I on E is cut out on s by lines through a fixed point O, and the pairs of I are in (1, 1) correspondence with the lines through O. So any ratio parameter (ξ, η) of the pencil of lines with vertex O is also a ratio parameter of the $R\infty^1$ formed by the pairs of I. But it has just been proved that two pairs $|P_1, P_3|$, $|P_2, P_4|$ of I are apolar on s IF the lines P_1P_3, P_2P_4 are conjugate relative to s, i.e. IF these two lines through O harmonically separate the tangents to s through O, i.e. IF the values of (ξ, η) which correspond to the pairs $|P_1, P_3|$, $|P_2, P_4|$ of I harmonically separate the values of (ξ, η) which correspond to the double points of I.

4·66. *Polar theory of a conic scroll*

Dually a theory of poles and polars relative to a conic scroll may be developed. Through any point L there pass two lines l, l' of a conic scroll S, which may be distinct or coincident. Two lines p_1, p_2 through L are said to be *conjugate* relative to S IF the pairs $|p_1, p_2|$ and $|l, l'|$ are apolar in the pencil of lines through L. A line is self-conjugate IF it belongs to S, and a line p of S is conjugate to all lines through the contact of p with S. The reader should go through the dual of the arguments of 4·61 to 4·65. The following is a summary of the dual results.

The lines which are conjugate to a given line p relative to a proper conic scroll S form a pencil of lines whose vertex P is called the pole of p relative to S. If p does not belong to S, P is the meet of the two lines of S whose contacts lie on p. If p belongs to S, P is the contact of p with S. If the pole of p_1 lies on p_2, then p_1 and p_2 are conjugate, so the pole of p_2 lies on p_1. Pairs of lines through any point P which are conjugate relative to S form an involution pencil, whose double lines are the lines of S through P.

Any point P is the pole of ONE *line p, which is called the polar of P relative to S. The poles of the lines through P form a related range of points on p, and the polars of the points of p form a related pencil of lines through P.*

A triangle is said to be self-polar relative to S IF *each vertex is the pole of the opposite side. If P is the pole of a line p which does not* *belong to S, and if $| q, r |$ is any pair of distinct conjugate lines through P, then pqr is a self-polar triangle relative to S.*

Two points are said to be conjugate relative to S if the polar line of each contains the other. Two distinct points P_1, P_2 are conjugate relative to S IF *they harmonically separate the contacts of S which lie on P_1P_2. A point is self-conjugate relative to S* IF *it is a contact. Also two points are conjugate relative to S* IF *their polars are conjugate relative to S or* IF *the pairs of lines of S through them are apolar in S.*

4·67. *Polar theory of a proper conic*

For a proper conic, whose points form a proper conic locus s and whose tangents form a proper conic scroll S, there are thus two dual definitions of conjugate points, conjugate lines, poles and polar lines. *These two sets of definitions are exactly equivalent.* For by 4·65 two lines are conjugate relative to s IF they are conjugate relative to S, and dually two points are conjugate relative to S IF they are conjugate relative to s.

The centre of an involution on s and the axis of an involution in S were defined in 4·52 and 4·54. Any point P is the centre of the involution I on s which is cut out by lines through P, and the tangents to s at the pairs of points of I form an involution I' in S, whose axis p is the polar line of P relative to the conic. Dually and conversely any line p is the axis of an involution I' in S, and the contacts of the pairs of lines of I' form an involution I on s, whose centre P is the pole of p relative to the conic. The double points of I are the points (p, s), while the double lines of I' are the lines $[P, S]$.

4·68. *Degenerate conics. Line-pairs*

The polar theory of degenerate conics requires separate consideration. It has been shown that in building up the theory for a proper conic we may either consider the locus s or the scroll S, both attitudes leading to the same result. But a line-pair must be regarded as a locus and a point-pair must be regarded as a scroll, so in a degenerate case only one attitude is possible.

The definition of conjugate points given in 4·61 applies to a line-pair, and the polar of a point P is defined to be the locus of points conjugate to P. Consider a proper line-pair $\| l, l' \|$ with vertex L. The locus of points conjugate to a general point P is the line through L which is the harmonic conjugate of LP relative to l and l'

in the pencil of lines through L; this line is the polar line of P relative to $\| l, l' \|$. Thus any point P other than L has ONE polar line p, which is also the polar line of all points of LP other than L. The polar line of a point of l other than L is l itself. The point L is conjugate to all points of the plane, so the polar of L is indeterminate.

If a line p does not contain L, the polars of points of p form a related pencil of lines through L, so L may be said to be the pole of p, though the polar of L is indeterminate. A line through L has no unique pole, but is the polar of all points other than L of a line through L. The tangents of $\| l, l' \|$ are the lines whose two intersections with $\| l, l' \|$ coincide, i.e. the lines through L. Any self-polar triangle has one vertex at L and the other vertices P, Q are such that LP, LQ harmonically separate l, l' in the pencil of lines through L.

Consider a repeated line $\| l, l \|$. Each point of l is conjugate to all points of the plane. Any point not on l has l as its unique polar line, while the polar of a point on l is indeterminate. No line has a unique pole, and no line other than l is the polar of any point. The tangents are all the lines of the plane. Any self-polar triangle has l as one side.

It follows that:

The polar of a point P relative to a conic locus s is indeterminate IF *s is a line-pair with vertex P.*

Also

If a point P is conjugate to three non-collinear points relative to a conic locus s, then s is a line-pair with vertex P.

4·69. *Point-pairs*

Dually consider a proper point-pair $\| L, L' \|$ with axis l. Two lines p, p' through a point A are conjugate relative to $\| L, L' \|$ IF they harmonically separate the lines AL, AL' of $\| L, L' \|$ which contain A, i.e. IF p and p' meet l in points which harmonically separate L and L'. The lines which are conjugate to a general line p are therefore the lines through the harmonic conjugate P of pl relative to L and L'; this point P is the pole of p relative to $\| L, L' \|$. Thus any line p other than l has ONE pole P, which is also the pole of all lines through pl other than l. The pole of any line through L other than l is L. The line l is conjugate to all lines of the plane, so the pole of l is indeterminate. If a point P does not lie on l, the poles of lines through P form a related range of points on l, so l may be said to be the polar of P, though the pole of l is indeterminate. A point of l has no unique polar, but is the pole of all lines other than l through a point of l. The contacts of $\| L, L' \|$ are the points such that the two lines of $\| L, L' \|$ through them coincide, i.e. the points of l. Any self-polar triangle has l as one side, and the other sides p, q are such that pl, ql harmonically separate L and L'.

Finally consider a repeated point $\| L, L \|$. Each line through L is conjugate to all lines of the plane. Any line which does not contain L has L as its unique pole, while the pole of any line through L is indeterminate. No point has a unique polar, and no point other than L is the pole of any line. The contacts are all the points of the plane. Any self-polar triangle has L as one vertex.

It follows that:

The pole of a line p relative to a conic scroll S is indeterminate IF *S is a point-pair with axis p.*

Also

If a line p is conjugate to three non-concurrent lines relative to a conic scroll S, then S is a point-pair with axis p.

4·7. Analytical theory of pole and polar. Reciprocation

4·71. *Conic locus*

In the notation of 4·11 the intersections of the line P_1P_2 with the conic locus $s = 0$ are given by

$$\lambda^2 s_{11} + 2\lambda\mu s_{12} + \mu^2 s_{22} = 0, \tag{·711}$$

and the values of the ratio parameter (λ, μ) of the line P_1P_2 which correspond to P_1 and P_2 are the roots of

$$\lambda\mu = 0. \tag{·712}$$

Thus P_1 and P_2 are conjugate relative to s IF the pairs of roots of (·711) and (·712) are apolar, i.e. IF $s_{12} = 0$. It follows that the locus of points which are conjugate to P_1 is the line $s_1 = 0$, so this is the equation of the polar line of P_1; it was shown in 4·12 that, if P_1 is on s, the equation of the tangent at P_1 is $s_1 = 0$. Also the line P_1P_2 is a tangent of s IF the roots of (·711) are equal, i.e. IF $s_{12}^2 = s_{11}s_{22}$, so the point equation of the pair of tangents to s through P_1 is

$$s_1^2 = s_{11}s. \tag{·713}$$

The polar line $s_1 = 0$ is indeterminate IF

$$ax_1+hy_1+gz_1 = hx_1+by_1+fz_1 = gx_1+fy_1+cz_1 = 0, \quad (\cdot 714)$$

i.e. IF s is a line-pair with vertex P_1, by 4·11.

If $[X_1, Y_1, Z_1]$ is a given line, and if $\delta \neq 0$, the equations

$$\left.\begin{aligned} ax_1+hy_1+gz_1 &= \rho X_1, \\ hx_1+by_1+fz_1 &= \rho Y_1, \\ gx_1+fy_1+cz_1 &= \rho Z_1, \end{aligned}\right\} \quad (\cdot 715)$$

determine ONE point (x_1, y_1, z_1), namely the point

$$(AX_1+HY_1+GZ_1, \quad HX_1+BY_1+FZ_1, \quad GX_1+FY_1+CZ_1), \quad (\cdot 716)$$

where $A, B, \ldots, H$ are the cofactors of $a, b, \ldots, h$ in δ. Thus, if s is proper, every line has ONE pole, as defined in 4·63, the pole of $[X_1, Y_1, Z_1]$ being the point (·716).

If P_1, P_2 are distinct points of a line p, the point P given by $s_1 = s_2 = 0$ is the pole of p relative to s, since it is conjugate to P_1 and to P_2. The polar of the point $\lambda P_1 + \mu P_2$ relative to s is the line $\lambda s_1 + \mu s_2 = 0$, so the polars of points of p form a related pencil of lines through P, as was shown in 4·63.

The lines p_1 and p_2 will be conjugate relative to s in the sense defined in 4·65 IF the pole of p_1 lies on p_2, i.e. IF the point (x_1, y_1, z_1) determined by (·715) also satisfies

$$X_2x_1+Y_2y_1+Z_2z_1 = 0. \quad (\cdot 717)$$

On eliminating x_1, y_1, z_1, ρ from these four equations it follows that p_1 and p_2 are conjugate relative to s IF

$$\begin{vmatrix} a & h & g & X_1 \\ h & b & f & Y_1 \\ g & f & c & Z_1 \\ X_2 & Y_2 & Z_2 & 0 \end{vmatrix} = 0, \quad (\cdot 718)$$

i.e. IF $S_{12} = 0$, where

$$S \equiv AX^2+BY^2+CZ^2+2FYZ+2GZX+2HXY. \quad (\cdot 719)$$

It was shown in 4·65 that a line is a tangent of s IF it is self-conjugate relative to s, so the line equation of the tangent scroll of s is $S = 0$. Also the pole of p_1 relative to s, the point (·716), is given by the line equation $S_1 = 0$.

4·72. *Conic scroll*

Dually in the notation of 4·15 the lines of the conic scroll $S = 0$ through the point $p_1 p_2$ are given by

$$\lambda^2 S_{11} + 2\lambda\mu S_{12} + \mu^2 S_{22} = 0, \qquad (\cdot 721)$$

so p_1 and p_2 are conjugate relative to S IF $S_{12} = 0$, and the lines conjugate to p_1 pass through the point $S_1 = 0$, which is therefore the pole of p_1 relative to S. Also the point $p_1 p_2$ is a contact of S IF $S_{12}^2 = S_{11} S_{22}$, so the line equation of the contacts of S which lie on p_1 is

$$S_1^2 = S_{11} S. \qquad (\cdot 722)$$

If S is proper, every point has ONE polar line, as defined in 4·66, the polar of (x_1, y_1, z_1) being the line

$$[\bar{a}x_1 + \bar{h}y_1 + \bar{g}z_1, \quad \bar{h}x_1 + \bar{b}y_1 + \bar{f}z_1, \quad \bar{g}x_1 + \bar{f}y_1 + \bar{c}z_1], \qquad (\cdot 723)$$

where $\bar{a}, \bar{b}, \ldots, \bar{h}$ are the cofactors of $A, B, \ldots, H$ in Δ. If p_1, p_2 are distinct lines through a point P, the line p given by $S_1 = S_2 = 0$ is the polar of P relative to S, since it is conjugate to p_1 and to p_2. The pole of the line $\lambda p_1 + \mu p_2$ relative to S is the point $\lambda S_1 + \mu S_2 = 0$, so the poles of lines through P form a related range of points on p.

The points P_1 and P_2 are conjugate relative to S in the sense defined in 4·66 IF $\bar{s}_{12} = 0$, where

$$\bar{s} \equiv \bar{a}x^2 + \bar{b}y^2 + \bar{c}z^2 + 2\bar{f}y^2 + 2\bar{g}zx + 2\bar{h}xy.$$

The equation of the contact curve of S is $\bar{s} = 0$, and the polar of P_1 relative to S is given by $\bar{s}_1 = 0$.

4·73. *Proper conic*

If, as in 4·71, the conic scroll $S = 0$ is the tangent scroll of the conic locus $s = 0$, then

$$\bar{a} = \delta a, \bar{b} = \delta b, \ldots, \bar{h} = \delta c,* \qquad (\cdot 731)$$

so $\bar{s} \equiv \delta s$. Thus, as was shown in 4·55, the contact curve of S is the conic locus s IF $\delta \neq 0$, i.e. IF s is proper. It also follows that the polar theory of a proper conic is the same whether the locus of points or the scroll of tangents is considered, for from either point of view two points P_1 and P_2 are conjugate IF $s_{12} = 0$ while two lines p_1 and p_2 are conjugate IF $S_{12} = 0$.

If $s = 0$ is a proper line-pair with vertex L, then $\delta = 0$, so $\bar{a} = \bar{b} = \ldots = \bar{h} = 0$ and consequently S is a perfect square; in fact

* See, for example, Durell and Robson, *Advanced Algebra*, 2 *and* 3 (1937), pp. 411–412.

$S = 0$ is the repeated point $\| L, L \|$. Also, if $s = 0$ is a repeated line, then $A = B = \ldots = H = 0$, and S is identically zero; in fact all lines of the plane are tangents of s.

4·74. *Particular triangle of reference*

When dealing with the properties of a conic it is often convenient to choose a co-ordinate system so that the equations of the conic take a simple form. The conic $s = 0$ contains the point $(1, 0, 0)$ IF $a = 0$ and touches the line $[1, 0, 0]$ IF $A = 0$. Thus

(i) If a conic locus contains the reference points, its point equation is of the form

$$fyz + gzx + hxy = 0, \qquad (\cdot 741)$$

and its line equation is of the form

$$f^2X^2 + g^2Y^2 + h^2Z^2 - 2ghYZ - 2hfZX - 2fgXY = 0. \qquad (\cdot 742)$$

Provided the conic locus is proper, when $fgh \neq 0$, the equations may be taken as

$$yz + zx + xy = 0, \qquad (\cdot 743)$$

$$X^2 + Y^2 + Z^2 - 2YZ - 2ZX - 2XY = 0, \qquad (\cdot 744)$$

without loss of generality.* Parametric equations of the conic (·743) are

$$x : y : z = \lambda : 1 - \lambda : \lambda(\lambda - 1), \qquad (\cdot 745)$$

the reference points being given by the values 1, 0, ∞ of the parameter λ; the tangents are given by

$$X : Y : Z = -(\lambda + 1)^2 : \lambda^2 : 1. \qquad (\cdot 746)$$

(ii) If a conic scroll contains the reference lines, its line equation is of the form

$$FYZ + GZX + HXY = 0 \qquad (\cdot 747)$$

and its point equation is

$$F^2x^2 + G^2y^2 + H^2z^2 - 2GHyz - 2HFzx - 2FGxy = 0. \qquad (\cdot 748)$$

Provided the conic scroll is proper, we may take $F = G = H = 1$, and obtain parametric equations as in (i).

4·75. *Self-polar triangle as triangle of reference*

The points $(0, 1, 0)$ and $(0, 0, 1)$ are conjugate relative to the conic locus $(abcfgh \between xyz)^2 = 0$ IF $f = 0$, so when the triangle of reference is

* For the transformation $x : y : z = fx' : gy' : hz'$ reduces (·741) to
$$y'z' + z'x' + x'y' = 0.$$

a self-polar triangle the point equation of a conic locus is of the form

$$ax^2+by^2+cz^2 = 0, \qquad (\cdot 751)$$

and its line equation is

$$bcX^2+caY^2+acZ^2 = 0. \qquad (\cdot 752)$$

Provided the conic locus is proper, when $abc \neq 0$, the equations may be taken as

$$x^2+y^2+z^2 = 0, \qquad (\cdot 753)$$

$$X^2+Y^2+Z^2 = 0 \qquad (\cdot 754)$$

without loss of generality.

4·76. *Parametric equations*

(i) It was shown in 4·43 that a proper conic s may be given by equations of the form

$$x:y:z = \lambda^2:\lambda:1,$$

the point equation being $y^2-zx = 0$. The tangent at the point λ is

$$x-2\lambda y+\lambda^2 z = 0,$$

or $[1, -2\lambda, \lambda^2]$, so the tangent scroll is given by

$$X:Y:Z = 1:-2\lambda:\lambda^2$$

and its line equation is $\qquad Y^2 = 4ZX.$

The polar of the point (x_1, y_1, z_1) is the line $[z_1, -2y_1, x_1]$ and the pole of the line $[X_1, Y_1, Z_1]$ is the point $[2Z_1, -Y_1, 2X_1]$.

(ii) It is sometimes more convenient to take the equations of a proper conic s in the form

$$x:y:z = \lambda^2:2\lambda:1,$$

when the point equation is

$$y^2 = 4zx.$$

The tangent at λ is $\qquad x-\lambda y+\lambda^2 z = 0,$

or $[1, -\lambda, \lambda^2]$, so the tangent scroll is given by

$$X:Y:Z = 1:-\lambda:\lambda^2$$

and its line equation is $\qquad Y^2 = ZX.$

The polar of the point (x_1, y_1, z_1) is the line $[2z_1, -y_1, 2x_1]$ and the pole of the line $[X_1, Y_1, Z_1]$ is the point $(Z_1, -2Y_1, X_1)$.

4·77. *Reciprocation*

If we have any geometrical construct K we may consider the construct k' obtained by replacing the points and lines of K by their polar lines and poles relative to a proper conic s. This process is known as *reciprocation* relative to s, and k' is called the reciprocal of K relative to s.

If the point equation of s is $s \equiv (abcfgh \between xyz)^2 = 0$ and its line equation $S \equiv (ABCFGH \between XYZ)^2 = 0$, in the usual notation, the reciprocal of the point P, (x, y, z), is the line p',

$$[ax+hy+gz, \quad hx+by+fz, \quad gx+fy+cz],$$

and the reciprocal of the line p, $[X, Y, Z]$, is the point P',

$$[AX+HY+GZ, \quad HX+BY+FZ, \quad GX+FY+CZ].$$

We have

$$(AX+HY+GZ)(ax+hy+gz)+(HX+BY+FZ)(hx+by+fz) \\ +(GX+FY+CZ)(gx+fy+cz) \equiv \delta(Xx+Yy+Zz),$$

so that if P lies on p, the CONDITION for which is $Xx+Yy+Zz = 0$, then p' contains P', a fundamental result which we have already proved in 4·62.

Since the co-ordinates of p' are linear in those of P and the co-ordinates of P' are linear in those of p, it follows that the reciprocal relative to s of a curve (or scroll) of order (or class) n is a scroll (or curve) of class (or order) n. The reciprocal of a curve and its tangents is a scroll and its contacts; in particular, s and S reciprocate into S and s.

4·8. Pencils of conics

4·81. *Intersections of two conic loci*

Consider a proper conic locus γ given by

$$x:y:z = \theta^2 : 2\theta : 1. \qquad (\cdot 811)$$

The points of γ which lie on another conic locus

$$s \equiv ax^2+by^2+cz^2+2fyz+2gzx+2hxy = 0 \qquad (\cdot 812)$$

are obtained by substituting $\theta^2 : 2\theta : 1$ for $x:y:z$ in (·812), and are therefore given by the quartic equation

$$\phi(\theta) \equiv a\theta^4+4h\theta^3+(4b+2g)\theta^2+4f\theta+c = 0. \qquad (\cdot 813)$$

In general this equation has distinct roots, so two general conic loci have four distinct intersections. If θ_0 is a root of multiplicity two, three or four, then s is said to have two, three or four intersections with γ at the point θ_0, and the two conic loci are said to have simple contact, three-point contact or four-point contact at this point. Any conic s which has more than four intersections with γ coincides with γ, for all the coefficients of (·813) must vanish.

On comparing (·813) with the general quartic equation

$$a_0\theta^4+4a_1\theta^3+6a_2\theta^2+4a_3\theta+a_4=0, \tag{·814}$$

it follows that all conic loci whose intersections with γ are given by (·814) have equations of the form

$$\lambda(a_0x^2+a_2y^2+a_4z^2+2a_3yz+2a_2zx+2a_1xy)+\mu(y^2-4zx). \tag{·815}$$

Similarly it follows that all conic loci which have the same intersections with γ as s have equations of the form

$$\lambda s+\mu(y^2-4zx)=0. \tag{·816}$$

4·82. *Definition of a point pencil of conics*

If $s\equiv(abcfgh\between xyz)^2=0$ and $s'\equiv(a'b'c'f'g'h'\between xyz)^2=0$ are any two distinct conic loci, the ∞^1 system of conic loci whose equations are

$$\lambda s+\mu s'=0,$$

is called a *point pencil of conics* and will be denoted by σ. Since the conic loci of σ may be identified with the values of the ratio variable (λ,μ), a point pencil of conics is an $R\infty^1$. In fact the ratio sets

$$(\lambda a+\mu a',\lambda b+\mu b',\ldots,\lambda h+\mu h')$$

formed by the coefficients in the point equations of the conic loci of σ form a linear ∞^1 system.*

The point pencil σ is uniquely determined by the two conic loci s and s' when a system of co-ordinates is assigned. The same system of conic loci is similarly defined with the same co-ordinate system by any two of its members; for if $\lambda_1s+\mu_1s'=0$ and $\lambda_2s+\mu_2s'=0$ are any two distinct conic loci of σ, the point equation of any conic locus of σ may be written in the form

$$\xi(\lambda_1s+\mu_1s')+\eta(\lambda_2s+\mu_2s')=0.$$

Also, if the co-ordinate system is changed by a non-singular linear transformation

$$\left.\begin{aligned} x&=a_1x'+b_1y'+c_1z',\\ y&=a_2x'+b_2y'+c_2z',\\ z&=a_3x'+b_3y'+c_3z',\end{aligned}\right\} \tag{·821}$$

and if the quadratic forms s and s' in x, y, z become quadratic forms $\bar{s}$ and $\bar{s}'$ in x', y', z', then the form $\lambda s+\mu s'$ becomes $\lambda\bar{s}+\mu\bar{s}'$, so in the

* See 2·23, p. 46.

new system of co-ordinates the point equations of the conic loci of σ are

$$\lambda\bar{s}+\mu\bar{s}' = 0.$$

Thus:

Any two distinct conic loci belong to ONE *point pencil. If in any co-ordinate system the point equations of the two conic loci are $s = 0$ and $s' = 0$, then the point equations of the conic loci of the pencil are*

$$\lambda s+\mu s' = 0.$$

4·83. *Base points*

Consider the point pencil σ of conics which is determined by two distinct conic loci $s = 0$ and $s' = 0$. The points of intersection of s and s' satisfy $\lambda s+\mu s' = 0$ and therefore lie on all the conic loci of σ; they are called the *base points* of σ.

Suppose that s' is proper, and that the co-ordinate system has been chosen so that s' is given by

$$y^2 = 4zx$$

or by

$$x:y:z = \theta^2:2\theta:1.$$

Then σ is given by

$$\lambda s+\mu(y^2-4zx) = 0.$$

It follows from 4·81 that the base points of σ are the points of s' given by $\phi(\theta) = 0$, and that the conic loci which have the same intersections with s' as s are the conic loci of σ. Thus:

A general point pencil σ of conics has four distinct base points A_1, A_2, A_3, A_4 and the conic loci of σ are the conic loci which contain A_1, A_2, A_3, A_4.

Conversely

The conic loci which contain four distinct points A_1, A_2, A_3, A_4, of which no three are collinear, form a point pencil of conics.

For if P is a general point there is no line-pair which contains the five points A_1, A_2, A_3, A_4, P, so the ONE conic locus s' which contains these five points is proper. It is possible to choose a conic locus s, distinct from s', which contains A_1, A_2, A_3, A_4 and it follows that the point pencil determined by s and s' consists of the conic loci which contain A_1, A_2, A_3, A_4.

4·84. *Line-pairs*

The conic locus $\lambda s+\mu s'=0$ is a line-pair IF

$$\begin{vmatrix} \lambda a+\mu a' & \lambda h+\mu h' & \lambda g+\mu g' \\ \lambda h+\mu h' & \lambda b+\mu b' & \lambda f+\mu f' \\ \lambda g+\mu g' & \lambda f+\mu f' & \lambda c+\mu c' \end{vmatrix} = 0. \qquad (\cdot 841)$$

In general this cubic equation in (λ, μ) has three distinct roots, but in particular cases it may have coincident roots or it may be an identity. Thus:

A general point pencil σ of conics contains three line-pairs. If A_1, A_2, A_3, A_4 are the four distinct base points of σ, the line-pairs are $\| A_1A_4, A_2A_3 \|$, $\| A_2A_4, A_3A_1 \|$ and $\| A_3A_4, A_1A_2 \|$.

A point pencil of conics is determined by two of its line-pairs, or by a proper conic locus and a line-pair.

4·85. *Involution theorem for a point pencil*

ONE conic locus of a point pencil σ contains any point P which is not a base point; for the CONDITION for $\lambda s+\mu s'=0$ to contain a point P gives ONE value of (λ, μ) unless it is identically satisfied by all values of (λ, μ). Thus, if a line p is not part of any conic locus of σ, the pairs of intersections of p with the conic loci of σ satisfy the conditions of 3·54, so

Any line p, which is not part of a conic locus of a point pencil σ, is met by the conic loci of σ in pairs of points of an involution. This involution, which is denoted by (p, σ), has a fixed point* IF *p contains a base point of σ.*

The involution (p, σ) is said to be *cut out* on p by the conics of σ, and the $R\infty^1$ formed by the pairs of (p, σ) is related to the $R\infty^1$ formed by the conics of σ.

Analytically the general point of p is

$$(\kappa_1 x_1+\kappa_2 x_2, \kappa_1 y_1+\kappa_2 y_2, \kappa_1 z_1+\kappa_2 z_2),$$

where P_1 and P_2 are two distinct points of p; this point lies on $\lambda s+\mu s'=0$ IF

$$\lambda(\kappa_1^2 s_{11}+2\kappa_1\kappa_2 s_{12}+\kappa_2^2 s_{22})+\mu(\kappa_1^2 s'_{11}+2\kappa_1\kappa_2 s'_{12}+\kappa_2^2 s'_{22}) = 0, \quad (\cdot 851)$$

* This is known as the involution theorem of Desargues-Sturm.

and by 3·61 the pairs of values of (κ_1, κ_2) which are given by this equation form an involution unless

$$s_{11} : s_{12} : s_{22} = s'_{11} : s'_{12} : s'_{22},$$

when one value of (λ, μ) makes (·851) an identity in (κ_1, κ_2) and consequently p is part of a conic locus of σ.

The two double points of the involution (p, σ) harmonically separate all the pairs of (p, σ) and are coincident IF (p, σ) has a fixed point.* Thus:

Any line p, which does not contain a base point of σ, is a tangent of two distinct conics of σ, and the two points of contact are conjugate relative to all conics of σ. If p contains a base point A of σ, and is not part of a conic locus of σ, then ONE *conic of σ touches p at A.*

4·86. *Common tangents of two conic scrolls*

Consider the proper conic scroll Γ given by

$$X : Y : Z = \theta^2 : 2\theta : 1. \qquad (\cdot 861)$$

The lines of Γ which belong to another conic scroll

$$S \equiv AX^2 + BY^2 + CZ^2 + 2FYZ + 2GZX + 2HXY = 0 \qquad (\cdot 862)$$

are given by the quartic equation

$$\Phi(\theta) \equiv A\theta^4 + 4H\theta^3 + (4B + 2G)\theta^2 + 4F\theta + C = 0. \qquad (\cdot 863)$$

In general this equation has distinct roots, so two general conic scrolls have four distinct common lines. On comparing (·863) with the general quartic equation

$$a_0\theta^4 + 4a_1\theta^3 + 6a_2\theta^2 + 4a_3\theta + a_4 = 0, \qquad (\cdot 864)$$

it follows that all conic scrolls whose common lines with Γ are given by (·864) have equations of the form

$$\lambda(a_0X^2 + a_2Y^2 + a_4Z^2 + 2a_3YZ + 2a_2ZX + 2a_1XY) + \mu(Y^2 - 4ZX) = 0. \qquad (\cdot 865)$$

Similarly all conic scrolls whose common tangents with Γ are those of S have equations of the form

$$\lambda S + \mu(Y^2 - 4ZX) = 0. \qquad (\cdot 866)$$

4·87. *Definition of a line pencil of conics*

If $S = 0$ and $S' = 0$ are line equations of two distinct conic scrolls, the ∞^1 system of conic scrolls whose equations are

$$\lambda S + \mu S' = 0,$$

* See 3·58, p. 94.

is called a *line pencil of conics*, and will be denoted by Σ. Thus a line pencil of conics is the dual of a point pencil of conics, and as in 4·82 it follows that:

Any two distinct conic scrolls belong to ONE *line pencil* Σ. *If in any co-ordinate system the line equations of the two conic scrolls are* $S = 0$ *and* $S' = 0$, *then the line equations of the conic scrolls of* Σ *are*

$$\lambda S + \mu S' = 0.$$

The conic scrolls of Σ *form an* $R\,\infty^1$ *of which* (λ, μ) *is a ratio parameter.*

4·88. *Base lines and point-pairs*

Consider the line pencil Σ given by

$$\lambda S + \mu S' = 0, \qquad (\cdot 881)$$

where
$$S \equiv (ABCFGH \between XYZ)^2,$$
$$S' \equiv (A'B'C'F'G'H' \between XYZ)^2.$$

The common lines of S and S' satisfy $\lambda S + \mu S' = 0$ and therefore belong to all the conic scrolls of Σ. They are called the *base lines* of Σ. It follows as in 4·83 that, if S' is proper, the conic scrolls which have the same common lines with S' as S are the conic scrolls of Σ. Also

A general line pencil Σ *of conics has four distinct base lines* a_1, a_2, a_3, a_4 *and the conic scrolls of* Σ *are the conic scrolls which contain* a_1, a_2, a_3, a_4. *Conversely, if* a_1, a_2, a_3, a_4 *are any four distinct lines no three of which are concurrent, the conic scrolls which contain* a_1, a_2, a_3, a_4 *form a line pencil of conics.*

The conic scroll $\lambda S + \mu S' = 0$ is a point-pair IF

$$\begin{vmatrix} \lambda A + \mu A' & \lambda H + \mu H' & \lambda G + \mu G' \\ \lambda H + \mu H' & \lambda B + \mu B' & \lambda F + \mu F' \\ \lambda G + \mu G' & \lambda F + \mu F' & \lambda C + \mu C' \end{vmatrix} = 0, \qquad (\cdot 882)$$

and in general this equation has three distinct roots. Thus:

A general line pencil Σ *of conics contains three point-pairs. If* a_1, a_2, a_3, a_4 *are the four distinct base lines of* Σ, *the point-pairs are* $\| a_1 a_4, a_2 a_3 \|$, $\| a_2 a_4, a_3 a_1 \|$, $\| a_3 a_4, a_1 a_2 \|$.

4·89. *Involution theorem for a line pencil*

The dual of 4·85 is as follows. Consider a line pencil Σ of conics. Any line p other than a base line of Σ belongs to ONE conic scroll of Σ. Thus:

If the pencil of lines with vertex P is not part of any conic scroll of Σ, the pairs of lines through P which belong to the conic scrolls of Σ form an involution. This involution, which is denoted by $[P,\Sigma]$ has a fixed line IF *P lies on a base line of Σ.*

The two double lines of the involution $[P,\Sigma]$ harmonically separate all the pairs of $[P,\Sigma]$ and are coincident IF $[P,\Sigma]$ has a fixed line. Also the two lines of a conic scroll which pass through a point P are coincident IF P is a contact of the scroll, so the double lines of $[P,\Sigma]$ arise from the conic scrolls of Σ which have P as a contact. Thus:

Any point P, which does not lie on a base line of Σ, is a contact of two distinct conic scrolls of Σ; the lines of these two conic scrolls through P are conjugate relative to all conic scrolls of Σ. If P lies on a base line a of Σ, and if the pencil of lines with vertex P is not part of any conic scroll of Σ, then for ONE *conic scroll of Σ P is the contact of a.*

4·9. Particular cases

4·91. *Pencils containing no proper conics*

If s' is a proper line-pair, its equation may be taken as $2yz = 0$, and the equation (·841) reduces to

$$\delta\lambda^3 + 2(hg - af)\lambda^2\mu - a\lambda\mu^2 = 0,$$

which is an identity in (λ, μ) IF

$$\delta = 0, \quad a = 0 \quad \text{and} \quad gh = 0.$$

If $\delta = 0$, s is a line-pair, and its equation is of the form

$$s \equiv (lx + my + nz)(l'x + m'y + n'z) = 0,$$

and $a = hg = 0$ gives

$$ll' = (nl' + n'l)(lm' + l'm) = 0.$$

This happens IF $l = l' = 0$, or $l = m = 0$, or $l = n = 0$, or $l' = m' = 0$, or $l' = n' = 0$, i.e. IF s is a line-pair with the same vertex as s' or a line-pair having a line in common with s'. Similarly, if s' is a repeated line $\| p, p \|$, it is easily shown that the equation (·841) is an identity in (λ, μ) IF s is either a line-pair with vertex on p or a line-pair containing p. Thus:

Point pencils of conics whose members are all line-pairs are of two types. In the first type the line-pairs have a common vertex and are formed by the pairs of lines of an involution pencil. In the second type one of the lines of the line-pairs is fixed, while the other passes through a fixed point.

Dually there are two types of line pencils of conics whose members are all point-pairs. In the first type the point-pairs have a common axis and are formed by the pairs of points of an involution on this axis. In the second type one of the points is fixed, while the other lies on a fixed line.

If s' is a proper conic locus, it has been shown that any conic locus s which has more than four intersections with s' must coincide with s'. Also, if s is distinct from s', it has been shown that the conic loci which have the same intersections with s' as s form a point pencil. These statements require modification if s' is a line-pair $\| p_1, p_2 \|$. A conic locus which has more than four intersections with s' need not coincide with s', but must contain either p_1 or p_2. Also, if s' is a proper line-pair, it is easily seen that the conic loci which have the same intersections with s' as s still form a point pencil, provided that s is not a line-pair with the same vertex as s'; this is no longer true if s' is a repeated line.

4·92. *Multiple contact*

In all cases except those just mentioned the general conic locus of a point pencil σ is proper. Let s and s' be two proper conic loci of σ and let A be one of their common points. Choose a co-ordinate system so that s' is given by

$$x : y : z = \theta^2 : 2\theta : 1,$$

and A is the point $\theta = 0$. Let s be given by

$$s \equiv (abcfgh \between xyz)^2 = 0,$$

so that the intersections of s and s' are given by

$$\phi(\theta) \equiv a\theta^4 + 4h\theta^3 + (4b + 2g)\,\theta^2 + 4f\theta + c = 0,$$

where $c = 0$ since s contains A. If $\theta = 0$ is a double root of $\phi(\theta) = 0$, two of the four intersections of s with s' are at A, and s is said to touch s' at A or to have simple contact with s' at A. If $\theta = 0$ is a 3-ple or a 4-ple root of $\phi(\theta) = 0$, s is said to have *three-point contact* or *four-point contact* with s' at A.

The tangents of s' are given by

$$X : Y : Z = 1 : -\theta : \theta^2,$$

so the common tangents of s and s' are given by

$$\Phi(\theta) \equiv C\theta^4 - 2F\theta^3 + (B+2G)\theta^2 - 2H\theta + A = 0.$$

If $\theta = 0$ is a root of $\Phi(\theta) = 0$, $A = bc - f^2 = 0$, and $c = 0$, so $f = 0$ and consequently $\theta = 0$ is a double root of $\phi(\theta) = 0$. Thus, if s and s' have the same tangent at A, at least two of the intersections of s and s' coincide in A.

If $\theta = 0$ is a double root of $\phi(\theta) = 0$, $c = f = 0$, so

$$A = bc - f^2 = 0, \quad H = fg - ch = 0,$$

and consequently $\theta = 0$ is also a double root of $\Phi(\theta) = 0$. Similarly, if $\theta = 0$ is a 3-ple or 4-ple root of $\phi(\theta) = 0$, it is easily shown that $\theta = 0$ is also a 3-ple or 4-ple root of $\Phi(\theta) = 0$. Thus:

If two proper conics have r point contact at A, where $r = 2$, 3 *or* 4, *then the conics have a common tangent at A which counts r times among the common tangents of the two conics.*

Dually, if a line p counts r times among the common tangents of two proper conics, where $r = 2$, 3 *or* 4, *then the points of contact of the conics with p coincide in a point at which the conics have r point contact.*

It follows as a corollary that:

Two proper conics whose four intersections are distinct have four distinct common tangents, and conversely.

4·93. *Particular cases*

It follows from 4·92 that the intersections and common tangents of two proper conics s and s' must be of one of the following types.

(i) *The general case*: Four distinct common points A_1, A_2, A_3, A_4 and four distinct common tangents a_1, a_2, a_3, a_4.

(ii) *Simple contact*: Two of the intersections coincide in A_1, and there are two other distinct intersections A_2, A_3. The conics have a common tangent a_1 at A_1, which counts twice among the common tangents, and there are two other distinct common tangents a_2 and a_3.

(iii) *Three-point contact*: Three of the intersections coincide in A_1, and there is one other distinct intersection A_2. The conics have a common tangent a_1 at A_1, which counts three times among the common tangents. There is one other common tangent a_2.

(iv) *Four-point contact*: All four intersections coincide in A_1. The conics have a common tangent a_1 at A_1, which counts four times among the common tangents. There is no other common tangent.

(v) *Double contact*: Two of the intersections coincide in A_1 and the other two coincide in A_2. The conics have common tangents a_1 and a_2 at A_1 and A_2, each of which counts twice among the common tangents. The line A_1A_2 is called the *chord of contact* and the point a_1a_2 is the pole of the chord of contact relative to both conics. Four-point contact is a particular case of double contact, when A_1 and A_2 coincide and the chord of contact is a tangent. When A_1 and A_2 are distinct the conics are said to have *proper double contact*.

4·94. *Line-pairs and point-pairs*

(i) In the general case the two proper conics s, s' determine a point pencil σ of conic loci $\lambda s+\mu s' = 0$, and a line pencil Σ of conic scrolls $\lambda S+\mu S' = 0$. These pencils σ and Σ have no common conics other than s, s'.* The pencil σ contains three distinct line-pairs, $\| A_1A_4, A_2A_3 \|$, $\| A_2A_4, A_3A_1 \|$, $\| A_3A_4, A_1A_2 \|$, corresponding to the roots of the equation (·841). The pencil Σ contains three distinct point-pairs, $\| a_1a_4, a_2a_3 \|$, $\| a_2a_4, a_3a_1 \|$, $\| a_3a_4, a_1a_2 \|$, corresponding to the roots of (·882).

In the particular cases the equations (·841), (·882) do not have distinct roots; the three line-pairs of σ are not distinct, and the three point-pairs of Σ are not distinct. The following statements can be established.

(ii) *Simple contact*: The point pencil σ contains only two line-pairs $\| A_1A_2, A_1A_3 \|$ and $\| a_1, A_2A_3 \|$, the first of which corresponds to a double root of (·841). Dually the line pencil Σ contains only two point-pairs $\| a_1a_2, a_1a_3 \|$ and $\| A_1, a_2a_3 \|$, the first of which corresponds to a double root of (·882).

(iii) *Three-point contact*: The point pencil σ contains only one line-pair $\| a_1, A_1A_2 \|$, corresponding to a triple root of (·841). Dually Σ contains ONE point-pair $\| A_1, a_1a_2 \|$, corresponding to a triple root of (·882).

(iv) *Four-point contact*: The point pencil σ contains ONE line-pair, the repeated line $\| a_1, a_1 \|$. Dually Σ contains ONE point-pair, the repeated point $\| A_1, A_1 \|$.

(v) *Proper double contact*: The point pencil σ contains TWO line-pairs, $\| a_1, a_2 \|$ and the repeated line $\| A_1A_2, A_1A_2 \|$, the latter corresponding to a double root of (·841). Dually Σ contains TWO

* A proof of this statement will be given in 4·98.

point-pairs, $\| A_1, A_2 \|$ and the repeated point $\| a_1 a_2, a_1 a_2 \|$, the latter corresponding to a double root of (·882).

4·95. *Method of proof*

It is not convenient to give a full proof of the statements of 4·94 at this stage. The effect of a change of the co-ordinate system on the equation (·841) which gives the line-pairs of the point pencil σ will be discussed in 9·6. It will then appear that, if the statements of 4·94 are established for any particular co-ordinate system, they hold for all co-ordinate systems.

Consider the case of simple contact, and choose a co-ordinate system so that s' is given by

$$s' \equiv y^2 - 4zx = 0,$$

or by

$$x:y:z = \theta^2 : 2\theta : 1,$$

and the points A_1, A_2, A_3 correspond to the values 0, 1, ∞ of θ. A_1, A_2, A_3 are then the points $(0,0,1)$, $(1,2,1)$, $(1,0,0)$, and the quartic equation in θ which gives the intersections of s' with s reduces to

$$\theta^3 - \theta^2 = 0.$$

One conic locus whose intersections with s' are given by this equation is

$$y^2 - 2xy = 0,$$

so the conic s must be given by an equation of the form

$$s \equiv (y^2 - 2xy) + \eta(y^2 - 4zx) = 0.$$

The cubic equation (·841) in (λ, μ), which is the CONDITION for $\lambda s + \mu s' = 0$ to be a line-pair, reduces to

$$(\eta\lambda + \mu)^2 \{(\xi + \eta)\lambda + \mu\} = 0.$$

The double root $(1, -\eta)$ of this equation corresponds to the conic locus $s - \eta s' = 0$ of σ, i.e. to

$$y^2 - 2xy = 0,$$

which is the line-pair $\| A_1 A_2, A_1 A_3 \|$. The simple root $(-1, \xi + \eta)$ corresponds to the conic locus $s - (\xi + \eta) s' = 0$, i.e. to

$$xy - 2zx = 0,$$

which is the line-pair $\| a_1, A_2 A_3 \|$. Thus 4·94 (ii) is established, and the other particular cases can be similarly treated.

4·96. *Equations of conics of a pencil*

Consider a proper conic locus s, whose point and line equations are $s = 0$, $S = 0$. It is now possible to write down the equations of conic loci or scrolls which have prescribed common points or common tangents with s. Let A_1, A_2, A_3, A_4 be distinct points of s, and let a_1, a_2, a_3, a_4 be distinct tangents of s. Let the point equations of the chord A_iA_j and of the tangent to s at A_i be $u_{ij} = 0$ and $u_{ii} = 0$. Let the line equations of the point a_ia_j and of the contact of a_i with s be $U_{ij} = 0$ and $U_{ii} = 0$. Then

(i) *The general case*: The conic loci whose intersections with s are A_1, A_2, A_3, A_4 form a point pencil which contains the line-pairs $\| A_1A_4, A_2A_3 \|$, $\| A_2A_4, A_3A_1 \|$, $\| A_3A_4, A_1A_2 \|$, and it was shown in 4·82 that this point pencil is determined by any two of its members. It follows that the point equation of any conic locus through A_1, A_2, A_3, A_4 may be expressed in any of the forms

$$\begin{aligned} \lambda_1 s + \mu_1 u_{14} u_{23} &= 0, & \lambda_4 u_{24} u_{31} + \mu_4 u_{34} u_{12} &= 0,\\ \lambda_2 s + \mu_2 u_{24} u_{31} &= 0, & \lambda_5 u_{34} u_{12} + \mu_5 u_{14} u_{23} &= 0,\\ \lambda_3 s + \mu_3 u_{34} u_{12} &= 0, & \lambda_6 u_{14} u_{23} + \mu_6 u_{24} u_{31} &= 0, \end{aligned}$$

by suitable choice of the ratio numbers $(\lambda_1, \mu_1), (\lambda_2, \mu_2), \ldots, (\lambda_6, \mu_6)$.

Dually the line equation of any conic scroll containing the lines a_1, a_2, a_3, a_4 may be expressed in any one of the forms

$$\begin{aligned} \lambda_1 S + \mu_1 U_{14} U_{23} &= 0, & \lambda_4 U_{24} U_{31} + \mu_4 U_{34} U_{12} &= 0,\\ \lambda_2 S + \mu_2 U_{24} U_{31} &= 0, & \lambda_5 U_{34} U_{12} + \mu_5 U_{14} U_{23} &= 0,\\ \lambda_3 S + \mu_3 U_{34} U_{12} &= 0, & \lambda_6 U_{14} U_{23} + \mu_6 U_{24} U_{31} &= 0. \end{aligned}$$

(ii) *Simple contact*: The conics whose intersections with s are A_1, A_1, A_2, A_3 form a point pencil containing the line-pairs

$$\| A_1A_2, A_1A_3 \|, \quad \| a_1, A_2A_3 \|,$$

where a_1 is the tangent to s at a_1. It follows that the point equation of any conic which touches s at A_1 and contains A_2, A_3 may be expressed in any of the forms

$$\lambda_1 s + \mu_1 u_{12} u_{13} = 0, \quad \lambda_2 s + \mu_2 u_{11} u_{23} = 0, \quad \lambda_3 u_{12} u_{13} + \mu_3 u_{11} u_{23} = 0.$$

Dually the conics whose common tangents with s are a_1, a_1, a_2, a_3 form a line pencil containing the point-pairs

$$\| a_1a_2, a_1a_3 \|, \quad \| A_1, a_2a_3 \|,$$

where A_1 is the contact of a_1 with s. Thus, remembering 4·92, the line equation of any conic which touches s at A_1 and contains the lines a_2, a_3 may be expressed in any of the forms

$$\lambda_1 S + \mu_1 U_{12} U_{13} = 0,$$
$$\lambda_2 S + \mu_2 U_{11} U_{23} = 0,$$
$$\lambda_3 U_{12} U_{13} + \mu_3 U_{11} U_{23} = 0.$$

(iii) *Three-point contact*: By a similar argument it follows that the point equation of any conic which has three-point contact with s at A_1 and meets s again in A_2 may be expressed in the form

$$\lambda s + \mu u_{11} u_{12} = 0.$$

Dually, if a_1 is the tangent to s at A_1 and a_2 is another tangent, the line equation of any conic which has three-point contact with s at A_1 and touches a_2 may be expressed in the form

$$\lambda S + \mu U_{11} U_{12} = 0.$$

(iv) *Four-point contact*: The point equation of any conic which has four-point contact with s at A_1 may be expressed in the form

$$\lambda s + \mu a_{11}^2 = 0,$$

and its line equation may be expressed in the form

$$\lambda S + \mu A_{11}^2 = 0.$$

(v) *Proper double contact*: The point equation of any conic which touches s at two distinct points A_1, A_2 may be expressed in any of the forms

$$\lambda_1 s + \mu_1 u_{11} u_{22} = 0, \quad \lambda_2 s + \mu_2 u_{12}^2 = 0, \quad \lambda_3 u_{11} u_{22} + \mu_3 u_{12}^2 = 0.$$

The line equation of such a conic may be expressed in any of the forms

$$\lambda_1 S + \mu_1 U_{11} U_{22} = 0, \quad \lambda_2 S + \mu_2 U_{12}^2 = 0, \quad \lambda_3 U_{11} U_{22} + \mu_3 U_{12}^2 = 0.$$

4·97. *An example*

The following theorem provides an example of the application of 4·96.

If P_1, P_2, P_3 are three general points which do not lie on a proper conic s, then there are six conics which contain P_1, P_2, P_3 and have three-point contact with s.

Choose a co-ordinate system so that s is given by

$$x:y:z = \theta^2:\theta:1$$

and let P_1, P_2, P_3 be (x_1, y_1, z_1), (x_2, y_2, z_2), (x_3, y_3, z_3). Any conic which has three-point contact with s at the point θ_1 and meets s again at the point θ_2 has a point equation of the form

$$\lambda(y^2-zx)+(x-2\theta_1 y+\theta_1^2 z)\{x-(\theta_1+\theta_2)y+\theta_1\theta_2 z\} = 0.$$

This conic contains P_1, P_2, P_3 IF

$$\lambda(y_i^2-z_i x_i)+(x_i-2\theta_1 y_i+\theta_1^2 z_i)(x_i-\theta_1 y_i) \\ -\theta_2(x_i-2\theta_1 y_i+\theta_1^2 z_i)(y_i-\theta_1 z_i) = 0,$$

for $i = 1, 2, 3$. On eliminating λ and θ_2 from these three equations we obtain the following sextic equation in θ_1, each of whose roots gives one conic having the required property:

$$\begin{vmatrix} y_1^2-z_1x_1, & (x_1-2\theta_1 y_1+\theta_1^2 z_1)(x_1-\theta_1 y_1), & (x_1-2\theta_1 y_1+\theta_1^2 z_1)(y_1-\theta_1 z_1) \\ y_2^2-z_2x_2, & (x_2-2\theta_1 y_2+\theta_1^2 z_2)(x_2-\theta_1 y_2), & (x_2-2\theta_1 y_2+\theta_1^2 z_2)(y_2-\theta_1 z_2) \\ y_3^2-z_3x_3, & (x_3-2\theta_1 y_3+\theta_1^2 z_3)(x_3-\theta_1 y_3), & (x_3-2\theta_1 y_3+\theta_1^2 z_3)(y_3-\theta_1 z_3) \end{vmatrix} = 0.$$

4·98. *Double contact*

Consider again the point and line pencils σ and Σ determined by two proper conics s and s'.

When s and s' have double contact, the proper conics of σ are also the proper conics of Σ. In all other cases σ and Σ have no conic in common other than s and s'.

For, in the notation of 4·93 (v), when $a_1 a_2$, A_1, A_2 are chosen as reference points, the line-pairs of σ are

$$y^2 = 0 \quad \text{and} \quad zx = 0,$$

while the point-pairs of Σ are

$$Y^2 = 0 \quad \text{and} \quad ZX = 0$$

by 4·94 (v). Consequently any proper conic locus of σ has a point equation of the form

$$y^2+2gzx = 0,$$

where $g \neq 0$, and its line equation is

$$gY^2+2ZX = 0,$$

which is the line equation of a conic of Σ. It is easily verified that the same is true in the case of four-point contact.

Also any common tangent a of s and s' is also a tangent of all conics of Σ. But by 4·85 only two conics of σ touch a, unless a is one of the lines of a line-pair of σ. Thus Σ can only contain a conic of σ other than s and s' if all the common tangents of s and s' are members of line-pairs of σ, which only happens in the cases of double contact, i.e. proper double contact and four-point contact.

CHAPTER V

CONFIGURATIONS

5·1. Quadrangle and quadrilateral

5·11. The figure determined by four points is called a *quadrangle*, the points being called the *vertices*, and the six joins of pairs of the points being called the *sides* of the quadrangle.* We shall assume that no three of the points A_1, A_2, A_3, A_4 are collinear, so that the six sides are distinct. In Fig. 7 the vertices are A_1, A_2, A_3, A_4, and the meets of pairs of opposite sides, i.e. the points (A_2A_3, A_1A_4),

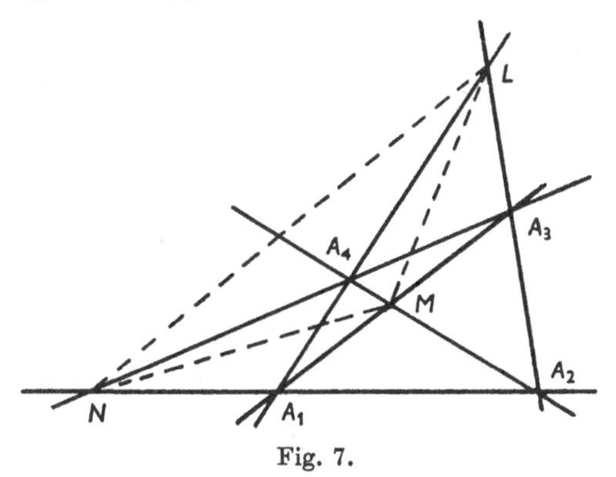

Fig. 7.

(A_3A_1, A_2A_4), (A_1A_2, A_3A_4) are denoted by L, M, N. These points L, M, N are called the *diagonal points* and the triangle LMN is called the *diagonal triangle* of the quadrangle.

The conic loci which contain A_1, A_2, A_3, A_4 form a point pencil σ; the line pairs of σ are $\| A_2A_3, A_1A_4 \|$, $\| A_3A_1, A_2A_4 \|$, $\| A_1A_2, A_3A_4 \|$ and the vertices of these line-pairs are L, M, N.

5·12. Let p be a line which is not a side of the quadrangle and therefore is not part of a conic of σ. Then, by 4·85, the conics of σ cut out an involution (p, σ) on p. In particular

The three pairs of opposite sides of the quadrangle meet p in three pairs of points of an involution.

* The figure determined by four points A, B, C, D, taken in that order, may be called the *ordered quadrangle* $ABCD$. It has the four ordered sides AB, BC, CD, DA, and so with equal propriety may be called an *ordered quadrilateral*. This figure is self-dual.

It was further proved in 4·85 that two conics of σ touch p, at the double points of (p, σ), that the double points are conjugate relative to all conics of σ, and that they coincide in a vertex of the quadrangle IF p contains this vertex.

5·13. The double points of the involution on the line MN are M and N, since the intersections of two of the line-pairs of σ coincide in these points. Thus M and N are conjugate relative to all conics of σ. Similarly L is conjugate to M and to N relative to all conics of σ. Thus:

The diagonal triangle of a quadrangle is self-polar relative to all conics through the vertices of the quadrangle.

5·14. In particular M and N harmonically separate the intersections of MN with A_2A_3 and A_1A_4, since these two lines form a line-pair of σ. This is known as the harmonic property of the quadrangle, and may be stated thus:

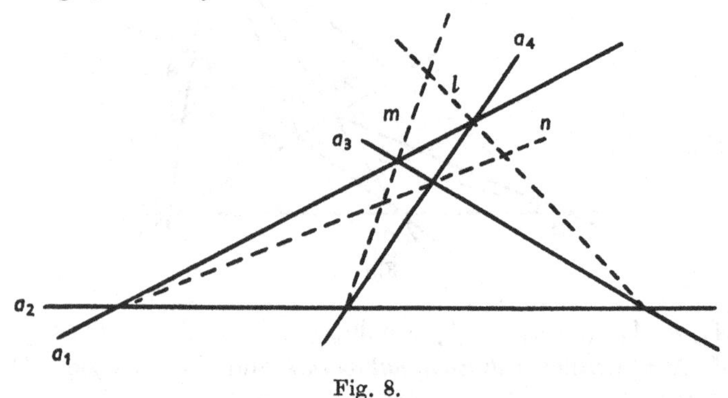

Fig. 8.

Each pair of opposite sides of a quadrangle meet at a diagonal point and harmonically separate the sides of the diagonal triangle through this diagonal point.

5·15. Dually the figure determined by four lines is called a *quadrilateral.* We shall assume that no three of the lines are concurrent. The lines are called the *sides* of the quadrilateral. A point of intersection of two sides is called a *vertex* and the meet of the other two sides is called the opposite vertex. There are six vertices, the three lines joining pairs of opposite vertices are called the diagonal lines of the quadrilateral, and the triangle formed by these lines is called the diagonal triangle. In Fig. 8 the sides are a_1, a_2, a_3, a_4 and the diagonal lines are l, m, n.

The conic scrolls which contain a_1, a_2, a_3, a_4 form a line pencil Σ, and any line p other than a_1, a_2, a_3, a_4 belongs to ONE conic scroll of Σ. Three conic scrolls of Σ are point-pairs, namely

$$\| a_2a_3, a_1a_4 \|, \quad \| a_3a_1, a_2a_4 \|, \quad \| a_1a_2, a_3a_4 \|,$$

and the axes of these point-pairs are the diagonal lines.

5·16. Let P be a point which is not a vertex of the quadrilateral and therefore is not part of a conic scroll of Σ. By 4·89 the pairs of lines through P which belong to scrolls of Σ are pairs of an involution pencil $[P, \Sigma]$. In particular

The three pairs of lines joining P to opposite vertices of the quadrilateral are pairs of an involution pencil.

Two scrolls of Σ contain P, their tangents at P being the double lines of $[P, \Sigma]$; these double lines are conjugate relative to all scrolls of Σ and they coincide in a side of the quadrilateral IF P lies on this side.

5·17. The double lines of the involution pencil $[mn, \Sigma]$ are m and n, so that m and n are conjugate relative to all scrolls of Σ. Thus:

The diagonal triangle of a quadrilateral is self-polar relative to all conic scrolls containing the sides of the quadrilateral.

5·18. In particular m and n harmonically separate the lines joining mn to a_2a_3 and a_1a_4. This is the harmonic property of the quadrilateral and may be stated thus:

Each pair of opposite vertices of a quadrilateral lie on a diagonal line and harmonically separate the meets of this diagonal line with the other two diagonal lines.

5·2. Desargues' Theorem. Related ranges and pencils. Pappus' Theorem

5·21. An immediate deduction from 5·12 is the following, usually known as Desargues' Theorem:

If the joins of corresponding vertices of two triangles ABC, $A'B'C'$ meet in a point U, then the meets of corresponding sides lie on a line u.

The triangles are then said to be *in perspective*; U is called the *centre of perspective* and u the *axis of perspective*.

For, denoting the points $(BC, B'C')$, $(CA, C'A')$, $(AB, A'B')$ by D, E, F, as in Fig. 9, we have to show that EF contains D. Let EF meet AA', BB', CC' in P, Q, R. Then the involutions cut out on EF by the conic loci through U, A, B, C and through U, A', B', C' both contain the pairs $| E, Q |$ and $| F, R |$ and are therefore the same involution. Thus BC and $B'C'$ both meet EF in the mate of P in this involution.

The converse of Desargues' Theorem, that if the meets of corresponding sides of two triangles are collinear, then the joins of corresponding vertices are concurrent, follows from the theorem itself. For, if D, E, F are collinear, the triangles ECC', FBB' are in perspective from D, wherefore (BB', CC') is on the join of $(C'E, B'F)$

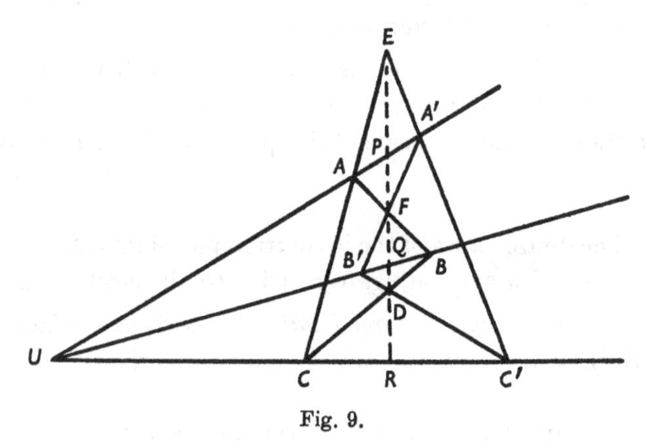

Fig. 9.

and (EC, FB), which is AA'. The converse of Desargues' Theorem is also the dual.

An alternative proof is as follows: With the same notation

$$(UCRC') = E(UCRC') = (UAPA') = F(UAPA') = (UBQB').$$

But U is a self-corresponding point of the subranges

$$\{U, C, R, C'\}, \quad \{U, B, Q, B'\},$$

so BC, $B'C'$, QR are concurrent.*

The proofs here given of Desargues' Theorem and its converse arise naturally in the present context, but they are not the simplest possible. An elementary algebraic proof is given in 5·29 and even simpler proofs will be found in 5·3.

* See 3·25 and 4·35, pp. 81, 111.

5·22. Consider related ranges $\{P, Q, R, \ldots\}$ and $\{P', Q', R', \ldots\}$ on two distinct lines l and l', whose meet is O. If O is not a self-corresponding point, but corresponds to points L and L' on l and l', then it follows from 4·34 that the joins PP', QQ', ... of pairs of corresponding points are the tangents of a proper conic S, which touches l and l' at L and L'. Thus LL' is the polar of O relative to S.

The points like $(PQ', P'Q)$, obtained from two pairs of corresponding points, are called the *cross-meets* of the related ranges. Since $(P, Q, O, L) = (P', Q', L', O')$, it follows that

$$P'(P, Q, O, L) = P(P', Q', L', O'),$$

and PP' is a common ray of these two subpencils of lines, so, by 4·25, the point $(P'Q, PQ')$ lies on LL'. Thus:

If O is not a self-corresponding point, all the cross-meets lie on the line LL', which is called the cross-axis of the related ranges. The cross-axis is the polar of O relative to the proper conic S whose tangents are the joins of pairs of corresponding points, and S touches l and l' at L and L'.

5·23. *If O is a self-corresponding point, the relation is a perspectivity, and all the cross-meets lie on a line through O which is called the cross-axis of the perspectivity. The cross-axis is the polar of the centre of perspective relative to the line-pair formed by l and l'.*

For, if H is the centre of the perspectivity, and if P, P' and Q, Q' are any two pairs of corresponding points other than O, the three points O, H and $(PQ', P'Q)$ are the diagonal points of the quadrangle $PP'QQ'$, and so, by 5·14, the line joining O to $(PQ', P'Q)$ is the harmonic conjugate of OH relative to l and l'.

5·24. Dually, if $p, q, r, \ldots$ and $p', q', r', \ldots$ are corresponding lines of two related pencils with distinct vertices L and L', the lines like $[pq', p'q]$ are called the *cross-joins* of the related pencils. Thus:

If LL' is not a self-corresponding line, but corresponds to distinct lines l and l' of the pencils, then all the cross-joins pass through the point ll', which is called the cross-pole of the related pencils. The cross-pole is the pole of LL' relative to the proper conic s whose points are the meets of corresponding lines of the pencils, and s touches l and l' at L and L'.

If LL' is a self-corresponding line, the meets of pairs of corresponding lines other than LL' lie on a fixed line h. The cross-joins pass through a point of LL' which is the pole of h relative to the point-pair formed by L and L'.

5·25. An immediate deduction from 5·22 and 5·23 is *Pappus' Theorem*, namely

If A, B, C are three points of a line l, and A', B', C' are three points of another line l', then the three points $(BC', B'C)$, $(CA', C'A)$, $(AB', A'B)$ lie on a line p.

For p is the cross-axis of the related ranges on l and l' in which $A \to A'$, $B \to B'$, $C \to C'$. The line p is called the *Pappus line* of the triads ABC, $A'B'C'$. Dually

If a, b, c are three lines through a point L and a', b', c' are three lines through another point L', then the three lines $[bc', b'c]$, $[ca', c'a]$, $[ab', a'b]$ meet in a point P.

5·26. *Six Pappus lines of two triads*

The Pappus line of 5·25 was obtained by relating the points A, B, C to the points A', B', C' in that order. By rearranging A', B', C', six Pappus lines are obtained. Thus, by taking A, B, C with A', B', C', with B', C', A' and with C', A', B' three Pappus lines p_1, p_2, p_3 are obtained, where

$$(BC', B'C),\ (CA', C'A),\ (AB', A'B) \text{ lie on } p_1,$$

$$(BA', CC'),\ (CB', AA'),\ (AC', BB') \text{ lie on } p_2,$$

and $$(BB', CA'),\ (CC', AB'),\ (AA', BC') \text{ lie on } p_3.$$

Similarly, by taking A, B, C with A', C', B', with C', B', A' and with B', A', C' three other Pappus lines p_1', p_2', p_3' are obtained, where

$$(BB', CC'),\ (CA', AB'),\ (AC', BA') \text{ lie on } p_1',$$

$$(BA', CB'),\ (CC', AA'),\ (AB', BC') \text{ lie on } p_2',$$

and $$(BC', CA'),\ (CB', AC'),\ (AA', BB') \text{ lie on } p_3'.$$

Consider the triangles formed by the lines BC', CB', AA' and by the lines CA', AC', BB'. The corresponding sides of these triangles meet in the points (BC', CA'), (CB', AC'), (AA', BB'), which lie on p_3'; so, by the converse of Desargues' Theorem, the joins of corresponding vertices are concurrent. But the corresponding vertices (CB', AA'), (AC', BB') lie on p_2, (AA', BC') and (BB', CA') lie on p_3, and (BC', CB') and (CA', AC') lie on p_1, so p_1, p_2, p_3 are concurrent. Similarly p_1', p_2', p_3' are concurrent. Thus

The six Pappus lines determined by two sets of three collinear points form two sets of three concurrent lines.

5·27. *If a triangle ABC is in perspective with $A'B'C'$ from P and with $B'C'A'$ from Q, then ABC is also in perspective with $C'A'B'$ from a point R.*

For Pappus' Theorem applied to the triads APA', BQC' shows that the point (AC', BA') is collinear with $(PC', A'Q)$ and (AQ, PB), which are C and B' (Fig. 10).

Alternatively the dual of Pappus' Theorem applied to the sets of concurrent lines PB, PC, PA and QC, QA, QB shows that the lines $C'A$, $A'B$, $B'C$ are concurrent.

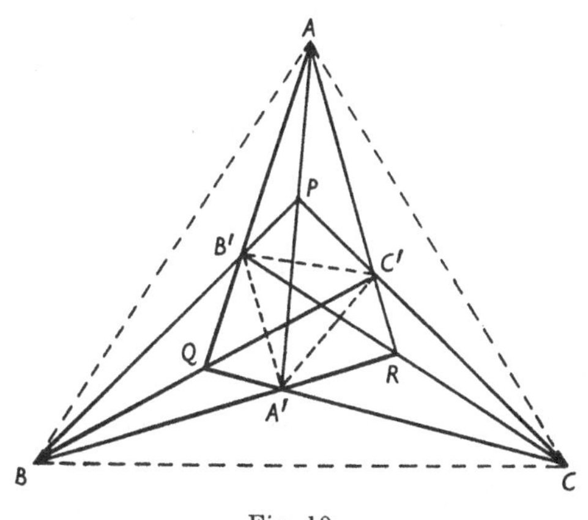

Fig. 10.

5·28. *If three triangles ABC, $A'B'C'$, $A''B''C''$ are in perspective from a point O, the three axes of perspective of the triangles taken in pairs are concurrent.*

For the sets of lines AB, $A'B'$, $A''B''$ and AC, $A'C'$, $A''C''$ form two triangles whose corresponding sides meet in the collinear points A, A', A''. The joins of corresponding vertices are therefore concurrent, and they are the three axes of perspective. Dually

If the corresponding sides of three triangles pass through three collinear points, the centres of perspective of the triangles taken in pairs are collinear.

5·29. Some of these theorems may be proved neatly by coordinates. Thus, for the theorem of 5·27, take ABC as triangle of reference and A', B', C' as (x_1, y_1, z_1), (x_2, y_2, z_3), (x_3, y_3, z_3). Then AA', BB', CC' are

$$z_1 y = y_1 z, \quad x_2 z = z_2 x, \quad y_3 x = x_3 y,$$

and these lines are concurrent IF

$$z_1 x_2 y_3 = y_1 z_2 x_3.$$

Similarly AB', BC', CA' are concurrent IF

$$z_2 x_3 y_1 = y_2 z_3 x_1.$$

But these two CONDITIONS imply

$$z_3 x_1 y_2 = y_3 z_1 x_2,$$

which is a CONDITION for AC', BA', CB' to be concurrent.

For Desargues' Theorem of 5·21 take ABC as triangle of reference and U as $(1, 1, 1)$. Then A', B', C' have co-ordinates of the form

$$(1+\lambda, 1, 1), \quad (1, 1+\mu, 1), \quad (1, 1, 1+\nu).$$

The point $(BC, B'C')$, or D, is then $(0, \mu, -\nu)$, and similarly E and F are $(-\lambda, 0, \nu)$ and $(\lambda, -\mu, 0)$.* These three points lie on the line

$$\frac{x}{\lambda}+\frac{y}{\mu}+\frac{z}{\nu} = 0.$$

Also A', B', C' are the poles of BC, CA, AB relative to the conic whose line equation is

$$\lambda X^2+\mu Y^2+\nu Z^2+(X+Y+Z)^2 = 0.$$

Thus

If two triangles are in perspective there is a conic relative to which the sides of each triangle are the polar lines of the corresponding vertices of the other triangle.

5·3. The Desargues figure

5·31. The figure of Desargues' Theorem deserves further study. In the notation of 5·21 the vertices and sides of the triangles ABC and $A'B'C'$, the points U, D, E, F and the lines u, AA', BB', CC' form a symmetrical figure of ten points and ten lines, each line containing three points and each point lying on three lines. A symmetrical notation for the points and lines is given in Fig. 11, from which it is evident that each line a_{ij} is the axis of perspective of the two triangles of the figure which are in perspective from P_{ij}.

There is a conic relative to which the ten lines a_{ij} are the polars of the ten points P_{ij}.

* The general point of the line P_1P_2 is $(\xi x_1+\eta x_2, \xi y_1+\eta y_2, \xi z_1+\eta z_2)$. The intersection of P_1P_2 with $x=0$ is therefore obtained by taking $\xi = x_2$, $\eta = -x_1$, and similarly for the intersections with $y=0$, $z=0$.

For, since the triangles $P_{23}P_{24}P_{25}$ and $P_{13}P_{14}P_{15}$ are in perspective from P_{12}, it follows from 5·29 that there is a conic S relative to which a_{13}, a_{14}, a_{15} are the polar lines of P_{13}, P_{14}, P_{15} and a_{23}, a_{24}, a_{25} are the polar lines of P_{23}, P_{24}, P_{25}. But P_{34} is $a_{15}a_{25}$ and is therefore the pole relative to S of the line $P_{15}P_{25}$, or a_{34}. Similarly the polar lines of P_{45}, P_{35} relative to S are a_{45}, a_{35}, which meet in P_{12}, so the polar line of P_{12} relative to S is $P_{45}P_{35}$, or a_{12}.

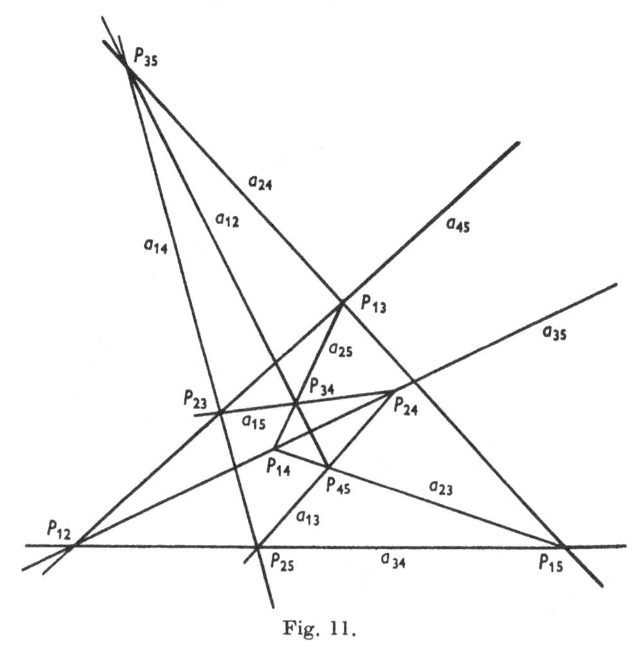

Fig. 11.

5·32. We now make a temporary excursion into space of three dimensions. If P_1, P_2, P_3, P_4, P_5 are five general points in space, there are ten lines P_iP_j joining pairs of the points, and ten planes π_{ij} joining sets of three of the points, where π_{12} for instance denotes the plane through P_3, P_4, P_5. Taking a section by a general plane π we obtain Desargues' figure in π; this is shown in Fig. 11, where the ten points P_{ij} and the ten lines a_{ij} are the intersections of π with the lines P_iP_j and the planes π_{ij}. For through each line P_iP_j there pass three of the planes and each of the ten planes contains three lines P_iP_j.

5·33. Conversely any Desargues figure in a plane may be obtained in this way from a set of five points in space. For, returning to the notation of 5·21, let π be the plane containing the figure, and let H and H' be any two points on a line through U which does not lie in π (Fig. 12). Then HA meets $H'A'$ in a point A'', for these two lines lie in the plane determined by U, A and H. Similarly HB meets $H'B'$ in a point B'', and HC meets $H'C'$ in a point C''. Further, A'', B'', C'' form a triangle, for, if they lay on a line l, then A, B, C would be collinear on the line of intersection of the planes π and $[Hl]$.

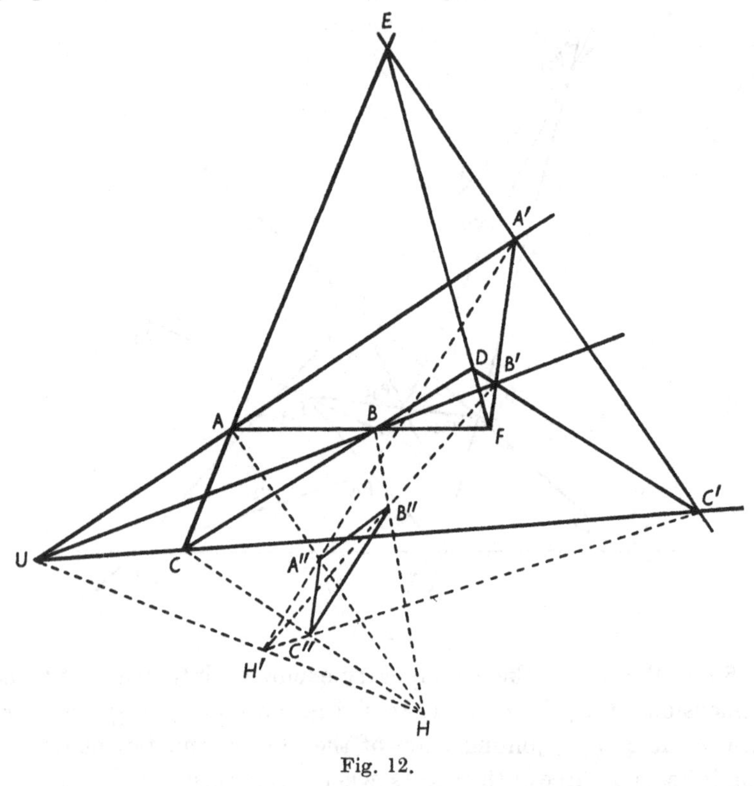

Fig. 12.

Now $B''C''$ meets BC, since these lines lie in the plane determined by H, B and C. Similarly $B''C''$ meets $B'C'$. But H does not lie in π, so $B''C''$ cannot lie in π and therefore meets π in a unique point which lies both on BC and on $B'C'$ and is therefore D. Similarly $C''A''$ contains E and $A''B''$ contains F. Thus the given Desargues' figure is obtained as in 5·32 from the five points H, H', A'', B'', C''.

5·34. Fig. 12 may be used to obtain a proof ofDesargues' Theorem which assumes only the incidence theorems of space of three dimensions.* First of all, if ABC and $A'B'C'$ are two triangles in different planes which are in perspective from a point U, then BC meets $B'C'$ in a point D since these two lines lie in the plane UBC. Similarly CA meets $C'A'$ in a point E and AB meets $A'B'$ in a point F. But the three points D, E, F lie both in the plane ABC and in the plane $A'B'C'$, so they are collinear on the line of intersection of these two planes. Thus Desargues' Theorem is established for two triangles in different planes.

5·35. Now let two triangles ABC and $A'B'C'$ in the same plane π be in perspective from a point U, and let the points $(BC, B'C')$ $(CA, C'A')$, $(AB, A'B')$ be denoted as before by D, E, F. We choose any two points H and H' on any line through U which does not lie in π, and as in 5·33 we construct the points A'', B'', C'' which do not lie in π. As in 5·33 we show that $B''C''$, $C''A''$, $A''B''$ contain D, E, F, appealing only to the incidence theorems. But the triangles ABC and $A''B''C''$ lie in different planes and are in perspective from H, so by 5·34 the pair of corresponding sides meet in collinear points. Thus D, E, F are collinear and Desargues' Theorem is proved for two triangles in the same plane.

5·36. A very simple proof of Desargues' Theorem in a plane is the following: $U, A, B, C, A', B', C', D, E, F$ and π are defined as above. Choose any two points A_1, A'_1 on any line through U which does not lie in π, and let AA_1, $A'A'_1$, meet in V. The triangles A_1BC and $A'_1B'C'$ are in perspective from U and therefore, by 5·34, the points $(BC, B'C')$, $(CA_1, C'A'_1)$, (A_1B, A'_1B') are collinear, say on a line u_1. If we now project the figure from V on to π then A_1 goes to A, A'_1 to A', and the three above points into D, E, F, which are therefore collinear on the line u in which π meets $[Vu_1]$.

5·37. In 5·32 we took five points in space, which determine ten lines joining pairs of the points and ten planes joining three of the points, and on taking a section by a general plane we obtained Desargues' figure.

Dually we may take five planes $\pi_1, \pi_2, \ldots, \pi_5$ of space. We then have ten lines p_{ij} of intersection of pairs of planes, and ten points P_{ij} of intersection of sets of three planes. Joining these lines and points to a general point O we obtain a figure of ten planes through

* In fact this proof might have been given at the end of Chapter I.

O and ten lines through O such that each plane contains three lines and each line lies in three planes. This is the three-dimensional dual of Desargues' figure in a plane. Taking a section of this figure of planes and lines through O by a general plane π we obtain a Desargues figure in π. This process of joining up to O and taking a section by π is equivalent to projecting the lines p_{ij} and the points P_{ij} from O on to π.*

5·38. *Proof of Desargues' Theorem by symbolic notation*

The symbolic notation of 1·9 affords another simple proof of Desargues' Theorem, whether in the plane or in [3]. In the plane case it is essentially the same as the algebraic proof given in 5·29. With the previous notation for the vertices of the triangles and the centre of perspective there is no loss of generality in writing, symbolically,

$$A' \equiv U + \alpha A, \quad B' \equiv U + \beta B, \quad C' \equiv U + \gamma C.$$

Hence

$$B' - C' \equiv \beta B - \gamma C \equiv D, \text{ say};$$

$$C' - A' \equiv \gamma C - \alpha A \equiv E; \quad A' - B' \equiv \alpha A - \beta B \equiv F.$$

The two symbolic expressions for D show that it is the intersection of $B'C'$, BC, and similarly for E and F. Further,

$$D + E + F \equiv B' - C' + C' - A' + A' - B' \equiv 0,$$

whence D, E and F are collinear.

5·4. Harmonic inversion. Quadrangles and quadrilaterals

5·41. We now return to geometry in a plane. If U is a fixed point and u a fixed line not containing U, then a general point P determines the line UP, which meets u in a point L. Let P' be the harmonic conjugate of P relative to U and L. P and P' coincide if P is on u; if P is at U the line UP is indeterminate but P' is also at U whatever line is taken through U. We thus have a (1, 1) correspondence between P and P' whose self-corresponding points are U and the points of u. The correspondence is clearly symmetrical (or involutory†), and it is known as *harmonic inversion* relative to U and u; the point U is called the *base point* and u the *base line*.

* Cf. H. F. Baker, *Principles of Geometry*, II (1922), p. 212.

† An extension of the use of the word involutory in 3·42, p. 87, in connection with an $R\infty^1$.

If p is a general line meeting u in M, then the points of p correspond to the points of the line p' which is the harmonic conjugate of p relative to MU and u. The correspondence between p and p' is also involutory, and p' coincides with p IF p passes through U or coincides with u.

If U is taken as the vertex X $(1, 0, 0)$ of the triangle of reference and u as $x = 0$, and if P is (x, y, z) or, symbolically, $xX + yY + zZ$, the symbolic expression of L is $yY + zZ$. The general point of UP is $\lambda X + yY + zZ$, where U, P, L correspond to the values $\infty, x, 0$ of the parameter λ; P' therefore corresponds to the value $\lambda = -x$ and is the point $-xX + yY + zZ$ or $(-x, y, z)$. The operation of harmonic inversion in this case thus merely has the effect of changing the sign of the x-co-ordinate. Similarly the line $[X, Y, Z]$ corresponds to $[-X, Y, Z]$.

It should be noticed that harmonic inversion is a self-dual transformation of points into points and lines into lines.

5·42. A pencil of lines with vertex P corresponds to a related pencil of lines with vertex P', so a conic locus, which is the locus of meets of corresponding lines of two related pencils, corresponds to another conic locus. Dually a conic scroll corresponds to another conic scroll.

If u is the polar of U relative to a proper conic C, then harmonic inversion relative to U and u leaves C unchanged, merely interchanging the pairs of intersections of C with lines through U and the pairs of tangents at these pairs of points.

5·43. Associated with any triangle LMN there are three harmonic inversions T_1, T_2, T_3 in which the base point and line are L and MN, M and NL, N and LM. If A_4 is any other point, and if A_1, A_2, A_3 are the points corresponding to A_4 in T_1, T_2, T_3, then the pairs (A_2, A_3), (A_3, A_1), (A_1, A_2) correspond in T_1, T_2, T_3. For if we take LMN as triangle of reference and A_4 as $(1, 1, 1)$, then A_1, A_2, A_3 are $(-1, 1, 1)$, $(1, -1, 1)$, $(1, 1, -1)$. Thus if any two of the inversions T_1, T_2, T_3 are applied in succession, the result is equivalent to the third inversion.

5·44. Starting with the triangle LMN and the point A_4 we have obtained Fig. 7,* in which LMN is the diagonal triangle of the quadrangle $A_1A_2A_3A_4$. Conversely, in Fig. 7, it follows from

* See 5·11, p. 147.

the harmonic property proved in 5·14 that A_1 and A_4 are corresponding points in T_1, and similarly that the other pairs of vertices correspond in one of the involutions. Thus, by taking the diagonal triangle as triangle of reference, the vertices of any quadrangle may be taken as $(1, 1, 1)$, $(-1, 1, 1)$, $(1, -1, 1)$, $(1, 1, -1)$. We saw in 5·13 that the diagonal triangle is self-polar relative to all conics through A_1, A_2, A_3, A_4, and we now see that the equation of these conics is

$$\lambda x^2 + \mu y^2 + \nu z^2 = 0,$$

where λ, μ, ν satisfy the equation

$$\lambda + \mu + \nu = 0.$$

We notice also that any conic through one vertex of a quadrangle, which has the diagonal triangle as a self-polar triangle, also contains the other three vertices of the quadrangle.

Dually, by taking the diagonal triangle as triangle of reference, the four lines of any quadrilateral may be taken as

$$[1, 1, 1], \quad [-1, 1, 1], \quad [1, -1, 1], \quad [1, 1, -1],$$

and the conics which touch the four lines have equation

$$\lambda X^2 + \mu Y^2 + \nu Z^2 = 0,$$

where $$\lambda + \mu + \nu = 0.$$

Also any conic, which touches one line of a quadrilateral and has the diagonal triangle as a self-polar triangle, also touches the other three sides of the quadrilateral.

5·45. *If two conic loci s and s' have four distinct intersections A_1, A_2, A_3, A_4 then the diagonal triangle of the quadrangle $A_1A_2A_3A_4$ is the only common self-polar triangle of the two conic loci.*

For suppose a point P_1 has the same polar line p_1 relative to s and s', and let Q_1 and Q_2 be two distinct points of p_1. Then the lines $s_1 = 0$ and $s_1' = 0$ coincide in p_1. P_1 does not lie on p_1, for this would imply that s and s' touch at P_1, which is contrary to the hypothesis that their four intersections are distinct. For some value λ_0 of λ the conic $\lambda_0 s + s' = 0$ passes through P_1. If this conic is not a line-pair with vertex P_1 the tangent to it at P_1 is $\lambda_0 s_1 + s_1' = 0$ or p_1, so that in this case P_1 would lie in p_1. It must therefore be a pair of lines through P_1, so that P_1 is a vertex of the diagonal triangle.

Dually, if two conic scrolls have four distinct common lines, the diagonal triangle of the quadrilateral formed by these lines is the only common self-polar triangle of the two conic scrolls.

5·46. In the quadrangle of Fig. 7,* determined by four points A_1, A_2, A_3, A_4, each side A_iA_j contains one diagonal point and meets the opposite side of the diagonal triangle in a point which we denote by P_{kl} or P_{lk}, where i, j, k, l are the numbers 1, 2, 3, 4 arranged in some order. We thus obtain six points P_{ij}. The triangles $A_1A_2A_3$ and

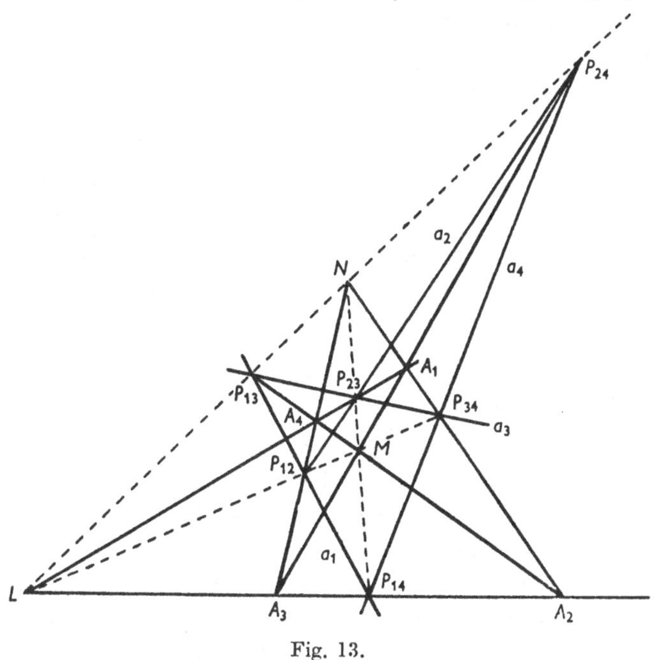

Fig. 13.

LMN are in perspective from A_4, so by Desargues' Theorem the points P_{14}, P_{24}, P_{34} lie on a line which we call a_4. Similarly we obtain three other lines a_1, a_2, a_3 as shown in Fig. 13. The quadrilateral formed by a_1, a_2, a_3, a_4 has LMN as its diagonal triangle.

The complete figure is self-dual, and may be obtained dually from the four lines a_1, a_2, a_3, a_4. The figure is also determined by the triangle LMN and the point A_4 or by the triangle LMN and the line a_4.

5·47. If A_1, A_2, A_3, A_4 are four distinct points of a proper conic C, then the diagonal triangle LMN is self-polar relative to C. Thus MN contains the meet of the tangents to C at A_1, A_4 and the meet of the tangents at A_2, A_3, so MN is a diagonal line of the quadrilateral formed by the tangents at A_1, A_2, A_3, A_4. Thus:

* See 5·11, p. 147.

The quadrangle formed by four distinct points of a proper conic and the quadrilateral formed by the tangents at these points have the same diagonal triangle.

Also, since MN is the polar of L relative to C, the double points of the involution on C determined by the pairs (A_1, A_4) and (A_2, A_3) are the meets of MN with C, and are therefore a pair of the involution determined by (A_1, A_2) and (A_3, A_4) and a pair of the involution determined by (A_1, A_3) and (A_2, A_4).

5·48. *If two quadrangles $A_1A_2A_3A_4$ and $B_1B_2B_3B_4$ have the same diagonal triangle LMN, their eight vertices lie on a conic.*

For we can draw a unique conic through A_1, A_2, A_3, A_4 and B_1, which will have LMN as self-polar triangle and must also contain B_2, B_3, B_4, by 5·44. Dually if two quadrilaterals have the same diagonal triangle, their sides touch a conic.

If a quadrangle A_1, A_2, A_3, A_4 and a quadrilateral $b_1b_2b_3b_4$ have the same diagonal triangle LMN, there are two conics which contain A_1, A_2, A_3, A_4 and touch b_1, b_2, b_3, b_4.

For, by 5·12, there are two conics through A_1, A_2, A_3, A_4 which touch b_1. These conics have LMN as self-polar triangle and therefore touch b_2, b_3, b_4. These two conics coincide IF the double points of the involution cut out on b_1 by conics through A_1, A_2, A_3, A_4 coincide, that is IF b_1 contains one of the points A_1, A_2, A_3, A_4. It follows that if b_1 contain A_1, then b_2, b_3, b_4 must also contain vertices of the quadrangle, and the conic touches the lines $b_1 \dots b_4$ at the point $A_1 \dots A_4$.

5·49. *If two proper conics have four distinct common tangents, the eight points of contact lie on another conic.*

For let A_1, A_2, A_3, A_4 be the four distinct common points of two conics C and C', and let b_1, b_2, b_3, b_4 be the four distinct common tangents.* Then by 5·45 C and C' have only one common self-polar triangle LMN which must be the diagonal triangle of the quadrangle $A_1A_2A_3A_4$ and of the quadrilateral $b_1b_2b_3b_4$. Since LMN is the diagonal triangle of $b_1b_2b_3b_4$ it follows from 5·47 that the quadrangles formed by the contacts of these four lines with C and with C' have LMN as a common diagonal triangle, so by 5·48 the

* It was proved in 4·92 that two proper conics which have four distinct common tangents also have four distinct common points.

eight contacts lie on a conic. Similarly the quadrilaterals formed by the tangents to C and to C' at A_1, A_2, A_3, A_4 have LMN as a common self-polar triangle, so we have the dual theorem:

If two proper conics have four distinct intersections, the eight tangents at these intersections touch a conic.

5·5. Pole and polar relative to a triangle

5·51. A particular case of Desargues' Theorem is the following theorem.

If the lines joining a point P_1 to the vertices of a triangle XYZ meet the opposite sides in X', Y', Z', then the meets L, M, N of corresponding sides of the triangles XYZ, $X'Y'Z'$ lie on a line p_1, which is called the polar of P_1 relative to the triangle XYZ (Fig. 14).

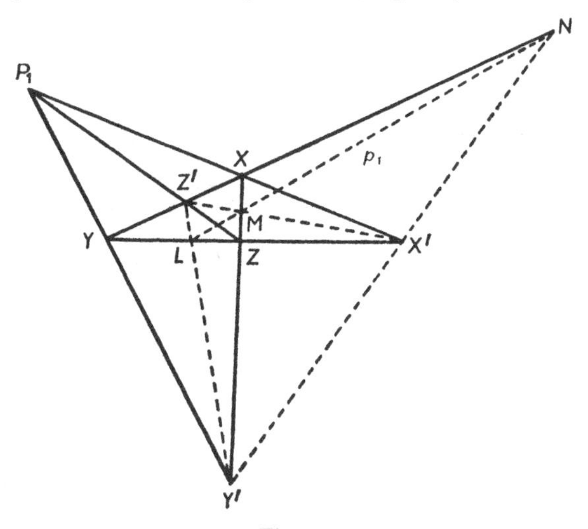

Fig. 14.

5·52. *Dually if a line p_1 meets the sides of a triangle XYZ in L, M, N, then the lines XL, YM, ZN form a triangle $X''Y''Z''$, and the lines XX'', YY'', ZZ'' meet in a point P_1 which is called the pole of p_1 relative to the triangle XYZ* (Fig. 15).

5·53. *If p_1 is the polar of P_1 relative to the triangle XYZ, then P_1 is the pole of p_1.*

For in Fig. 14 the line LMN is the polar line p_1 of P_1. If the points (YM, ZN), (ZN, XL), (XL, YM) are X'', Y'', Z'', then the pole of p_1 is the point of concurrence of the lines XX'', YY'', ZZ''.

But Pappus' Theorem applied to the triads $YZ'N$ and $ZY'M$ shows that X', X'', P_1 are collinear, so P_1 lies on XX'' similarly on YY'' and ZZ''.

Dually if P_1 is the pole of p_1 relative to the triangle XYZ, then p_1 is the polar of P_1.

It should be noticed that P_1 and p_1 are also pole and polar relative to the triangle $X'Y'Z'$ and relative to the triangle $X''Y''Z''$.

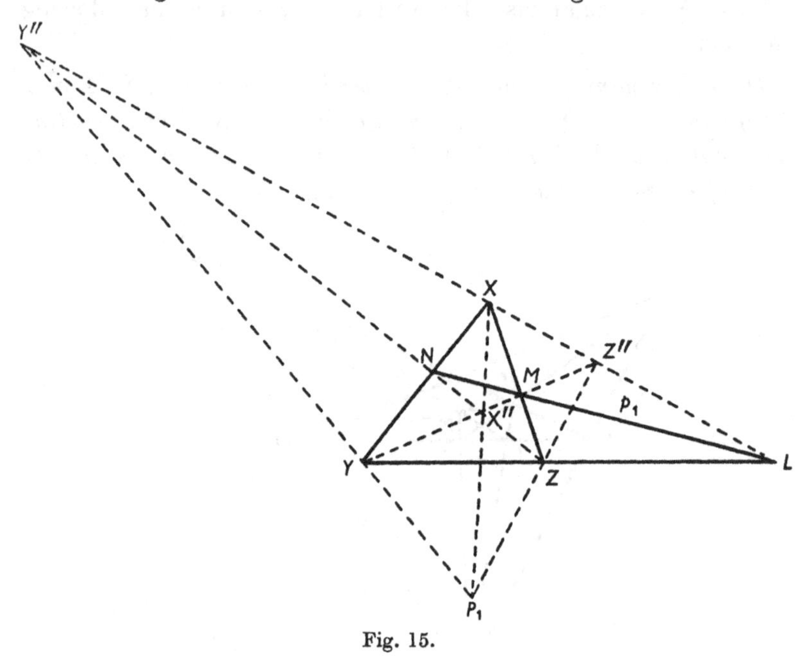

Fig. 15.

5·54. The theorems of 5·53 also follow from Fig. 13,* which as we saw is determined by the triangle LMN and the point A_4 or by the triangle LMN and the line a_4. For the polar of A_4 relative to the triangle LMN is a_4 and the pole of a_4 is A_4.

5·55. Analytically, if XYZ is the triangle of reference, and P_1 is (x_1, y_1, z_1), then in 5·51 the points X', Y', Z' are $(0, y_1, z_1)$, $(x_1, 0, z_1)$, $(x_1, y_1, 0)$, so L, M, N are $(0, y_1, -z_1)$, $(-x_1, 0, z_1)$, $(x_1, -y_1, 0)$. The points L, M, N lie on the line

$$\frac{x}{x_1}+\frac{y}{y_1}+\frac{z}{z_1}=0,$$

which is therefore the polar p_1 of P_1.

* See p. 161.

Similarly if the line p_1 of 5·52 is $\left[\frac{1}{x_1}, \frac{1}{y_1}, \frac{1}{z_1}\right]$, the lines XL, YM, ZN are $\left[0, \frac{1}{y_1}, \frac{1}{z_1}\right]$, $\left[\frac{1}{x_1}, 0, \frac{1}{z_1}\right]$, $\left[\frac{1}{x_1}, \frac{1}{y_1}, 0\right]$, so the lines XX'', YY'', ZZ'' are $\left[0, \frac{1}{y_1}, -\frac{1}{z_1}\right]$, $\left[-\frac{1}{x_1}, 0, \frac{1}{z_1}\right]$, $\left[\frac{1}{x_1}, -\frac{1}{y_1}, 0\right]$. The lines XX'', YY'', ZZ'' contain the point

$$x_1 X + y_1 Y + z_1 Z = 0$$

or (x_1, y_1, z_1), which is therefore the pole of p_1. Thus we have also proved 5·53.

Polarity relative to a triangle will be discussed in greater detail in 12·6.

5·6. A conic and two points, and a conic and two lines

5·61. Let Γ be a proper conic and A and B two distinct points, and consider the locus of points P such that PA and PB are conjugate lines relative to Γ. Suppose first that A and B are not conjugate relative to Γ, so that each line l through A has a unique pole L relative to Γ which is distinct from B. Then BL is the only line through B which is conjugate to l relative to Γ. Similarly each line m through B is conjugate to a unique line through A. Thus the correspondence between lines l through A and lines m through B, in which two lines correspond when they are conjugate relative to Γ, is a (1, 1) correspondence T, and it follows from 4·24 that the locus of points P of intersection of corresponding lines is a conic locus f which contains A and B.

The line AB is self-corresponding in T IF it is self-conjugate relative to Γ, that is IF it is a tangent of Γ. Suppose that this is not the case, so that the pole O of AB does not lie on AB. Then f is a proper conic locus and the tangents to f at A and B are the lines corresponding to AB in T, which are AO and BO. Thus O is the pole of AB relative to f as well as to Γ.

If a line l through A touches Γ at L, the corresponding line through B in T is BL, so L lies on f. Similarly the points of contact of tangents to Γ through B lie on f. Conversely, if a point P of f lies on Γ, since PA and PB are conjugate relative to Γ either PA or PB must be a tangent of Γ, so the intersections of f with Γ are the points of contact of the tangents of Γ through A and through B. If A (or B) lies on Γ, then f touches Γ at A (or B). Thus:

If A and B are distinct points which are not conjugate relative to a proper conic Γ, and if AB is not a tangent of Γ, then the locus of points P such that PA and PB are conjugate relative to Γ is a proper conic locus f through A and B, which is called the harmonic conic locus of Γ and the point-pair A, B. The line AB has the same pole relative to f and Γ, and the intersections of f and Γ are the points of contact of tangents to Γ through A and through B.

5·62. The lines PA and PB are conjugate relative to Γ IF they harmonically separate the tangents of Γ through P. Thus:

The conic locus f is also the locus of points P such that the pair of tangents to Γ through P are harmonically separated by A and B.

5·63. We have assumed that A and B are distinct points which are not conjugate relative to Γ and that AB is not a tangent of Γ. We now inquire what happens to the conic locus f if these restrictions are removed.

If AB is a tangent of Γ, and if E and D are the points of contact of the other tangents of Γ through A and B, then f consists of AB and DE.

For AB is a self-corresponding line in T, so the locus of meets of other pairs of corresponding lines is a line, which contains D and E since AD is conjugate to BD and AE to BE relative to Γ.

If A and B are conjugate relative to Γ, and if O is the pole of AB, then f consists of AO and BO.

For BO is the polar of A relative to Γ so the pole of every line through A lies on BO, the pole of AO being B itself. Thus in T every line through A corresponds to the same line BO, except the line AO which corresponds to all the lines through B.

If B coincides with A, then f consists of the pair of tangents from A to Γ.

For if P is a point of f, PA is self-conjugate relative to Γ and is therefore a tangent of Γ.

5·64. From the first theorem of 5·63 we deduce that, if the sides of a triangle ABC touch a proper conic Γ at the points D, E, F, and if P, Q, R are any three points lying respectively on the lines EF, FD, DE, then BP is conjugate to CP, CQ to AQ, and AR to BR relative to Γ.

The second theorem of 5·63 may be stated thus: If A and B are conjugate points relative to Γ and if a and b are conjugate lines relative to Γ passing respectively through A and B, then either a is the polar of B relative to Γ or b is the polar of A.

5·65. The argument of 5·61 still holds if Γ is a degenerate conic scroll consisting of a pair of points H, K. In fact the conic locus f is then the locus of points P such that the cross-ratio $P(AHBK)$ has the constant value -1 (see § 4·29, p. 108).

5·66. The dual of 5·61 is as follows:

If a and b are distinct lines which are not conjugate relative to a proper conic γ, and if ab is not a point of γ, then the scroll formed by lines p such that the points pa and pb are conjugate relative to γ is a proper conic scroll G containing a and b, which is called the harmonic conic scroll of γ and the line-pair $\|a, b\|$. The contacts of G with a and b lie on the polar of ab relative to γ, so ab has the same polar relative to γ and to G. The common tangents of G and γ are the tangents to γ at the points of intersection of γ with a and b.

This is still true if γ is a line-pair h, k, though the last statement needs careful interpretation. The scroll G consists of the lines p such that $p(ahbk) = -1$.

The reader should write out the dual of §§ 5·62 to 5·64.

5·67. If *Q and Q' are a pair of points of an involution on a line with double points H, K, and if Γ is a proper conic, then the four intersections of the tangents to Γ through Q with the tangents to Γ through Q' lie on the harmonic conic locus f of Γ and the point-pair H, K.*

For, if P is one of the intersections, the tangents to Γ through P are PQ and PQ', which are harmonically separated by H and K. Thus by allowing the pair Q, Q' to vary in the involution we obtain ∞^1 sets of four points on f, and each point of f belongs to one set.

Dually if q and q' are a pair of lines of an involution pencil of lines through a point with double lines h, k, and if γ is a proper conic, then the four lines joining the meets of q with γ to the meets of q' with γ belong to the harmonic conic scroll G of γ and the line-pair h, k. By varying the pair q, q' in the involution pencil we obtain ∞^1 sets of four lines of G, and each line of G belongs to one of these sets.

5·68. *Ordered quadrangles inscribed in one conic and circumscribed to another.*

If γ and Γ are a proper conic locus and a proper conic scroll, and if there is one set of four points A, B, C, D on γ such that the lines AB, BC, CD, DA belong to Γ, then there are ∞^1 such sets of four points of γ, and each point of γ belongs to one of these sets.

For, with the notation of Fig. 16, the diagonal triangle LMN of the quadrangle $ABCD$ is self-polar relative to γ, by 5·13, so L and N harmonically separate the two intersections H and K of LN with γ. In fact L and N are a pair of the involution on HK which has H and K as double points, so by 5·67 the harmonic conic locus of Γ and the point-pair H, K contains the six points A, B, C, D, H, K

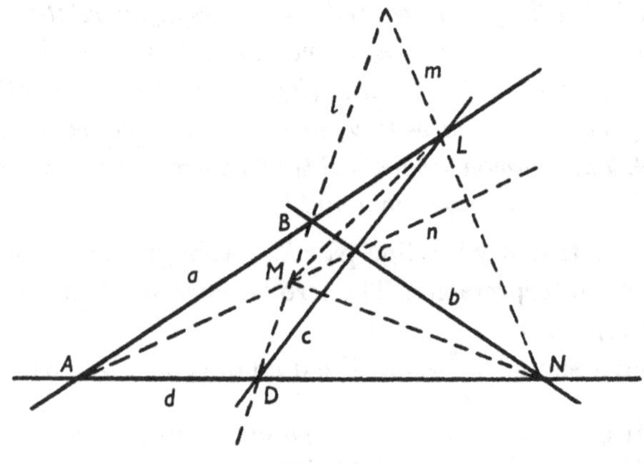

Fig. 16.

and therefore coincides with γ. It then follows from 5·67 that, if L', N' is any pair of the involution on HK, the tangents to Γ through L' meet the tangents to Γ through N' in four points of γ which have the same property as A, B, C, D.

Alternatively, if A' is any point of γ the tangents from A' to Γ meet HK in a pair of points L' and N' which harmonically separate H and K, since we have shown that γ is the harmonic conic locus of Γ and the point-pair H, K. Then, if $A'L'$ meets γ again in B', $B'N'$ is a tangent to Γ, since $B'L'$ is a tangent and the lines $B'L'$, $B'N'$ are harmonically separated by H and K. Similarly, if $B'N'$ meets γ again in C', then $C'L'$ is a tangent of Γ, and if $C'L'$ meets $A'N'$ in D', then D' lies on γ. We thus have four points A', B', C', D' of γ such that $A'B'$, $B'C'$, $C'D'$, $D'A'$ touch Γ.

Also, if $A'C'$ meets $B'D'$ in M', $L'M'N'$ is the diagonal triangle of the quadrangle $A'B'C'D'$, so that M' is the pole of HK relative to γ and therefore coincides with M. Thus, as A' varies on γ, the pairs of points A', C' and B', D' belong to an involution on γ with centre M.

The theorem is self-dual, for, if we denote the lines AB, BC, CD, DA of Fig. 16 by a, b, c, d, then the points ab, bc, cd, da are B, C, D, A. Thus a, b, c, d are four lines of Γ such that the points ab, bc, cd, da lie on γ, and we have shown that if there is one such set there are ∞^1.

It is perhaps of interest to go through the dual proof. The diagonal triangle of the quadrilateral $abcd$ is formed by the lines l, m, n in Fig. 16, so l and n are conjugate relative to Γ and harmonically separate the two tangents of Γ through M, which we will call h and k. Thus l, n is a pair of the involution pencil which has h and k as double lines, and by 5·67 the harmonic conic scroll of γ and the line-pair h, k contains a, b, c, d, h, k and therefore coincides with Γ. Thus each pair l', n' of the involution pencil gives a set of four tangents a', b', c', d' of Γ such that the points $a'b'$, $b'c'$, $c'd'$, $d'a'$ lie on γ. The pairs of lines a', c' form an involution among the tangents of Γ with m as axis.

It should be noted that it is not correct to say that the quadrangle $ABCD$ is circumscribed to Γ, for the quadrangle has six sides and only four of these touch Γ; it is the *ordered* quadrangle $ABCD$ which is circumscribed to Γ.

5·7. Point pencil of conics. Eleven-point conic

5·71. Consider a pencil of conic loci, as defined in 4·82, and denote the pencil by σ. We showed in 4·85 that a general line p meets the conics of σ in pairs of points of an involution, which we denote by (p, σ). Since the involution (p, σ) has two double points, P and P' say, there are two conics of σ which touch p, and their points of contact are P and P'. Since the double points of (p, σ) harmonically separate all the pairs of the involution, the points P and P' are conjugate relative to all the conics of σ. We shall say that two points are conjugate relative to the pencil σ IF they are conjugate relative to all conics of σ.

5·72. If two points P and P' are conjugate relative to two distinct conics of σ, there are three possibilities. If the two conics meet PP'

in distinct pairs of points, these pairs are both harmonically separated by P and P', so P and P' are the double points of the involution (PP', σ), and are therefore conjugate relative to σ. If the two conics meet PP' in the same pair of points, these must lie on all conics of σ, so again P and P' are conjugate relative to σ. If one of the conics contains the line PP', the intersections of the other conic with PP' lie on all conics of σ, so again P and P' are conjugate relative to σ. Thus any two points which are conjugate relative to two distinct conics of σ are conjugate relative to σ.

5·73. If a point P has the same polar line p relative to two distinct conics $s = 0$ and $s' = 0$ of σ, then all the points of p are conjugate to P relative to σ. But if Q is any point not lying on p, we can choose a value λ_0 of λ so that P shall be conjugate to Q relative to the conic locus $\lambda_0 s + s' = 0$, and since P is then conjugate to Q and to all the points of p, it follows from 4·68 that this conic $\lambda_0 s + s' = 0$ is a line-pair with a vertex at P. Thus the polar lines of P relative to any two conics of σ are distinct unless P is a vertex of a line-pair of σ. Conversely if P is a vertex of a line-pair of σ and if p is the polar of P relative to another conic s of σ, then the points of p are conjugate to P relative to two conics of σ, s and the line-pair, so l is the polar line of P relative to all conics of σ except the line-pair.

5·74. Thus the polars of a general point P relative to any two conics $s = 0$ and $s' = 0$ of σ meet in a unique point P', which is conjugate to P relative to σ. The polar lines of P relative to the conics of σ pass through P', and each line through P' is the polar of P relative to one conic of σ. For if P is (x_1, y_1, z_1), its polar lines relative to the conics $\lambda s + s' = 0$ are $\lambda s_1 + s_1' = 0$, which pass through the point $s_1 = s_1' = 0$ and are related to the values of λ. Thus:

Any point P, which is not a vertex of a line-pair of σ, determines a unique point P' which is conjugate to P relative to σ. The polars of P relative to the conics of σ pass through P' and form a pencil of lines related to the conics of σ. The points P and P' are the double points of the involution (PP', σ), and PP' is the tangent at P to the unique conic of σ through P. The relation between P and P' is symmetrical, and P' coincides with P IF *P is a base point of σ.*

5·75. *The locus of poles of a general line p relative to the conics of σ is a conic locus γ, which is also the locus of points whose conjugate points relative to σ, lie on p.*

For, if P_1 and P_2 are any two points of p, the pole of p relative to a conic s of σ is the meet of the polar lines s_1 and s_2 of P_1 and P_2 relative to s. But, by 5·74, as s varies in σ the lines s_1 and s_2 form related pencils, whose vertices P_1' and P_2' are the points conjugate to P_1 and P_2 relative to σ. Thus the locus of the poles $s_1 s_2$ of p relative to the conics of σ is a conic locus γ through P_1' and P_2'. But this locus is determined by p and σ and is therefore independent of the positions of P_1 and P_2 on p, so by allowing P_1 to vary on p we see that γ is also the locus of the points P_1' which are conjugate relative to σ to the points P_1 of p.

Algebraically, a general member of σ is $\lambda s + s' = 0$ and the pole of p relative to this is given by the equations

$$\lambda s_1 + s_1' = 0, \quad \lambda s_2 + s_2' = 0,$$

and the locus of the pole is

$$\gamma \equiv \begin{vmatrix} s_1 & s_1' \\ s_2 & s_2' \end{vmatrix} = 0.$$

Alternatively, a general point of p is $P_1 + \mu P_2$ and the conjugate of this relative to σ is given by the intersection of its polars relative to s and s', namely

$$s_1 + \mu s_2 = 0, \quad s_1' + \mu s_2' = 0.$$

Eliminating μ, we again obtain γ as the locus of these points.

5·76. We confine our attention to a general point pencil σ which has four distinct base points A_1, A_2, A_3, A_4 forming a quadrangle with a diagonal triangle LMN, as in Fig. 7.* If p is a general line the conic γ of 5·75 contains the two points of contact with p of the conics of σ which touch p, and the three points L, M, N, since these are the vertices of the line-pairs of σ. Also each side $A_i A_j$ of the quadrangle meets p in a point P_{ij}, and, if P_{ij}' is the harmonic conjugate of P_{ij} relative to A_i and A_j, the points P_{ij} and P_{ij}' are conjugate relative to σ. Thus γ also contains the six points P_{ij}', and is therefore known as the *eleven-point conic* of p and the quadrangle A_1, A_2, A_3, A_4, or of p and the point pencil σ.

* See p. 147.

5·77. It is interesting to see what happens to the eleven-point conic when p has particular positions.

(i) If p contains one of the vertices of the quadrangle, say A_1, the points P'_{12}, P'_{13}, P'_{14} and the two double points of the involution (p, σ) all coincide in A_1, so the conic γ contains L, M, N, P'_{34}, P'_{42}, P'_{23} and touches p at A_1.

(ii) If p contains one of the vertices of the diagonal triangle, say L, and if p' is the harmonic conjugate of p relative to A_1A_4 and A_2A_3, then γ consists of the two lines p' and MN. For the pole of p relative to every conic of σ except the line-pair $\| A_1A_4, A_2A_3 \|$ lies on MN, while p is the polar of every point of p' relative to this line-pair. Alternatively, each point of p other than L has a unique conjugate point relative to σ which lies on p', while the point L is conjugate to all points of MN relative to σ.

(iii) If p is a side of the quadrangle, say A_1A_4, γ consists of A_1A_4 and MN.

(iv) If p is a side of the diagonal triangle, say MN, γ consists of NL and LM.

5·8. Analytical treatment of general point pencils of conics

5·81. If the diagonal triangle of the quadrangle of base points of a general point pencil σ is taken as triangle of reference, the co-ordinate system may be chosen so that the base points are $(1, \pm 1, \pm 1)$, as we saw in 5·43, and the equations of the conics of σ are

$$s \equiv \lambda x^2 + \mu y^2 + \nu z^2 = 0,$$

where the coefficients λ, μ, ν satisfy the equation

$$u \equiv \lambda + \mu + \nu = 0.$$

The polar of a general point P_1, (x_1, y_1, z_1), relative to s is the line $[\lambda x_1, \mu y_1, \nu z_1]$, and for all conics s of σ this line contains the point $\left(\frac{1}{x_1}, \frac{1}{y_1}, \frac{1}{z_1}\right)$, since $u = 0$. Thus the point P'_1 which is conjugate to P_1 relative to σ is $\left(\frac{1}{x_1}, \frac{1}{y_1}, \frac{1}{z_1}\right)$.

The CONDITION for P_1 to lie on the line p_1, $[X_1, Y_1, Z_1]$, is

$$X_1x_1 + Y_1y_1 + Z_1z_1 = 0,$$

which is also the CONDITION for P'_1 to lie on the conic

$$\gamma \equiv \frac{X_1}{x} + \frac{Y_1}{y} + \frac{Z_1}{z} = 0,$$

which is therefore the eleven-point conic of p_1 and the quadrangle. Also the pole of p_1 relative to s is $\left(\frac{X_1}{\lambda}, \frac{Y_1}{\mu}, \frac{Z_1}{\nu}\right)$, which lies on γ IF $u = 0$, thus again verifying that γ is also the locus of poles of p_1 relative to the conics of σ.

5·82. The conic
$$\begin{vmatrix} x^2 & y^2 & z^2 \\ x_1^2 & y_1^2 & z_1^2 \\ 1 & 1 & 1 \end{vmatrix} = 0$$
contains P_1 and belongs to σ since the coefficients satisfy $u = 0$. It is therefore the unique conic of σ through P_1, and its tangent at P_1 is
$$\begin{vmatrix} x_1 x & y_1 y & z_1 z \\ x_1^2 & y_1^2 & z_1^2 \\ 1 & 1 & 1 \end{vmatrix} = 0,$$
or
$$\begin{vmatrix} x & y & z \\ x_1 & y_1 & z_1 \\ \frac{1}{x_1} & \frac{1}{y_1} & \frac{1}{z_1} \end{vmatrix} = 0,$$
which is of course the line $P_1 P_1'$.

Thus P_1 lies on the tangent at P_2 to the conic of σ through P_2 IF
$$\begin{vmatrix} x_1 & y_1 & z_1 \\ x_2 & y_2 & z_2 \\ \frac{1}{x_2} & \frac{1}{y_2} & \frac{1}{z_2} \end{vmatrix} = 0,$$
so the locus of points of contact of tangents through P_1 to the conics of σ is the cubic curve
$$\begin{vmatrix} yz & zx & xy \\ x & y & z \\ x_1 & y_1 & z_1 \end{vmatrix} = 0.$$

5·83. The line p_1, $[X_1, Y_1, Z_1]$, is the tangent to $s = 0$ at the meet of p_1 with another line p_2, $[X_2, Y_2, Z_2]$, IF the pole $\left(\frac{X_1}{\lambda}, \frac{Y_1}{\mu}, \frac{Z_1}{\nu}\right)$ of p_1 relative to s lies on p_1 and on p_2, that is IF
$$\frac{X_1^2}{\lambda} + \frac{Y_1^2}{\mu} + \frac{Z_1^2}{\nu} = 0$$
and
$$\frac{X_1 X_2}{\lambda} + \frac{Y_1 Y_2}{\mu} + \frac{Z_1 Z_2}{\nu} = 0.$$

Solving these equations for $\frac{1}{\lambda}, \frac{1}{\mu}, \frac{1}{\nu}$ we obtain

$$\frac{1}{\lambda}:\frac{1}{\mu}:\frac{1}{\nu} = Y_1 Z_1(Y_1 Z_2 - Y_2 Z_1) : Z_1 X_1(Z_1 X_2 - Z_2 X_1) : X_1 Y_1(X_1 Y_2 - X_2 Y_1).$$

But s is a conic of σ IF λ, μ, ν satisfy $u = 0$, so the CONDITION for p_1 to be the tangent to a conic of σ at the point $p_1 p_2$ is

$$\frac{X_1}{Y_1 Z_2 - Y_2 Z_1} + \frac{Y_1}{Z_1 X_2 - Z_2 X_1} + \frac{Z_1}{X_1 Y_2 - X_2 Y_1} = 0.$$

Regarding p_1 as fixed, we deduce that the line equation of the point-pair formed by the double points of the involution (p_1, σ) is

$$\frac{X_1}{Y_1 Z - Z_1 Y} + \frac{Y_1}{Z_1 X - X_1 Z} + \frac{Z_1}{X_1 Y - Y_1 X} = 0.$$

Regarding p_2 as fixed we deduce that the tangents to conics of σ at their intersections with p_2 form the cubic envelope

$$\frac{X}{Y_2 Z - Z_2 Y} + \frac{Y}{Z_2 X - X_2 Z} + \frac{Z}{X_2 Y - Y_2 X} = 0.$$

5·84. Suppose $s \equiv ax^2 + by^2 + cz^2 = 0$, $s' \equiv a'x^2 + b'y^2 + c'z^2 = 0$ are the point equations of two proper conics referred to the diagonal triangle of their quadrangle of base points. The pole of a general line p_1, $[X_1, Y_1, Z_1]$, relative to s is the point $\left(\frac{X_1}{a}, \frac{Y_1}{b}, \frac{Z_1}{c}\right)$ and therefore, by 4·71, the equation of the pair of tangents to s at its intersections with p_1 is

$$\left(\frac{X_1^2}{a} + \frac{Y_1^2}{b} + \frac{Z_1^2}{c}\right)(ax^2 + by^2 + cz^2) = (X_1 x + Y_1 y + Z_1 z)^2.$$

Similarly the pair of tangents to s' at its intersections with p_1 has the equation

$$\left(\frac{X_1^2}{a'} + \frac{Y_1^2}{b'} + \frac{Z_1^2}{c'}\right)(a'x^2 + b'y^2 + c'z^2) = (X_1 x + Y_1 y + Z_1 z)^2.$$

The four intersections of these two pairs of tangents lie on the conic

$$\left(\frac{X_1^2}{a} + \frac{Y_1^2}{b} + \frac{Z_1^2}{c}\right)(ax^2 + by^2 + cz^2) = \left(\frac{X_1^2}{a'} + \frac{Y_1^2}{b'} + \frac{Z_1^2}{c'}\right)(a'x^2 + b'y^2 + c'z^2),$$

which belongs to the point pencil $\lambda s + s' = 0$, the corresponding value of λ being given by

$$\frac{X_1^2}{a} + \frac{Y_1^2}{b} + \frac{Z_1^2}{c} = \lambda\left(\frac{X_1^2}{a'} + \frac{Y_1^2}{b'} + \frac{Z_1^2}{c'}\right).$$

But this is the CONDITION for the line p_1 to touch

$$S - \lambda S' = 0,$$

where $S = 0$, $S' = 0$ are the line equations of s, s'.

This result may be stated in various ways. For instance we can deduce the following theorem:

If q and q' are tangents to s and s' at Q and Q', the point qq' lies on a conic $\lambda s + s' = 0$ of the point pencil determined by s and s' IF *QQ' touches the conic $S - \lambda S' = 0$ of the line pencil determined by s and s'.*

5·9. Line pencil of conics. Eleven-line conic

5·91. Consider a line pencil of conics, as defined in 4·87, and denote the pencil by Σ. We showed in 4·89 that the pairs of tangents through a point P to conics of Σ form an involution pencil, which we denote by $[P, \Sigma]$.

Since $[P, \Sigma]$ has two double lines, p and p' say, there are two conics of Σ which pass through P, and their tangents at P are p and p'. Since the double lines of $[P, \Sigma]$ harmonically separate all the pairs of the involution pencil, the lines p and p' are conjugate relative to all the conics of Σ. We shall say that two lines are conjugate relative to Σ IF they are conjugate relative to all the conics of Σ.

5·92. By the principle of duality we deduce from 5·72 that any two lines which are conjugate relative to two distinct conics of Σ are conjugate relative to Σ. From 5·73 we deduce that the poles of a line p relative to two conics of Σ are distinct unless p is an axis of a point-pair of Σ, and conversely that the axis of a point-pair of Σ has the same pole relative to all the conics of Σ except this point-pair.

Then, from 5·74 we deduce:

Any line p, which is not an axis of a point-pair of Σ, determines a unique line p' which is conjugate to p relative to Σ. The poles of p relative to the conics of Σ lie on p' and form a range of points related to the conics of Σ. The lines p and p' are the double lines of the involution pencil $[pp', \Sigma]$, and pp' is the point of contact with p of the unique conic of Σ which touches p. The relation between p and p' is symmetrical, and p' coincides with p IF *p is a base line of Σ.*

5·93. From 5·75 we deduce:

The polar lines of a general point P relative to the conics of Σ form a conic scroll Γ, which is also the scroll formed by lines whose conjugate lines relative to Σ contain P. The conic scroll Γ contains the double lines of the involution pencil $[P, \Sigma]$ and the axes of the point-pairs of Σ.

5·94. A general line pencil Σ has four distinct base lines a_1, a_2, a_3, a_4 forming a quadrilateral with a diagonal triangle lmn as in Fig. 8.* If P is a general point, the conic scroll Γ contains the two tangents at P to the two conics of Σ through P, and the three lines l, m, n. Also each vertex $a_i a_j$ of the quadrilateral is joined to P by a line p_{ij}, and, if p'_{ij} is the harmonic conjugate of p_{ij} relative to a_i and a_j, then p'_{ij} is the line conjugate to p_{ij} relative to Σ. Thus Γ also contains the six lines p'_{ij} and is called the *eleven-line conic* of P and the quadrilateral a_1, a_2, a_3, a_4, or of P and the line pencil Σ. The dual of 5·77 is left to the reader.

5·95. By taking lmn as triangle of reference, the line equations of the conics of the general tangential pencil Σ may be taken as

$$S \equiv \lambda X^2 + \mu Y^2 + \nu Z^2 = 0,$$

where

$$U \equiv \lambda + \mu + \nu = 0,$$

the sides of the base quadrilateral being $[1, \pm 1, \pm 1]$.

Since $U = 0$ the point $[\lambda X_1, \mu Y_1, \nu Z_1]$, which is the pole of $[X_1, Y_1, Z_1]$ relative to S, lies on the line $\left[\frac{1}{X_1}, \frac{1}{Y_1}, \frac{1}{Z_1}\right]$, which is therefore the line conjugate to $[X_1, Y_1, Z_1]$ relative to Σ.

If P is the point (x_1, y_1, z_1) the CONDITION for the line $[X_1, Y_1, Z_1]$ to contain P is

$$x_1 X_1 + y_1 Y_1 + z_1 Z_1 = 0,$$

which is also the CONDITION for the conjugate line $\left[\frac{1}{X_1}, \frac{1}{Y_1}, \frac{1}{Z_1}\right]$ of $[X_1, Y_1, Z_1]$ relative to Σ to belong to the conic scroll

$$\frac{x_1}{X} + \frac{y_1}{Y} + \frac{z_1}{Z} = 0.$$

Since $U = 0$ this scroll contains the polar line $\left[\frac{x_1}{\lambda}, \frac{y_1}{\mu}, \frac{z_1}{\nu}\right]$ of P relative to S, and is the eleven-line scroll of P and the quadrilateral.

* See 5·15, p. 148.

5·96. If p_1 is the line $[X_1, Y_1, Z_1]$ and if p_1' is the conjugate line $\left[\frac{1}{X_1}, \frac{1}{Y_1}, \frac{1}{Z_1}\right]$ of p_1 relative to Σ, then the conic scroll of Σ which contains p_1 is

$$\begin{vmatrix} X^2 & Y^2 & Z^2 \\ X_1^2 & Y_1^2 & Z_1^2 \\ 1 & 1 & 1 \end{vmatrix} = 0,$$

and the contact of p_1 with this conic scroll is the point

$$\begin{vmatrix} X_1X & Y_1Y & Z_1Z \\ X_1^2 & Y_1^2 & Z_1^2 \\ 1 & 1 & 1 \end{vmatrix} = 0, \quad \text{or} \quad \begin{vmatrix} X & Y & Z \\ X_1 & Y_1 & Z_1 \\ \frac{1}{X_1} & \frac{1}{Y_1} & \frac{1}{Z_1} \end{vmatrix} = 0,$$

which is the point p_1p_1'. Also the scroll formed by the tangents to the conics of Σ at their intersections with p_1 is the cubic scroll

$$\begin{vmatrix} X_1 & Y_1 & Z_1 \\ X & Y & Z \\ YZ & ZX & XY \end{vmatrix} = 0.$$

5·97. The point P_1, (x_1, y_1, z_1), is the contact with S of one of the tangents of S through the point P_2, (x_2, y_2, z_2), IF

$$\frac{1}{y_1z_1(y_1z_2 - y_2z_1)} + \frac{1}{z_1x_1(z_1x_2 - z_2x_1)} + \frac{1}{x_1y_1(x_1y_2 - x_2y_1)} = 0$$

by the dual of 5·83. So the equation of the double lines of the involution $[P_1, \Sigma]$ is

$$\frac{x_1}{y_1z - z_1y} + \frac{y_1}{z_1x - x_1z} + \frac{z_1}{x_1y - y_1x} = 0,$$

while the locus of points of contacts of tangents through P_1 to conics of Σ is the cubic curve

$$\frac{x}{y_1z - z_1y} + \frac{y}{z_1x - x_1z} + \frac{z}{x_1y - y_1x} = 0.$$

CHAPTER VI

METRICAL GEOMETRY

6·1. Metrical Geometry. Types of conics

6·11. In Chapter I we briefly considered various types of Geometry $G_1, G_2, \ldots, G_6$, showing how they could be regarded as a sequence leading to the Algebraic Geometry G_a which is our subject. In this book we are thinking of progress in a definite direction, namely towards the general study of algebraic loci in linear space of any number of dimensions. Although our study is limited to space of one, two and three dimensions, rational systems, linear transformations, curves and surfaces of order two, and the like, it should be regarded as an introduction to a much wider field which includes, for example, the theory of algebraic curves which are not rational and the general theory of algebraic transformations.

Nevertheless, although we are primarily looking in this direction, it is well to look back at intervals to observe the connection of what we have been doing with the Euclidean Geometry of rectangular Cartesian co-ordinates. It is of interest to find that some familiar Euclidean property is a particular case of a theorem that we have proved in G_a. Also, since the language of Euclidean Geometry is so well developed, we shall often find that the statement of a theorem in metrical form brings it more clearly before the mind's eye.

In this short chapter we give some examples of theorems of metrical geometry which are deducible from G_a. They must necessarily appear a rather haphazard collection. This is because G_a, unlike metrical geometry, is restricted to algebraic processes and because in this book we are dealing with only elementary properties of G_a. For example, although plane cubic curves are hardly touched on in this book, certain properties of such curves can easily be obtained in Cartesian Geometry. Again, certain properties of conics are essentially tied up with the theory of irrational ∞^1 systems, but such properties can be dealt with to a certain extent in Cartesian Geometry. It is in fact the case that many properties which are quite easy to prove in a specialized form in Cartesian Geometry are connected with general conceptions in G_a which are quite advanced. But we are of course not greatly interested in using G_a to prove metrical theorems, our real field of study lying elsewhere. Consequently in this chapter, and in the brief metrical illustrations which occur elsewhere in the book, we do not attempt to force G_a to deal with metrical properties of an unsuitable type, but only mention cases in which metrical results can be readily deduced.

We must first remind ourselves of the projective meanings of metrical terms. What follows is little more than a brief statement of facts.*

6·12. If co-ordinates are so chosen that the circular points I and J are given by $x^2+y^2 = z = 0$, the line equation of the point-pair $\|I, J\|$ is $X^2+Y^2 = 0$. A *direction* is a point at infinity, that is a point on the line $[0, 0, 1]$, and two directions $(x_1, y_1, 0)$, $(x_2, y_2, 0)$ are *perpendicular* IF they harmonically separate I and J, that is IF $x_1x_2+y_1y_2 = 0$. The direction of a line $[X, Y, Z]$ is the point at infinity on the line, namely $(Y, -X, 0)$. Two lines $[X_1, Y_1, Z_1]$ and $[X_2, Y_2, Z_2]$ are *parallel* or in the same direction IF they have the same point at infinity, that is IF $X_1Y_2-X_2Y_1 = 0$. Two lines $[X_1, Y_1, Z_1]$ and $[X_2, Y_2, Z_2]$ are perpendicular IF their directions are perpendicular, that is IF $X_1X_2+Y_1Y_2 = 0$, that is IF the lines are conjugate relative to the point-pair $\|I, J\|$. Thus a line $[X, Y, Z]$ is perpendicular to itself IF $X^2+Y^2 = 0$, that is IF it contains I or J; it is then called an *isotropic* line. The line at infinity, $z = 0$, has no unique direction, but contains all directions; it is both parallel and perpendicular to all lines.

A point which does not lie on the line at infinity is called a finite point, or an accessible point. If P_1 and P_2 are two finite points whose ordinary Cartesian co-ordinates are (x_1, y_1) and (x_2, y_2), a point on the line P_1P_2 such that $P_1P : PP_2 = k_2 : k_1$ has co-ordinates

$$\left(\frac{k_1x_1+k_2x_2}{k_1+k_2}, \frac{k_1y_1+k_2y_2}{k_1+k_2}\right).$$

If we use instead the homogeneous Cartesian co-ordinates $(x_1, y_1, 1)$ and $(x_2, y_2, 1)$, which is permissible since P_1 and P_2 are finite points, the point P is $(k_1x_1+k_2x_2, k_1y_1+k_2y_2, k_1+k_2)$ or, symbolically, $k_1P_1+k_2P_2$. A general point of P_1P_2 is $\lambda P_1+\mu P_2$, and (λ, μ) has the values $(1, 0)$, (k_1, k_2), $(0, 1)$ for P_1, P, P_2 respectively. Let

* For a fuller discussion the reader may refer to $R(1)$, Chapter 8, and to $R(2)$, Chapters 23 and 24.

T be the point at infinity on P_1P_2, the corresponding value of (λ, μ) being $(1, -1)$. The cross-ratio (P_1PP_2T) is therefore equal to $(\infty, k_1/k_2, 0, -1)$, or $-k_2/k_1$. The metrical statement about the ratio of the two distances P_1P and PP_2 has thus been turned into an 'algebraic projective' statement that the cross-ratio of four collinear points has a certain value. In particular the mid-point of P_1P_2 is the harmonic conjugate relative to P_1 and P_2 of the point at infinity on P_1P_2. If A is a point on a line a, then pairs of points L, L' on a, which are equidistant (in opposite senses) from A, form an involution of which the double points are A and the point at infinity on a.

If AB and CD are two equal segments of a line a and T is the point at infinity on a, the metrical statement '$AB = CD$' can be replaced by the projective statement '$(ABCT) = (DCBT)$'. If AB and CD are unequal, with $AB = kCD$, take the unique point A' such that $(A'BCT) = (DCBT)$. Then

$$k = \frac{AB}{CD} = \frac{AB}{A'B} = (ABA'T).$$

If OA and OB are non-collinear segments of equal length we may, anticipating 6·18, say that A and B lie on a circle with centre O, or, in projective language, that the unique conic through A which touches OI at I and OJ at J also contains B.

If two lines AB and CD are parallel and the segments AB, CD are of equal length, then AC and BD are parallel. Thus, if we are given the segment AB and a point C, the point D can be constructed as the intersection of the lines CT and BT', where T and T' are the points at infinity on AB and AC respectively. The reader will have no difficulty in seeing how, by a combination of the preceding results, we can express in algebraic projective language the fact that the lengths of any two linear segments are in a certain proportion.

6·13. Consider the two lines

$$l_1 \equiv x\sin\theta_1 - y\cos\theta_1 + k_1 z = 0,$$
$$l_2 \equiv x\sin\theta_2 - y\cos\theta_2 + k_2 z = 0.$$

Their directions are the points

$$P_1(\cos\theta_1, \sin\theta_1, 0) \quad \text{and} \quad P_2(\cos\theta_2, \sin\theta_2, 0)$$

at infinity. If I is taken as the point $(1, i, 0)$ and J as $(1, -i, 0)$ we have

$$(P_1 I P_2 J) = (\cot\theta_1, -i, \cot\theta_2, i)$$
$$= \frac{(\cos\theta_1 + i\sin\theta_1)(\cos\theta_2 - i\sin\theta_2)}{(\cos\theta_1 - i\sin\theta_1)(\cos\theta_2 + i\sin\theta_2)}$$
$$= e^{2i(\theta_1-\theta_2)}.$$

Similarly
$$(P_1 J P_2 I) = e^{-2i(\theta_1-\theta_2)}$$
$$= e^{2i\{\pi-(\theta_1-\theta_2)\}}.$$

In the Euclidean plane the angles between l_1 and l_2 are, within integral multiples of 2π and with a suitable convention for the sense in which the angle is measured, $\theta_1-\theta_2$ and $\pi-(\theta_1-\theta_2)$; the above equations, which are due to Laguerre, enable us to interpret the idea of an angle in terms of the value of a cross-ratio.*

If the lines l_1 and l_2 are perpendicular, the above cross-ratios each have the value -1, and P_1, P_2 harmonically separate I, J.

Consider involution pencils of pairs of lines through a finite point U. Two lines through U are perpendicular IF they harmonically separate the isotropic lines UI and UJ; hence the involution pencil with UI and UJ as double lines, which is called the orthogonal† involution pencil with vertex U, consists of all pairs of perpendicular lines through U. Since two different involutions of pairs of an $R\infty^1$ have one pair in common it follows that a general involution pencil with vertex U contains one pair of perpendicular lines, while an involution pencil which contains two pairs of perpendicular lines is an orthogonal involution pencil. If UI and UJ are a pair of an involution pencil, the double lines are perpendicular and bisect the angles between all the pairs of lines of the pencil; for if the double lines are taken to be $xy = 0$ the pairs of the involution pencil are the line-pairs of the form $\lambda x^2 - \mu y^2 = 0$ which are, in metrical language, equally inclined to the x-axis in opposite senses.

6·14. A conic locus γ meets the line at infinity in two points T and T'. The *asymptotes* of γ are the tangents at T and T', and the *centre* of γ is the pole of the line at infinity, that is the meet of the asymptotes. A *diameter* is a line through the centre, and, if A and B are the intersections of a diameter with γ, then the mid-point of AB is the centre.

* Cf. $R(2)$, p. 35.

† Cf. $R(2)$, p. 34.

If T and T' coincide in T then γ is a parabola, its centre T is at infinity, and its diameters are parallel. A degenerate parabola is a line-pair with its vertex at infinity or a repeated line.

If T and T' are I and J, γ is a circle. A degenerate circle consists either of the line at infinity and another line or of two lines OI and OJ. The latter is called the point circle O.

If T and T' harmonically separate I and J, γ is a rectangular hyperbola, which may degenerate into two perpendicular lines.

The distinction between ellipse and hyperbola only occurs in real geometry.

If the centre O of γ is not at infinity, and if a diameter p meets the line at infinity in P, then the pole of p relative to γ is the harmonic conjugate P' of P relative to T and T'; hence OP' is the diameter p' which is conjugate to p. The locus of mid-points of chords parallel to p is the polar of P, or p', and similarly p is the locus of mid-points of chords parallel to p'. The pairs of conjugate diameters form an involution pencil with OT and OT' as double lines. Hence, by 6·13, there is one pair of perpendicular conjugate diameters, called the *principal axes*, unless OT and OT' are OI and OJ, in which case γ is a circle and all pairs of conjugate diameters are perpendicular.

6·15. A *focus* of γ is a point P such that the tangents of γ through P are PI and PJ. A *directrix* is the polar line of a focus.* If γ is a proper conic locus, γ and the point-pair $\|I, J\|$ determine a line pencil of conic scrolls, which is called a *confocal system* since all proper conics of the system have the same foci. As a confocal system is a line pencil, its members should be regarded as scrolls, and the degenerate members are point-pairs, not line-pairs. In fact a confocal system is any line pencil of conic scrolls which contains the point-pair $\|I, J\|$. A confocal system is determined by any one of its members, other than $\|I, J\|$, and, if $S = 0$ is the line equation of a conic scroll, the conic scrolls of the confocal system determined by S have line equations of the form $S + \lambda(X^2 + Y^2) = 0$.

6·16. If γ is a proper conic locus, and is not a parabola or a circle, its foci are the two pairs of opposite vertices other than I, J of the quadrilateral formed by the tangents to γ through I and through J.

* Cf. $R(2)$, p. 144.

In Fig. 17 the foci are S, S', H and H', and the triangle OXY is self-polar relative to γ since it is the diagonal triangle of the quadrilateral. Thus O is the centre of γ and OX, OY are conjugate diameters; since, by the harmonic property of the quadrilateral, they harmonically separate OI, OJ, they are the principal axes of γ. If XYO is taken as triangle of reference the equation of γ is of the form

$$ax^2+by^2+cz^2 = 0.$$

The confocal system determined by γ consists of the conic scrolls inscribed in the quadrilateral of tangents, and the point-pairs of the system are $\|I, J\|$, $\|S, S'\|$ and $\|H, H'\|$.

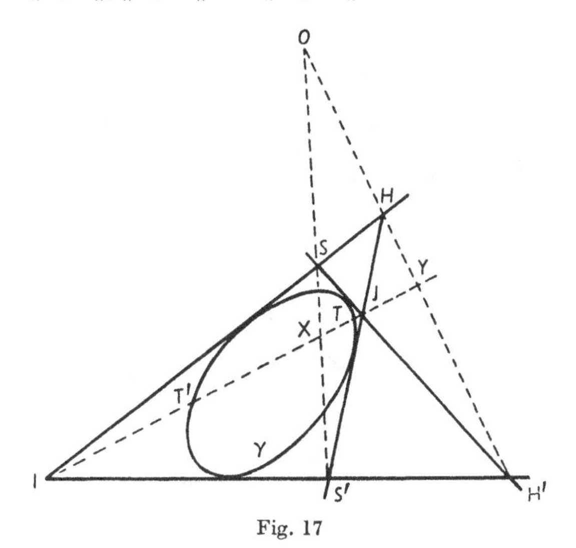

Fig. 17

6·17. If γ is a proper parabola touching the line at infinity at O, then O is the centre of γ, and the other tangents through I and J meet in the focus S (Fig. 18). The other intersection V of OS with γ is the *vertex*, and OS is the *axis* of γ. The polar of S is the directrix; if this meets the line at infinity in U, the polar of U is SO and hence the tangent at V contains U. SU and SO harmonically separate SI and SJ, since they are conjugate lines relative to γ; hence U and O are perpendicular directions and the tangent at the vertex is perpendicular to the axis. If OUV is taken as triangle of reference the equation of γ is of the form $y^2 = 4azx$.

When we talk about the focus of γ we always mean the point S, but O is also a focus and there are no others. The confocal system determined by γ consists of the conic scrolls whose common tangents with γ are IS, JS and IJ counted twice. Thus the proper confocals are the proper conics which touch γ at O and also touch IS and JS; the point-pairs of the system are $\| I, J \|$ and $\| S, O \|$.

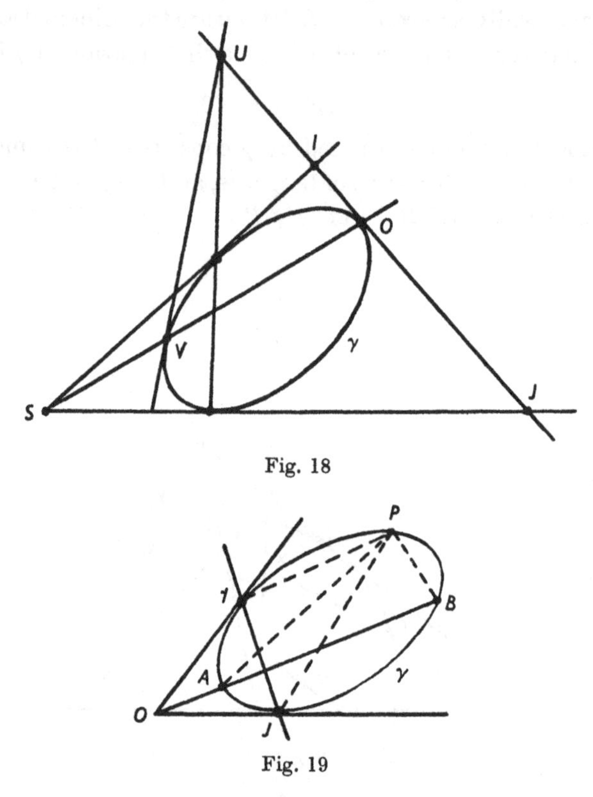

Fig. 18

Fig. 19

6·18. If γ is a circle, the tangents at I and J meet in the centre O, which is also the only focus (Fig. 19). If we take the triangle formed by any two perpendicular diameters and the line at infinity as triangle of reference, and I and J as $(1, \pm i, 0)$, the equation of γ will be of the form $x^2 + y^2 = a^2z^2$.

The proper conics of the confocal system determined by γ are the proper conics which touch γ at I and J, that is the circles concentric with γ; the point-pairs of the system are $| I, J |$ and $| O, O |$.

If A, B are the points of γ on any diameter, A and B harmonically separate I and J on γ, and therefore, if P is any point of γ, the lines PA, PB harmonically separate PI, PJ; in metrical language, a diameter subtends a right angle at any point of the circumference.

Taking P at A we see that the tangent to a circle at any point is perpendicular to the diameter through the point.

If A and B are any two points, H the point at infinity on AB, and K the harmonic conjugate of H relative to I and J, then perpendicular lines through A and B form related pencils in which AB corresponds to AK and BK, AI corresponds to BI and AJ to BJ. Thus the locus of points P such that PA is perpendicular to PB is a circle through A and B, whose centre is the mid-point of AB. More generally the locus of points P for which the lines AP and BP contain a fixed angle, that is for which the cross-ratio $P(AIBJ)$ is constant, is a circle through A and B.

6·19. It may have been noticed that the present section is somewhat unsymmetrical. In the definition of a parabola and a circle we were thinking of a conic as a locus, while as soon as we mentioned confocal systems we were driven to think of the conics as scrolls. This is unavoidable as the complete duality of projective geometry is lost when we pass to metrical geometry, because a plane with two fixed points is not self-dual. We could in fact develop a dual metrical geometry with two lines instead of two points, but this would be of little interest. Much more interesting is the self-dual metrical geometry which can be developed by taking a proper conic instead of the point-pair $\|I, J\|$; this is beyond the scope of this book, but for an account of such a geometry the reader is referred to H. F. Baker, *Principles of Geometry*, vol. II, chapter V. The same book also contains a much deeper analysis of the ideas of distance and angle in the Euclidean plane than is necessary for our present purpose.

6·2. Metrical applications

6·21. A general point pencil σ of conics cuts an involution, Λ say, of pairs of points on the line at infinity. Two conics of σ touch the line at infinity; hence:

A general point pencil of conics contains two parabolas.

The involution whose double points are I, J has in general one pair common with Λ; hence:

A general point pencil of conics contains one rectangular hyperbola.

If two conics of σ are rectangular hyperbolas the double points of Λ are I and J. In this case all the conics are rectangular hyperbolas.

In particular the three line-pairs of σ are perpendicular, and the base points of σ form the vertices of a triangle and its orthocentre. Thus:

All rectangular hyperbolas through the vertices of a triangle contain the orthocentre.

If the double points of Λ harmonically separate I, J they are the points at infinity on the principal axes of all conics of σ, and in particular they are the points at infinity on the angle bisectors of the line-pairs of σ. Thus:

If the principal axes of two conics s and s' are parallel they are parallel to the principal axes of any conic of the point pencil determined by s and s'.

One member of σ is a circle. Thus:

The common chords of a conic and a circle are equally inclined to the principal axes of the conic.

6·22. A pencil σ of circles through I, J and two finite points A, B is called a *coaxal system* of circles and the line AB is the *radical axis*. One degenerate member of σ is the line-pair $\|AB, IJ\|$. The others are, with the notation of Fig. 20, the line-pairs $\|LI, LJ\|$ and $\|MI, MJ\|$; these are the point-circles of σ, and L

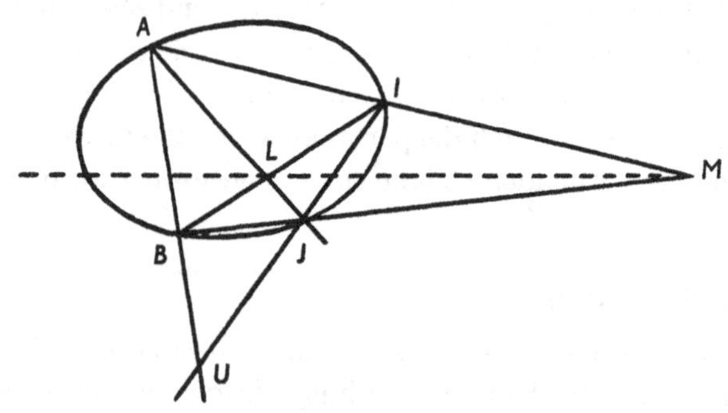

Fig. 20

and M are known as the *limiting points*. The triangle ULM is self-polar to each circle of σ. The poles of IJ with respect to the coaxal circles all lie on LM, which is thus the *line of centres* of the system; it is perpendicular to the radical axis. Further, the points L, M, being the double points of the involution (LM, σ), are conjugate relative to every circle of σ.

6·23. *Nine-point circle and Apollonius hyperbola*

Consider the eleven-point conic γ of the line at infinity with respect to a general point pencil σ of conics. Following 5·7 we see that γ may be regarded as the locus of the poles of the line at infinity, that is the locus of centres of conics of σ, and as the locus of points conjugate, relative to all conics of σ, to points at infinity. If A_1, A_2, A_3, A_4 are the base points of σ, LMN the diagonal triangle of the quadrangle $A_1A_2A_3A_4$, and U, V the points at infinity on the two parabolas of σ, the conic γ contains L, M, N, U, V and the mid-points of the sides A_iA_j.

In the case where U, V are I, J the conics of σ are, as we have seen, all rectangular hyperbolas, and each of the base points is the orthocentre of the triangle formed by the other three. The eleven-point conic γ becomes the *nine-point circle* of the quadrangle $A_1A_2A_3A_4$ or, as is more commonly said, of the triangle formed by any three of the points A_i. Thus:

The nine-point circle of a triangle is the locus of centres of rectangular hyperbolas through the vertices.

If U, V harmonically separate I, J, the base points of σ are concyclic; the conic γ is a rectangular hyperbola and its asymptotes are parallel to the principal axes of every conic of σ. Let C be the circle of σ and O its centre, s any other conic of σ and P a point of intersection of s with γ. P is conjugate to some point P' at infinity relative to all conics of σ, and P' must therefore lie on the tangent to s at P. The polar of P' with respect to the circle C contains P and the centre O of C and is therefore the line OP. OP meets the line at infinity in the harmonic conjugate of P' relative to I and J. It follows that OP is perpendicular to the tangent to s at P, that is that P is the foot of a normal to s from O. Conversely, it is easily shown that if P is the foot of a normal from O it lies on γ. The rectangular hyperbola γ, which thus contains the feet of the four normals from O to s, is known as the *Apollonius hyperbola* of O with respect to s; it is equally the Apollonius hyperbola of O with respect to any conic of the pencil determined by s and any circle with centre O.

6·24. *Director circle*

Let Γ be a proper conic scroll and f the harmonic conic locus of Γ and the point-pair $\| I, J \|$. As in 5·6 f is the locus of the point of intersection P of conjugate lines to Γ through I and J, or of points

P whose tangents to Γ harmonically separate PI, PJ and are therefore perpendicular. This locus of points of intersection of perpendicular tangents is the *director circle* of Γ; its centre coincides with the centre O of Γ. The director circle meets Γ in the contacts with Γ of the tangents to Γ from I and J, that is in the intersections of Γ with its directrices.

If Γ degenerates into a point-pair $\|A, B\|$ the director circle f becomes the circle on AB as diameter. If Γ is a parabola f degenerates into the line at infinity and another line; this other line contains the contacts of the tangents to Γ, other than IJ, from I and J, and is therefore the directrix of Γ, the polar of its proper focus. If Γ is a rectangular hyperbola I and J are conjugate to Γ and f becomes the line-pair $\|OI, OJ\|$, i.e. the point circle O.

In the general case if Q, Q' is a pair of points on the line at infinity harmonically separating I, J, it follows, as in 5·6 (5·67 and 5·68), that the four intersections of the tangents to Γ from Q with the tangents from Q' lie on f; these two pairs of tangents are thus the two pairs of opposite sides of a rectangle inscribed in f and circumscribed to Γ, and there are ∞^1 such rectangles, each with centre O. Thus the figure of a conic and its director circle is projectively equivalent to that of two conics for which there exist ∞^1 ordered quadrangles circumscribed to one and inscribed in the other.

6·25. Let Γ and Γ' be two conic scrolls and A and B the points of intersection, other than I and J, of their director circles. The lines AI, AJ harmonically separate the tangents from A to both Γ and Γ', and are thus the double lines of the involution pencil of tangents from A to conic scrolls of the line pencil Σ determined by Γ and Γ'. Hence A, and similarly B, lie on the director circle of each conic of Σ; the director circles form a coaxal system with radical axis AB. In particular, considering the degenerate members of Σ, we have the theorem:

The three circles described on the joins of pairs of opposite vertices of a complete quadrilateral as diameters have two common points.

If the base lines of the pencil Σ are three lines a_1, a_2, a_3 and the line at infinity the scrolls are parabolas and their directrices all pass through a fixed point A. The directrix of the degenerate parabola consisting of the point a_2a_3 and the point at infinity on a_1 is the line through a_2a_3 perpendicular to a_1. Thus:

The directrices of parabolas inscribed in a triangle pass through the orthocentre.

6·26. Let Σ be a line pencil of confocal conics, the point-pairs of the system being $\|I, J\|$, $\|S, S'\|$ and $\|H, H'\|$, as in Fig. 17.* The involution pencil Λ of pairs of tangents from a general point P contains the pair PI, PJ, and hence the double lines of Λ are perpendicular. These are the tangents at P to the two confocals which pass through P. Hence:

Two confocal conics intersect orthogonally at each of their four intersections.

By 6·13 each pair of Λ has the double lines as angle bisectors. One of these pairs is $|PS, PS'|$. Thus:

If P is a point of a conic γ, and S, S' two foci on the same principal axis, the tangent and normal at P are the bisectors of the angle SPS'.

A special case of this is:

If P is a point of a parabola γ with focus S, the tangent and normal at P bisect the angles between PS and the line through P parallel to the axis.

6·27. *Auxiliary circle*

With the notation of Fig. 21 let the tangent t to a conic γ at the point P meet the line at infinity in Q, let Q' be the harmonic conjugate of Q relative to I, J, and let SQ' meet t in N. The point N is thus the foot of the perpendicular from the focus S to the tangent t. We shall prove that the locus of N as t varies is a circle.

Let t meet SI in L and SJ in M, let IN meet SJ in V, and let JN meet SI in W. We have

$$(LWSI) = J(LNMQ) = S(LNMQ) = (IQ'JQ) = -1;$$

similarly

$$(MVSJ) = I(MNLQ) = S(MNLQ) = (JQ'IQ) = -1.$$

Suppose that we had started with a general line through I meeting SJ in V. This determines the point M such that $(MVSJ) = -1$. From M we can draw ONE tangent t to γ apart from MJ, touching γ at P and meeting IV, IS, IJ in N, L, Q respectively. If SN meets IJ in Q' we have

$$(JQ'IQ) = S(MNLQ) = I(MVSJ) = -1,$$

* See p. 183.

from which it follows that SN is perpendicular to t. Thus a general line through I contains ONE point N which is the foot of the perpendicular from S to a tangent of γ. We have in fact

$$I\{N\} \equiv \{V\} \equiv \{M\} \equiv \{P\} \equiv \{L\} \equiv \{W\} \equiv J\{N\},$$

where the ranges $\{V\}$ and $\{M\}$ are on SJ, the ranges $\{L\}$ and $\{W\}$ on SI, and the range $\{P\}$ on γ. The locus of N is therefore a circle, δ say, generated by related pencils with vertices I and J. If V and J

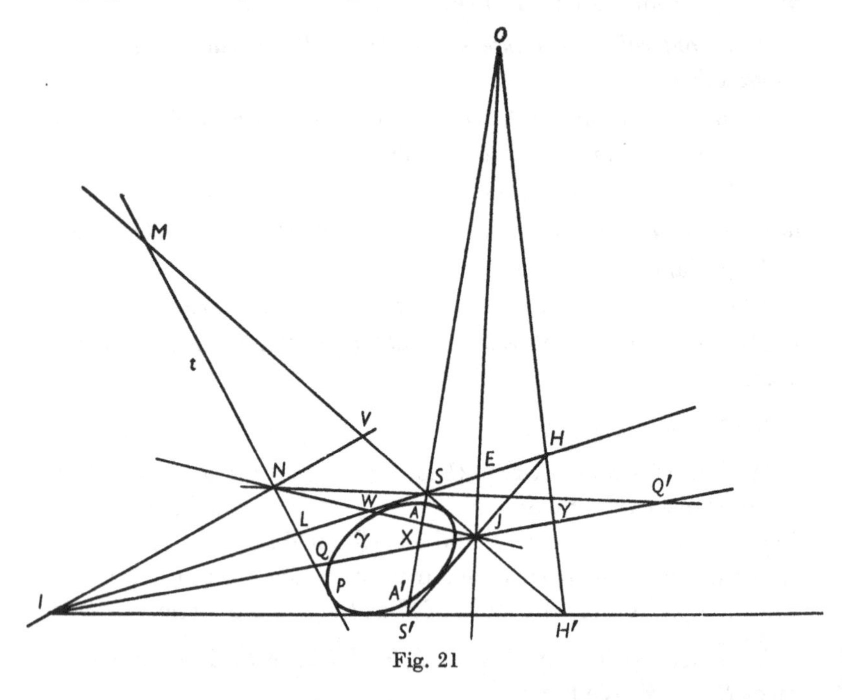

Fig. 21

coincide, M and J also coincide, and L and W become the points H and E of Fig. 21, since $(HESI) = -1$. Thus to the line IJ in the pencil through I there corresponds the line JEO in the pencil through J. Hence JO is the tangent to δ at J, and similarly IO is the tangent at I; thus the circle δ is concentric with γ. Further, δ contains the intersections A, A' of the principal axis SS' with γ; for the tangent to γ at A goes through Y, and X is the harmonic conjugate of Y relative to I, J. It follows that δ is the circle on AA' as diameter. The circle δ is known as the *auxiliary circle*, and there is also another auxiliary circle having as ends of a diameter the points (γ, HH').

From the symmetry of the above result it is clear that δ must also be the locus of the foot of the perpendicular from S' to a variable tangent of γ. A direct proof can be given as follows. Let the other tangent t' to γ from the point N of Fig. 21 touch γ at P'. The involution pencil of pairs of tangents from N to conics confocal with γ contains $| NP', NP |$, $| NI, NJ |$, $| NS', NS |$, so, by 3·43,

$$N(P'IS'J) = N(PJSI) = (QJQ'I) = -1,$$

and thus N is the foot of the perpendicular from S' to t'.

When γ is a parabola and S its proper focus, essentially the same arguments can be applied, but in this case the line IJ is self-corresponding in the two related pencils and δ degenerates into IJ and the tangent at the vertex of γ. Thus:

The foot of the perpendicular from the focus of a parabola to a variable tangent lies on the tangent at the vertex.

6·28. *Pedal circle and pedal line*

If ABC is a triangle and S a general point there is ONE conic γ touching BC, CA, AB, SI, SJ. The circle through the feet of the perpendiculars from S to the sides of the triangle ABC, which is known as the *pedal circle* of S with respect to the triangle, is thus the auxiliary circle, associated with the focus S, of the conic γ.

Consider the line pencil Σ of parabolas touching BC, CA, AB, IJ. There is ONE tangent, apart from IJ, to every member of Σ from each of the points I and J. The locus of the points of intersection of these tangents, that is of the foci of members of Σ, is thus a circle; one member of Σ is the point-pair consisting of A and the point (BC, IJ), so that A, and similarly B and C, lie on this circle. We have therefore the theorem:

The foci of parabolas inscribed in a triangle lie on the circumcircle.

If the point S above lies on the circumcircle of ABC the conic γ is therefore a parabola, whose auxiliary circle degenerates into IJ and the tangent at the vertex. Thus:

The feet of the perpendiculars to the sides of a triangle ABC from a point S of the circumcircle are collinear.

The line of collinearity is the *pedal line* or *Simson line* of S with respect to the triangle.

6·29. Consider again a line pencil Σ of confocal conics with centre O and foci S, S', H, H'. The poles of a line p relative to the members of Σ are collinear on a line p', the conjugate of p relative to Σ. Since p' contains the pole of p relative to $\| I, J \|$ it is perpendicular to p; it is the normal at the point of contact to the confocal which touches p.

Let L be a general point of the plane and Γ the eleven-line conic of L with respect to Σ; as in 5·94, Γ is the envelope of the polar lines of L relative to members of Σ and also the envelope of lines p' which are conjugate, relative to Σ, to lines p through L. The polars of L relative to the point-pairs of Σ are SS', HH', IJ, so that Γ is a parabola touching the common principal axes of the confocals. SS' and HH' meet at right angles in O, so that O lies on the directrix of Γ. The tangents at L to the two confocals through it also belong to Γ and are perpendicular. The directrix of Γ is thus the line OL.

If M is the focus of Γ, the line MI belongs to Γ and is therefore the polar of L relative to some conic of Σ. IL, IM are therefore conjugate relative to this conic and harmonically separate the tangents IS, IS'; similarly JL, JM harmonically separate JS, JS'. The points L and M are thus symmetrically situated with respect to Σ, and L is the focus of the eleven-line parabola of M relative to Σ.

Let T, T' be the points of contact of tangents from L to a general conic C of Σ, and let the normals to C at T and T' meet in N. The conjugate of the line LT relative to Σ contains T and is therefore the line TN. Thus TN, $T'N$, TT' all belong to Γ. By 6·28 the circumcircle of the triangle TNT', which is the circle on LN as diameter, contains the focus M of Γ. As C varies in Σ the locus of the point N is the line through M perpendicular to LM; it contains, as particular positions of N, the *centres of curvature* at L of the two confocals through L.

If P_i $(i = 1, 2, 3, 4)$ are the feet of the normals from L to a conic C of Σ the tangents t_i to C at these points all belong to Γ, since they are conjugate relative to Σ to the normals. It follows that the reciprocal γ of Γ relative to C passes through the points P_i. Further, since Γ contains the principal axes of C, γ passes through the points at infinity on these principal axes and is therefore a rectangular hyperbola, in fact the Apollonius hyperbola* of Γ with respect to C.

* See 6·23, p. 187.

CHAPTER VII

HOMOGRAPHIC RANGES ON A CONIC

7·1. Homography on a conic

7·11. Suppose there is a general homography, or (1, 1) correspondence, T between the points P, P' of a proper conic s, so that we have the relation

$$\{P\} \equiv \{P'\},$$

where P, P' are regarded as elements of the $R\infty^1$ of points of s. This may also be expressed in the notation

$$P' = TP, \quad P = T^{-1}P',$$

where T^{-1} is the inverse of T, as defined in 3·32. We shall denote the line PP' by p and prove that the ∞^1 lines p belong to a conic scroll.

Let Q be a general point of the plane and let QP meet s again in P''. Then P, P'' are pairs of an involution on s and we have

$$\{P''\} \equiv \{P\} \equiv \{P'\}.$$

There is thus a homography between P'' and P', and this has two self-corresponding points. These are in general distinct. For, if $\lambda, \lambda', \lambda''$ are the values at P, P', P'' of a parameter on the conic, the relation between P and P' is given by an equation of the form

$$a\lambda\lambda' + b\lambda + c\lambda' + d = 0, \tag{·111}$$

and that between P and P'' by an equation

$$A\lambda\lambda'' + B(\lambda + \lambda'') + D = 0;$$

hence, eliminating λ, the relation between P' and P'' is

$$(aB - cA)\lambda'\lambda'' + (aD - cB)\lambda' + (bB - dA)\lambda'' + (bD - dB) = 0,$$

and the CONDITION for this is to have coincident self-corresponding points is

$$(aD - cB + bB - dA)^2 = 4(aB - cA)(bD - dB),$$

which, as is easily seen, cannot be identically true for general A, B, D unless $a = b = c = d = 0$.

Thus, if Q is a general point of the plane, there are two distinct lines p passing through it, whence it follows that the lines p belong to a conic scroll. In general this is a proper conic scroll, and not a point-pair or repeated point. For suppose that for a particular position of Q all the lines through Q are lines p. There are two cases. If Q does not lie on s then T must be an involution with centre Q and not a general homography. If Q lies on s all the lines through it are lines p and therefore Q corresponds to every point of s either in T or in T^{-1}: but in this case T is improper, the left-hand side of equation (·111) breaking up into two factors of the form $(\alpha\lambda-\beta)(\alpha'\lambda'-\beta')$. Hence, if T is not an involution and is not improper, the lines p are tangents of a proper conic Γ. Further, Γ touches s at the self-corresponding points of T. For, if Q is a point of intersection of Γ and s, the two tangents to Γ from Q coincide; but these are the lines joining Q to the points TQ and $T^{-1}Q$, and they coincide IF Q is a self-corresponding point of T. Thus:

The lines joining corresponding points of a general homography T on a proper conic s envelop a proper conic Γ which touches s at the self-corresponding points H and K of T; if H and K coincide in H, then Γ has four-point contact with s at H.

7·12. If T is an involution with centre A the lines p all pass through A, and Γ, regarded as a conic scroll, is the repeated point A. If T is improper, being given by the equation $(\alpha\lambda-\beta)(\alpha'\lambda'-\beta')=0$, the lines p form two pencils with vertices at the points $\lambda=\beta/\alpha$, $\lambda=\beta'/\alpha'$, and Γ is a point-pair; if $\beta/\alpha=\beta'/\alpha'$, T is an improper or parabolic involution and Γ is the repeated point $\lambda=\beta/\alpha$.

7·13. The preceding results are easily proved algebraically. In the general case, when H and K are distinct, we can take as triangle of reference that formed by the line HK and the tangents to s at H and K. With a suitable choice of co-ordinate system we can then, by 4·14, take the equation of s as

$$s \equiv y^2 - zx = 0,$$

with parametric expression

$$x:y:z = \lambda^2:\lambda:1.$$

The homography T is given by an equation of the form

$$b\lambda + c\lambda' = 0, \qquad (\cdot 131)$$

the self-corresponding points being $\lambda=0$ and $\lambda=\infty$.

From (·131) we deduce

$$(b\lambda+c\lambda')(c\lambda+b\lambda') = 0, \qquad (\cdot 132)$$

or

$$bc(\lambda+\lambda')^2 + (b-c)^2\lambda\lambda' = 0,$$

which is the CONDITION for the chord

$$x-(\lambda+\lambda')y+\lambda\lambda'z = 0,$$

joining the corresponding points λ and λ' of s to touch the conic whose line equation is

$$\Gamma \equiv bcY^2+(b-c)^2 ZX = 0,$$

and this conic Γ touches s at the points $(1, 0, 0)$, $(0, 0, 1)$, which are the self-corresponding points of T.

7·14. The introduction of the factor $c\lambda+b\lambda'$ in (·132) is not so artificial as it may seem. For Γ, besides being the envelope of the joins of corresponding points of T, is clearly also the envelope of the joins of corresponding points of T^{-1}, and the equation of T^{-1} is

$$c\lambda+b\lambda' = 0.$$

Similarly if we have a homography on s whose equation is

$$\phi \equiv a\lambda\lambda'+b\lambda+c\lambda'+d = 0,$$

the reverse homography is given by

$$\phi' \equiv a\lambda\lambda'+c\lambda+b\lambda'+d = 0,$$

and

$$\phi\phi' = d^2+bc(\lambda+\lambda')^2+a^2\lambda^2\lambda'^2+a(b+c)\lambda\lambda'(\lambda+\lambda')$$
$$+\{(b-c)^2+2ad\}\lambda\lambda'+d(b+c)(\lambda+\lambda').$$

So $\phi\phi' = 0$ is a CONDITION for the chord λ, λ' of s to touch the conic whose line equation is

$$C \equiv d^2X^2+bcY^2+a^2Z^2-a(b+c)YZ$$
$$+\{(b-c)^2+2ad\}ZX-d(b+c)XY = 0,$$

which is therefore the conic enveloped by the joins of pairs of corresponding points of the homography $\phi = 0$.

The self-corresponding points of $\phi = 0$ are given by

$$a\lambda^2+(b+c)\lambda+d = 0,$$

so they are the meets H, K of s with the line

$$ax+(b+c)y+dz = 0.$$

The pole of this line relative to s is the point

$$2dX-(b+c)Y+2aZ = 0,$$

the line equation of s is $Y^2-4ZX = 0$, and

$$4C \equiv \{2dX-(b+c)Y+2aZ\}^2-(b-c)^2(Y^2-4ZX),$$

which verifies the fact that C touches s at H and K.

7·15. If the self-corresponding points of T coincide in a point H, we may take s as

$$x:y:z = \lambda^2:\lambda:1$$

in such a way that H is the point $\lambda = 0$. The equation of T is then of the form

$$\phi \equiv a\lambda\lambda' + b(\lambda - \lambda') = 0.$$

Putting $d = 0$, $c = -b$ in 7·14 we see that the joins of corresponding points of T envelop the conic

$$a^2Z^2 - b^2(Y^2 - 4ZX) = 0,$$

which has four-point contact with s at H.

7·16. If P, Q, R ... and P', Q', R' ... are corresponding points in T, then the tangents $p, q, r, \ldots$ and $p', q', r', \ldots$ are corresponding lines in a homography among the tangents of s, so the dual of the theorem of 7·11 tells us that the locus of points of intersection of corresponding tangents is a conic γ which touches s at the self-corresponding points of T. Since the point pp' is the pole of the line PP' relative to s, this conic γ is the reciprocal of Γ relative to s.

7·2. The cross-axis. Pascal's Theorem

7·21. If P_1, P_1' and P_2, P_2' are two pairs of corresponding points of a homography T on a proper conic s, the point $(P_1P_2', P_1'P_2)$ is called the *cross-meet* of the two pairs. If T has distinct self-corresponding points, we may, as in 7·13, take its equation as

$$b\lambda + c\lambda' = 0,$$

and, since
$$b\lambda_1 + c\lambda_1' = b\lambda_2 + c\lambda_2' = 0,$$

we have
$$\lambda_1\lambda_2' = \lambda_1'\lambda_2.$$

But the chords P_1P_2' and $P_1'P_2$ are

$$x - (\lambda_1 + \lambda_2')y + \lambda_1\lambda_2'z = 0,$$
$$x - (\lambda_1' + \lambda_2)y + \lambda_1'\lambda_2z = 0;$$

by subtraction we see that the cross-meet lies on the line $y = 0$. Thus:

The cross-meets $(P_1P_2', P_1'P_2)$, obtained from any two pairs P_1, P_1' and P_2, P_2' of a general homography T on a proper conic s lie on the line joining the self-corresponding points of T, which is called the cross-axis of T.

By reciprocating with respect to s we see that the *cross-joins* $[p_1p_2', p_1'p_2]$ obtained from the tangents p_1, p_1' and p_2, p_2' at two pairs of corresponding points pass through the pole of the cross-axis relative to s.

7·22. *The theorem of* 7·21 *still holds for a homography T with coincident self-corresponding points, the cross-axis being the tangent at the unique self-corresponding point.*

For, with the notation of 7·15, we have

$$a\lambda_1\lambda_1' + b(\lambda_1 - \lambda_1') = a\lambda_2\lambda_2' + b(\lambda_2 - \lambda_2') = 0,$$

whence $$\lambda_1\lambda_1'(\lambda_2 - \lambda_2') = \lambda_2\lambda_2'(\lambda_1 - \lambda_1')$$

or $$\lambda_1'\lambda_2(\lambda_1 + \lambda_2') = \lambda_1\lambda_2'(\lambda_1' + \lambda_2),$$

which is a CONDITION for the chords P_1P_2' and $P_1'P_2$ to meet on the line $x = 0$.

7·23. The theorem of 7·21 and 7·22 is easily proved descriptively. If the self-corresponding points H and K of T are distinct, we have

$$(HKP_1P_2) = (HKP_1'P_2').$$

Hence $$P_1'(HKP_1P_2) = P_1(HKP_1'P_2'),$$

and, since P_1P_1' is a common self-corresponding ray, the remaining three pairs of corresponding rays meet in collinear points, so that $(P_1P_2', P_1'P_2)$ lies on HK.

7·24. If the self-corresponding points of T coincide in H, let Q be the point $(P_1P_2', P_1'P_2)$ and let QH meet s again in a point L supposed distinct from H. Then we have

$$(HLP_1P_2) = (LHP_2'P_1'),$$

since the corresponding points are pairs in the involution with centre Q. It follows that

$$(HLP_1P_2) = (HLP_1'P_2'),$$

whence L is also a self-corresponding point in the unique homography T which has H as a self-corresponding point and P_1, P_1' and P_2, P_2' as corresponding pairs. L cannot therefore be distinct from H, and the theorem is proved.

7·25. Homographic ranges on a conic are determined by three pairs of corresponding points, so from 7·21 and 7·22 we deduce:

If P_1, P_2, P_3, P_1', P_2', P_3' are any six points of a proper conic the three points $(P_2P_3', P_2'P_3)$, $(P_3P_1', P_3'P_1)$, $(P_1P_2', P_1'P_2)$ are collinear.

Calling the points C, A, E, F, D, B we have Pascal's Theorem:

If $ABCDEF$ is a hexagon inscribed in a proper conic, the points of intersection of opposite sides, namely (AB, DE), (BC, EF), (CD, FA), are collinear.*

The dual of this is Brianchon's Theorem:

If abcdef is a hexagon circumscribed to a proper conic, the lines joining opposite vertices, namely $[ab, de]$, $[bc, ef]$, $[cd, fa]$, are concurrent.*

Pascal's Theorem may also be stated as follows:

If A, B, C, D, E, F are six points of a proper conic, the three involutions determined by the two pairs $|A, B|$, $|D, E|$, by the two pairs $|B, C|$, $|E, F|$, and by the two pairs $|C, D|$, $|F, A|$ have a pair in common.

For three involutions on the conic have a common pair IF their centres are collinear.

In this form the theorem is essentially one about values of a parameter, and is therefore true when A, B, C, D, E, F represent elements of any $R\infty^1$. A reader who attempts to prove the theorem for six points on a line by a direct attack will appreciate the merits of descriptive methods.

We can deduce further results by supposing the six points of Pascal's Theorem not to be distinct, provided that we interpret the join of two coincident points as the tangent at that point. Thus, by considering the hexagon $AABBCC$, we have the following:

If a triangle is inscribed in a proper conic the tangents at the vertices meet the opposite sides in three collinear points.

The dual result, which is a special case of Brianchon's Theorem, is:

If a triangle is circumscribed to a proper conic the three lines joining the vertices to the points of contact of the opposite sides are concurrent.

It should be noted that though these and similar results can conveniently be deduced by limiting arguments, their proof does not require such arguments.

7·26. Pascal's Theorem gives us a method of constructing the homography in which three given points P_1, P_2, P_3 of a proper conic s correspond to three given points P'_1, P'_2, P'_3. For the line a joining

* Strictly speaking we should say an *ordered* hexagon.

the points $(P_1P'_2, P'_1P_2)$ and $(P_3P'_1, P'_3P_1)$ is the cross-axis of the homography, and its intersections with s are the self-corresponding points. If P is any point of s and if PP'_1 meets a in L, then the further intersection of P_1L with s is the point corresponding to P.

7·3. The product of two involutions

7·31. Consider two distinct involutions on a proper conic s with centres L and M. If P is a variable point of s let LP meet s again in Q and let MQ meet s again in R. Since P and Q are pairs in the involution with centre L we have on s

$$\{P\} \equiv \{Q\},$$

and, since Q and R are pairs in the involution with centre M,

$$\{Q\} \equiv \{R\}.$$

Hence $$\{P\} \equiv \{R\},$$

and P, R are corresponding points in a homography T, which is called the *product* of the two involutions.

If RL meets s again in F, and if FM meets s again in G, then G is the point corresponding to R in T, and G will not in general coincide with P. We now show that G coincides with P and that T is an involution IF L and M are conjugate relative to s.

7·32. Suppose first that neither L nor M lies on s and that LM meets s in distinct points H, K. Then when P is H, Q is K and R is H, so that H and similarly K are the self-corresponding points of T. But, for general P,

$$(HPKR) = Q(HPKR) = (HLKM) = \rho, \text{ a constant.}$$

In fact ρ is the constant cross-ratio of T, and T will be an involution

IF $$(HLKM) = -1,$$

that is IF L and M are conjugate relative to s.

If LM touches s at H the self-corresponding points of T coincide in H, so that if T is an involution it must be an improper* one with fixed point H. This is impossible if both L and M are distinct from H, for in that case, if P is a general point of s, Q and therefore also R is distinct from H. If, however, either L or M coincides with H, T is an improper involution with fixed point H.

Similarly if either L or M lies on s the homography T is degenerate and is an (improper) involution IF LM touches s. Thus:

* See 3·58, p. 94.

The product of two involutions on s is itself an involution IF *the centres of the two involutions are conjugate relative to s.*

7·33. If L, M are general points, the lines PR envelop a conic, two of whose tangents pass through a general point N. Thus:

There are in general two triangles PQR inscribed in a conic s whose sides PQ, QR, RP pass through three given points L, M, N.

But if L, M are conjugate relative to s, all the lines PR pass through the point of intersection of the tangents to s at H and K, which is the pole of LM relative to s. Thus:

There are ∞^1 triangles PQR inscribed in s whose sides PQ, QR, RP pass through three given points L, M, N IF *LMN is a self-polar triangle relative to s.**

7·34. *Any homography on s may be obtained as the product of two involutions in ∞^1 ways.*

For denote the homography by T, and let any point P_1 of s correspond to a point R_1. Choose any point L on the cross-axis of T, let LP_1 meet s again in Q_1, and let Q_1R_1 meet the cross-axis in M. Then if T has distinct self-corresponding points H and K, the product of the involutions with centres L and M is a homography which has three pairs $|H, H|$, $|K, K|$, and $|P_1, R_1|$ in common with T and which therefore coincides with T.

If the self-corresponding points of T coincide in a point H, the cross-axis of T is the tangent at H. There is then a point F_1 of s such that $F_1 \to Q_1$ in T, and by 7·22 the point (P_1Q_1, F_1R_1) lies on the cross-axis, so F_1R_1 contains L. Thus the homography which is the product of the involutions with centres L and M contains the pairs $|P_1, R_1|$, $|F_1, Q_1|$ and $|H, H|$, and therefore coincides with T.

7·4. Two homographies

7·41. Let T_1 and T_2 be two homographies on a proper conic s,† and let $P \to P'$ in T_1 and $P' \to P''$ in T_2. Denote the reverse correspondences by T_1^{-1} and T_2^{-1}, as in 3·32. Then in T_2^{-1} each $P'' \to$ a unique P', and in T_1^{-1} each $P' \to$ a unique P, so we have a (1, 1)

* Cf. $R(2)$, pp. 88, 89.

† Many of the results of this and other sections of this chapter can, with slight modification, be stated as theorems about homographies in the elements of any $R\infty^1$. In some cases, as in Pascal's Theorem, it is convenient to take the $R\infty^1$ as a proper conic; in others it is really irrelevant to specify the particular $R\infty^1$.

correspondence between P and P''. Symbolically we may say $P' = T_1 P$, $P'' = T_2 P'$, whence $P'' = T_2 T_1 P$. The homography $P \to P''$ is called the *product* of T_1 and T_2 and is denoted by $T_2 T_1$. Since $P'' \to P'$ in T_2^{-1} and $P' \to P$ in T_1^{-1}, the reverse of $T_2 T_1$ is $T_1^{-1} T_2^{-1}$.

7·42. If T_1 and T_2 are given by

$$a_1 \lambda\lambda' + b_1 \lambda + c_1 \lambda' + d_1 = 0,$$

$$a_2 \lambda'\lambda'' + b_2 \lambda' + c_2 \lambda'' + d_2 = 0,$$

then, eliminating λ', the homography $T_2 T_1$ is given by

$$\begin{vmatrix} a_1\lambda + c_1 & b_1\lambda + d_1 \\ a_2\lambda'' + b_2 & c_2\lambda'' + d_2 \end{vmatrix} = 0,$$

or

$$(a_1 c_2 - a_2 b_1)\lambda\lambda'' + (a_1 d_2 - b_1 b_2)\lambda + (c_1 c_2 - a_2 d_1)\lambda'' + c_1 d_2 - b_2 d_1 = 0.$$

7·43. Thus $T_1 T_2$ is not in general the same as $T_2 T_1$. Also $T_2 T_1$ is an involution IF $a_1 d_2 + a_2 d_1 = b_1 b_2 + c_1 c_2$, which is also a CONDITION for $T_1 T_2$ to be an involution, since it is symmetrical in the suffices. When T_1 and T_2 are both involutions this CONDITION expresses the fact that the centres of the involutions are conjugate relative to s, in agreement with 7·32.

7·44. We now consider what pairs of points F and G exist on a conic such that they are pairs of corresponding points in each of two given homographies T_1 and T_2. We are thinking of the pair of points rather than the individual points, so for each homography we may have either $F \to G$ or $G \to F$. If then such a pair exists we have the following possibilities:

(*a*) $G = T_1 F$ and $G = T_2 F$, giving $G = T_1 T_2^{-1} G$.

(*b*) $G = T_1 F$ and $G = T_2^{-1} F$, giving $G = T_1 T_2 G$.

(*c*) $F = T_1 G$ and $F = T_2 G$, giving $F = T_1 T_2^{-1} F$.

(*d*) $F = T_1 G$ and $F = T_2^{-1} G$, giving $F = T_1 T_2 F$.

Since (*c*) and (*d*) are the same as (*a*) and (*b*) with F and G interchanged, they will only give the same pairs and we need not consider them. From (*a*) we see that G must be one of the self-corresponding points of $T_1 T_2^{-1}$, and we thus obtain two pairs F, G. Similarly we obtain two pairs from (*b*) by taking G at one of the self-corresponding points of $T_1 T_2$. Thus:

In general, two homographies on a proper conic have four common (unordered) pairs.

This result is also evident when we remember that the joins of corresponding points of T_1 and T_2 are the lines of two conic scrolls Γ_1 and Γ_2, which have four common lines in general. If one of the homographies, say T_1, is an involution, Γ_1 is a pencil of lines counted twice, so that there are only two distinct common pairs. If both T_1 and T_2 are involutions there is only one common pair.

7·45. *Pencil of homographies*

If the equations of two homographies T_1 and T_2 on a proper conic s are

$$\phi_1 \equiv a_1\lambda\lambda' + b_1\lambda + c_1\lambda' + d_1 = 0, \quad \phi_2 \equiv a_2\lambda\lambda' + b_2\lambda + c_2\lambda' + d_2 = 0,$$

we may introduce the idea of a $R\infty^1$ or *pencil* of homographies determined by the equation $\phi_1 + \rho\phi_2 = 0$, where ρ is a variable parameter, with the usual convention that T_2 is the homography of the pencil corresponding to the value $\rho = \infty$. In general the pencil contains ONE involution corresponding to the value of ρ for which

$$b_1 + \rho b_2 = c_1 + \rho c_2,$$

but if two members of the pencil are involutions then they all are.

The homography $\phi_1 + \rho\phi_2 = 0$ has coincident self-corresponding points IF

$$(b_1 + c_1 + \rho b_2 + \rho c_2)^2 = 4(a_1 + \rho a_2)(d_1 + \rho d_2),$$

so that in general two members of the pencil have coincident self-corresponding points. In fact, if the parametric equation of s is

$$x : y : z = \lambda^2 : \lambda : 1,$$

we see by 7·14 that the line joining the self-corresponding points of $\phi_1 + \rho\phi_2 = 0$ is

$$(a_1 + \rho a_2)x + (b_1 + \rho b_2 + c_1 + \rho c_2)y + (d_1 + \rho d_2)z = 0;$$

as ρ varies this line describes a pencil and two of its members are tangents of s.

7·5. Cyclic homographies

7·51. Let T be a homography with distinct self-corresponding points H, K on a proper conic s,* so that, if P, P_1 is a pair of corresponding points, we have $P_1 = TP$ and

$$(HPKP_1) = \rho,$$

* See the footnote to 7·41, p. 200.

where ρ is a constant. Choose a parameter λ so that H and K are given by $\lambda = \infty$ and $\lambda = 0$. Then, if P, P_1 are given by λ, λ_1, we have

$$(\infty, \lambda, 0, \lambda_1) = \rho$$

or

$$\lambda_1 = \rho\lambda.$$

Now suppose that, in T, $P \to P_1$, $P_1 \to P_2$, ..., and $P_{n-1} \to P_n$, where n is some integer. Then if P_r is given by $\lambda = \lambda_r$, we have $\lambda_r = \rho\lambda_{r-1}$, so that $\lambda_n = \rho^n \lambda$. In fact

$$(HPKP_n) = \rho^n.$$

The relation $P \to P_n$ is thus a homography with cross-ratio ρ^n and with self-corresponding points H and K. It is denoted by T^n; since it is the product of T taken with itself n times.

7·52. If $\rho = 1$ the homography T is the trivial one of *identity*, which we shall denote by I. If $T = I$ then $T^n = I$, and for any homography T we have $T^{-1}T = TT^{-1} = I$.

If $P \to P_r$ in T^r then $P_r \to P$ in the rth power of T^{-1}, which we denote by T^{-r}. Thus T^{-r} is the reverse (or inverse) of T^r, and $T^rT^{-r} = I = T^{-r}T^r$.

7·53. If T is not identity, and if P is not a self-corresponding point, P_1 is distinct from P, P_2 from P_1, and so on. There is no reason why P_r should coincide with P for any value of r, but suppose that this does happen and that h is the least value of r for which P_r coincides with P. Then we must have $\rho^h = 1$, so that, for *any* position of P, P_h coincides with P, and T^h is identity. When this happens we say that T is a *cyclic homography of order* h.

Clearly T^mT^n is T^{m+n}, T^{m+nh} is T^m, and T^{-m} is T^{nh-m}, where m and n are integers. It follows that any power of T is one of the homographies $I, T, T^2, \ldots, T^{h-1}$; further, these homographies are all distinct, since h is the least value of r for which $T^h = I$. These h distinct homographies therefore form a *group* with the property that the product of any two members of the group is itself a member of the group.

7·54. Alternatively we may proceed as follows. If T is a homography with distinct self-corresponding points H and K, and if P is a particular point distinct from H and K such that $T^hP = P$ and $T^rP \neq P$ for $r < h$, then we have

$$T^hH = H, \quad T^hK = K \quad \text{and} \quad T^hP = P.$$

The homography T^h thus has three distinct self-corresponding points and is therefore identity. It follows that $T^h P = P$ for general P, and T is cyclic of order h.

7·55. *A cyclic homography cannot have coincident self-corresponding points.*

For a homography T, with coincident self-corresponding points at the point H with parameter $\lambda = 0$, is given by an equation of the form

$$a\lambda\lambda_1 + \lambda - \lambda_1 = 0,$$

where λ, λ_1 are the parameters of corresponding points P, P_1, and $a \neq 0$, if we exclude the trivial case when T is identity.

We have
$$\lambda_1 = \frac{\lambda}{1 - a\lambda}.$$

If $P_2 = TP_1 = T^2P$ and corresponds to the parameter λ_2 we have

$$\lambda_2 = \frac{\lambda_1}{1 - a\lambda_1} = \frac{\lambda}{1 - 2a\lambda},$$

and finally P_n or T^nP corresponds to the parameter

$$\lambda_n = \frac{\lambda}{1 - na\lambda}.$$

For no value of the integer n can $\lambda_n = \lambda$ identically for all λ, and $\lambda = 0$ is the only particular value for which the equation can be true.

7·56. The homography I may be regarded as cyclic of any order. Disregarding this we may say that a cyclic homography of order two is an involution and that three distinct points A, B, C of s determine ONE cyclic homography of order three in which $A \to B \to C \to A$, namely the unique homography with corresponding pairs A, B and B, C and C, A. This homography will be discussed in 7·6. If A, B, C, D are four distinct points of s there is in general no cyclic homography of order four in which

$$A \to B \to C \to D \to A.$$

We prove:

There is a homography on s in which $A \to B \to C \to D \to A$ IF *the lines AC, BD are conjugate relative to s.*

For if there exists such a homography, we have

$$(ABCD) = (BCDA) = (ADCB),*$$

* See 2·73, p. 68.

from which it follows that $(ABCD)$ is harmonic and AC, BD are conjugate relative to s. Conversely, if $(ABCD)$ is harmonic, we have

$$(ABCD) = (ADCB) = (BCDA),$$

and therefore the homography which has A, B; B, C; C, D as corresponding pairs also has D, A as a corresponding pair.

7·6. The Hessian of a triad

7·61. Let A, B, C be three points of a proper conic s and let the tangents to s at A, B, C meet the opposite sides of the triangle ABC in L, M, N. Pascal's Theorem applied to the hexagon $AABBCC$ shows that L, M, N lie on a line u, which will meet s in two points H and K. In fact u, as containing the points (AA, BC), (BB, CA), (CC, AB), is the cross-axis of the homography T determined by the pairs A, B and B, C and C, A. T is thus the cyclic homography of order three in which $A \to B \to C \to A$, and H and K are its self-corresponding points, necessarily distinct by 7·55. The points H and K are called the *Hessian points* of the triad A, B, C. Since L, M, N are collinear, the three involutions determined by the pairs $|A, A|$ and $|B, C|$, $|B, B|$ and $|C, A|$, $|C, C|$ and $|A, B|$ have a common pair $|H, K|$.

7·62. Let A' be the point of s which is the harmonic conjugate of A relative to B and C, and let B' and C' be similarly defined. The line AA' is the polar of L relative to s, so AA' contains the pole U of u relative to s. Similarly BB' and CC' contain U. Thus the pairs $|A, A'|$, $|B, B'|$, $|C, C'|$ belong to an involution whose double points are the points of contact of tangents through U, namely H and K.

7·63. By 7·53 the cross-ratio ρ of the homography T is given by $\rho^3 = 1, \rho \neq 1$. Thus $\rho = \omega$ or ω^2, where ω is an imaginary cube root of unity. We have therefore

$$(HBKC) = (HCKA) = (HAKB) = \omega \quad \text{or} \quad \omega^2.$$

Also, since $H, A, B, C \to H, B, C, A$ in T, we have

$$\sigma = (HABC) = (HBCA) = 1 - \frac{1}{\sigma},$$

whence $\sigma^2 - \sigma + 1 = 0$, giving $\sigma = -\omega$ or $-\omega^2$.

Thus $$(HABC) = -\omega \quad \text{or} \quad -\omega^2$$

and $$(KABC) = -\omega^2 \quad \text{or} \quad \omega.$$

There is only one other cyclic homography of order three determined by A, B, C, namely that in which $A \to C \to B \to A$, and this is merely the inverse T^{-1} of T.

7·64. The theorems we have just proved are essentially theorems about values of a parameter, and are therefore applicable to any $R\infty^1$. For convenience we shall restate the results for the points of a line:

If λ is a parameter of points of a line, and if $\lambda_1, \lambda_2, \lambda_3$ are the values of λ corresponding to three given points A_1, A_2, A_3 of the line, then the points H and K, whose parameters are given by

$$(\lambda\lambda_1\lambda_2\lambda_3) = -\omega \quad \text{and} \quad (\lambda\lambda_1\lambda_2\lambda_3) = -\omega^2,$$

are called the Hessian points of the triad A_1, A_2, A_3 and have the following properties:

(i) If A_i' is the harmonic conjugate of A_i relative to the other two points of the triad, then the pairs $|A_1, A_1'|, |A_2, A_2'|, |A_3, A_3'|$ belong to an involution whose double points are H and K.*

(ii) The three involutions which have one point of the triad as a double point and the other two points of the triad as a pair have the pair $|H, K|$ in common.

(iii) The two cyclic homographies in which $A_1 \to A_2 \to A_3 \to A_1$ and $A_3 \to A_2 \to A_1 \to A_3$ have H and K as self-corresponding points.

(iv) $(H, A_i, K, A_j) = \omega$ or ω^2, when $i \neq j$.

We showed in 2·74 that if the cross-ratio of four points taken in a given order is $-\omega$, then any cross-ratio obtained by permuting the order of the points has the value $-\omega$ or $-\omega^2$, the four points being then said to be equianharmonic. We can now say that

Four points are equianharmonic IF *any one is a Hessian point of the other three.*

7·7. Polar sets

7·71. The algebra connected with Hessian points is simplified if we use homogeneous co-ordinates. For convenience we continue to consider points of a line, though the theorems proved are applicable to any $R\infty^1$.

Let (x, y) be homogeneous co-ordinates of points of the line. We showed in 2·76 that if A_1, A_2 are the two points given by

$$ax^2 + 2bxy + cy^2 = 0$$

* Cf. $R(2)$, pp. 32, 33.

then the point given by

$$\left(x_1\frac{\partial}{\partial x}+y_1\frac{\partial}{\partial y}\right)(ax^2+2bxy+cy^2) = 0$$

is the harmonic conjugate of the point (x_1, y_1) relative to A_1 and A_2.

7·72. We now define the *first polar* of a point P_1, (x_1, y_1), relative to the set of three points A_1, A_2, A_3 given by

$$\phi(x,y) \equiv ax^3+3bx^2y+3cxy^2+dy^3 = 0$$

to be the pair of points given by

$$\left(x_1\frac{\partial}{\partial x}+y_1\frac{\partial}{\partial y}\right)\phi(x,y) = 0$$

or
$$\phi_1(x,y) \equiv x_1(ax^2+2bxy+cy^2)+y_1(bx^2+2cxy+dy^2) = 0.$$

The *second polar* of P_1 relative to the set A_1, A_2, A_3 is defined to be the point given by

$$\left(x_1\frac{\partial}{\partial x}+y_1\frac{\partial}{\partial y}\right)^2\phi(x,y) = 0$$

or by
$$\left(x_1\frac{\partial}{\partial x}+y_1\frac{\partial}{\partial y}\right)\phi_1(x,y) = 0,$$

and this is the harmonic conjugate of P_1 relative to the first polar of P_1.

7·73. It is not obvious that the first polar set is determined by P_1 and the three points A_1, A_2, A_3, for it might depend on the choice of a co-ordinate system. But any other system of homogeneous co-ordinates (x', y') is connected with (x, y) by a transformation of the form

$$x = px'+qy', \quad y = rx'+sy',$$

where
$$ps-qr \neq 0.$$

Thus if $\phi(x,y) = \psi(x',y')$ and P_1 is the point (x_1', y_1') in the new co-ordinates, we have

$$\begin{aligned}\frac{\partial}{\partial x'}\psi(x',y') &= \frac{\partial\phi(x,y)}{\partial x}\frac{\partial x}{\partial x'}+\frac{\partial\phi(x,y)}{\partial y}\frac{\partial y}{\partial x'}\\ &= \left(p\frac{\partial}{\partial x}+r\frac{\partial}{\partial y}\right)\phi(x,y)\end{aligned}$$

and
$$\frac{\partial}{\partial y'}\psi(x',y') = \left(q\frac{\partial}{\partial x}+s\frac{\partial}{\partial y}\right)\phi(x,y).$$

so

$$\left(x_1' \frac{\partial}{\partial x'} + y_1' \frac{\partial}{\partial y'}\right) \psi(x', y') = \left\{(px_1' + qy_1') \frac{\partial}{\partial x} + (rx_1' + sy_1') \frac{\partial}{\partial y}\right\} \phi(x, y)$$
$$= \left(x_1 \frac{\partial}{\partial x} + y_1 \frac{\partial}{\partial y}\right) \phi(x, y).$$

Thus $\left(x_1 \frac{\partial}{\partial x} + y_1 \frac{\partial}{\partial y}\right) \phi(x, y)$ and $\left(x_1' \frac{\partial}{\partial x'} + y_1' \frac{\partial}{\partial y'}\right) \psi(x', y')$

vanish for the same two points, and consequently the first polar set is determined by P_1 and the set A_1, A_2, A_3, and is independent of the choice of co-ordinates. We shall in Chapter XII meet other examples where algebraic processes are used to define a new geometrical construct from a given one, but where it can be shown that the result of these processes is independent of the co-ordinate system employed.

7·74. If A_1 is the point (α, β) we have

$$\phi(x, y) \equiv (\beta x - \alpha y) f(x, y),$$

where $f(x, y)$ is a homogeneous quadratic and A_2, A_3 are given by $f(x, y) = 0$. Then

$$\left(\alpha \frac{\partial}{\partial x} + \beta \frac{\partial}{\partial y}\right) \phi(x, y) = (\beta x - \alpha y) \left(\alpha \frac{\partial}{\partial x} + \beta \frac{\partial}{\partial y}\right) f(x, y),$$

so the first polar of A_1 relative to the set A_1, A_2, A_3 consists of A_1 and the harmonic conjugate A_1' of A_1 relative to A_2 and A_3. Similarly, if A_2', A_3' are the harmonic conjugates of A_2 and A_3 relative to A_3, A_1 and A_1, A_2, respectively, then the first polars of A_2 and A_3 are the pairs $|A_2, A_2'|$ and $|A_3, A_3'|$.

7·75. Now as P_1 varies, the first polars are given by

$$x_1 \frac{\partial \phi}{\partial x} + y_1 \frac{\partial \phi}{\partial y} = 0$$

or
$$x_1(ax^2 + 2bxy + cy^2) + y_1(bx^2 + 2cxy + dy^2) = 0,$$

and therefore form an involution of pairs of points, with double points given by

$$\begin{vmatrix} \dfrac{\partial^2 \phi}{\partial x^2} & \dfrac{\partial^2 \phi}{\partial x \, \partial y} \\ \dfrac{\partial^2 \phi}{\partial x \, \partial y} & \dfrac{\partial^2 \phi}{\partial y^2} \end{vmatrix} = 0$$

or by
$$\begin{vmatrix} ax+by, & bx+cy \\ bx+cy, & cx+dy \end{vmatrix} = 0.$$

But we have just shown in 7·74 that three particular pairs of this involution, obtained by taking P_1 at A_1, A_2, A_3 are $|A_1, A_1'|$, $|A_2, A_2'|$, $|A_3, A_3'|$; hence this involution is the one obtained in 7·64, (i), and the double points are the Hessian points of A_1, A_2, A_3. Thus:

The Hessian points of the three points given by
$$ax^3+3bx^2y+3cxy^2+dy^3 = 0$$
are the pair of points given by
$$(ac-b^2)\,x^2+(ad-bc)\,xy+(bd-c^2)\,y^2 = 0.$$

7·76. We note that the first polar of (x_1, y_1) is the point (x_2, y_2) counted twice IF
$$x_1(ax_2+by_2)+y_1(bx_2+cy_2) = 0$$
and
$$x_1(bx_2+cy_2)+y_1(cx_2+dy_2) = 0.$$
Since these equations are both symmetrical in (x_1, y_1) and (x_2, y_2) it follows that the only points whose first polars are repeated points are the Hessian points, and further that the first polar of each Hessian point is the other Hessian point counted twice.

7·77. From 7·75 it follows that, for a non-homogeneous parameter λ, the Hessian of the roots of
$$a\lambda^3+3b\lambda^2+3c\lambda+d = 0$$
is given by
$$\begin{vmatrix} a\lambda+b & b\lambda+c \\ b\lambda+c & c\lambda+d \end{vmatrix} = 0$$
or by
$$(ac-b^2)\,\lambda^2+(ad-bc)\,\lambda+(bd-c^2) = 0.$$

Thus, returning to the notation of 7·61, if s is the conic
$$x:y:z = \lambda^2:\lambda:1,$$
and A, B, C are given by
$$a\lambda^3+3b\lambda^2+3c\lambda+d = 0,$$
then H and K are given by
$$(ac-b^2)\,\lambda^2+(ad-bc)\,\lambda+bd-c^2 = 0.$$
Thus the equation of the line u is
$$(ac-b^2)\,x+(ad-bc)\,y+(bd-c^2)\,z = 0$$

and U is the point

$$\{2(bd-c^2), \quad -(ad-bc), \quad 2(ac-b^2)\}.$$

7·78. Again, if ABC is taken as triangle of reference, s will be of the form

$$fyz+gzx+hxy = 0,$$

and U is easily seen to be the point (f,g,h).

But a general point P_1, (x_1, y_1, z_1) has a polar line p_1 relative to the triangle ABC, whose equation is

$$\frac{x}{x_1}+\frac{y}{y_1}+\frac{z}{z_1} = 0,*$$

and if P_1 lies on s, p_1 contains (f,g,h), or U. Moreover as P_1 varies on s, p_1 forms the pencil of lines through U, and as P_1 tends to A, B or C, p_1 tends to AU, BU, or CU. The chord, q_1 say, joining the points of the first polar of P_1 relative to the triad A, B, C also passes through U, and as P_1 varies on s the lines p_1 and q_1 form related pencils of lines through U. But p_1 and q_1 coincide for three positions of P, namely A, B, C, so they coincide for all positions of p. Thus:

If A, B, C, P are four points of a proper conic, the first polar set of P relative to the triad A, B, C is the pair of points in which s is met by the polar line of P relative to the triangle ABC.

7·79. *Solution of the general cubic equation*

The notion of the Hessian pair of a triad is important in the theory of the solution of the general cubic equation

$$f(\lambda) \equiv a\lambda^3+3b\lambda^2+3c\lambda+d = 0. \qquad (\cdot 791)$$

Suppose α and β are the Hessian pair of the roots of (·791), namely the roots of the quadratic equation

$$\begin{vmatrix} a\lambda+b & b\lambda+c \\ b\lambda+c & c\lambda+d \end{vmatrix} = 0. \qquad (\cdot 792)$$

Then, if λ_1, λ_2, λ_3 are the roots of (·791), we have, as in 7·64 (iv),

$$(\alpha, \lambda_i, \beta, \lambda_j) = \omega \quad \text{or} \quad \omega^2, \text{ when } i \neq j.$$

Thus
$$\frac{\lambda_j-\alpha}{\lambda_j-\beta} = \omega\frac{\lambda_i-\alpha}{\lambda_i-\beta} \quad \text{or} \quad \omega^2\frac{\lambda_i-\alpha}{\lambda_i-\beta},$$

* See 5·55, p. 164.

whence it follows that λ_1, λ_2, λ_3 are the roots of an equation of the form

$$\left(\frac{\lambda-\alpha}{\lambda-\beta}\right)^3 = \rho.$$

In other words, if α and β are the roots of (·792), we can find numbers p and q such that

$$a\lambda^3+3b\lambda^2+3c\lambda+d \equiv p(\lambda-\alpha)^3+q(\lambda-\beta)^3.$$

It is merely necessary to choose p and q to satisfy any two of the equations

$$p+q = a, \quad p\alpha+q\beta = -b, \quad p\alpha^2+q\beta^2 = c, \quad p\alpha^3+q\beta^3 = -d,$$

and the other two will be automatically satisfied.

Thus the solution of the general cubic (·791) is made to depend on the solution of the quadratic (·792).

CHAPTER VIII

TWO CONICS. RECIPROCATION OF ONE CONIC INTO ANOTHER. PARTICULAR CASES

8·1. Standard forms for the equations of two proper conics

8·11. The intersections and common tangents of two proper conics were discussed in 4·8 and 4·9, when it was shown that five cases arise, namely:

(1) The general case: The conics have four distinct common points A_1, A_2, A_3, A_4 and four distinct common tangents a_1, a_2, a_3, a_4.

(2) Simple contact: The conics have a common tangent a at a common point A, TWO other distinct common points A_1, A_2, and TWO other distinct common tangents a_1, a_2.

(3) Three-point contact: The conics have a common tangent a at a common point A, ONE other common point A_1, and ONE other common tangent a_1.

(4) Four-point contact: The conics have a common tangent a at a common point A, and no other common point or tangent.

(5) Proper double contact:* The conics have two distinct common tangents a_1 and a_2 at two distinct common points A_1 and A_2, and no other common point or tangent. A_1A_2 is called the chord of contact.

The improper conics of the point pencil and line pencil determined by the two proper conics were also discussed in 4·8 and 4·9, but we shall now make a more detailed investigation.

The work involved in this chapter is rather tedious but it illustrates well the complications that can arise in theory when we admit variations from the general case.

8·12. Let the point equations of two proper conics be

$$s \equiv ax^2 + by^2 + cz^2 + 2fyz + 2gzx + 2hxy = 0,$$
$$s' \equiv a'x^2 + b'y^2 + c'z^2 + 2f'yz + 2g'zx + 2h'xy = 0,$$

and let their line equations be

$$S \equiv AX^2 + BY^2 + CZ^2 + 2FYZ + 2GZX + 2HXY = 0,$$
$$S' \equiv A'X^2 + B'Y^2 + C'Z^2 + 2F'YZ + 2G'ZX + 2H'XY = 0,$$

* 'Double contact' includes (4) and (5), four-point contact being regarded as improper double contact, the chord of contact being a tangent.

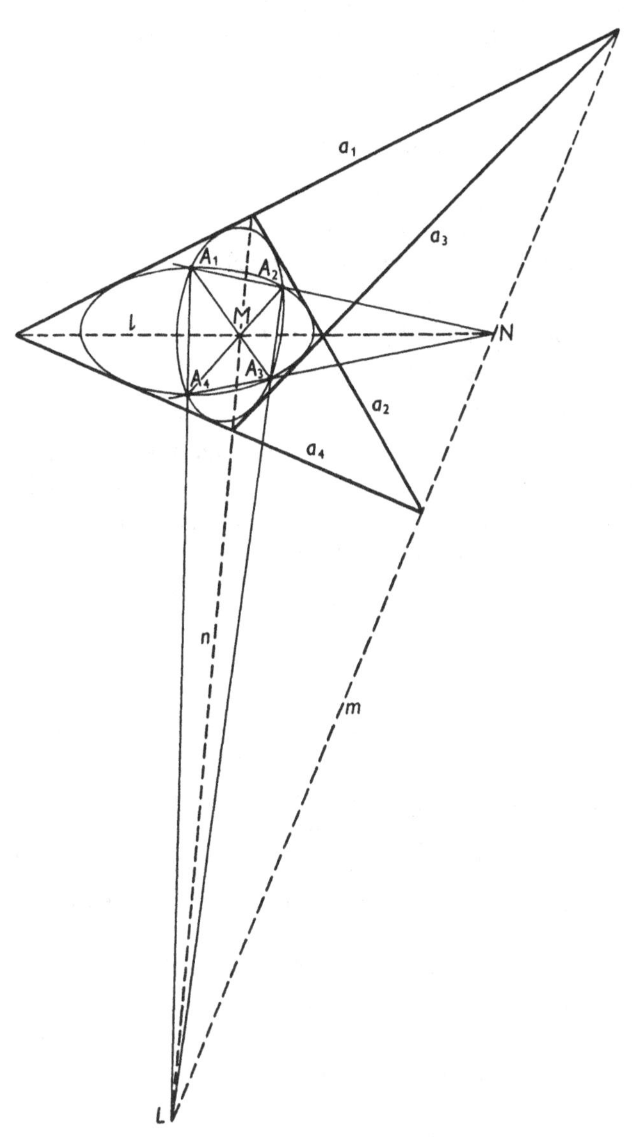

Fig. 22.

where $A, B, \ldots$ are the cofactors of the corresponding small letters in the determinants δ, δ' of s, s'. By 4·1 a point conic is a line-pair (in particular a repeated line) IF its determinant vanishes. Hence the line-pairs of the point pencil σ of conics $\lambda s + \mu s' = 0$ are given by the cubic equation

$$\phi(\lambda, \mu) \equiv \begin{vmatrix} a\lambda + a'\mu & h\lambda + h'\mu & g\lambda + g'\mu \\ h\lambda + h'\mu & b\lambda + b'\mu & f\lambda + f'\mu \\ g\lambda + g'\mu & f\lambda + f'\mu & c\lambda + c'\mu \end{vmatrix} = 0.$$

Dually the point-pairs of the line pencil Σ of conics $\lambda s + \mu s' = 0$ are given by

$$\Phi(\lambda, \mu) \equiv \begin{vmatrix} A\lambda + A'\mu & H\lambda + H'\mu & G\lambda + G'\mu \\ H\lambda + H'\mu & B\lambda + B'\mu & F\lambda + F'\mu \\ G\lambda + G'\mu & F\lambda + F'\mu & C\lambda + C'\mu \end{vmatrix} = 0.$$

Further, we recall from 5·73 that a point P_1 has the same polar line p_1 relative to s and s' IF it is the vertex of a line-pair of σ, and that p_1 is the polar of P_1 relative to every conic of σ except the line-pair with vertex P_1, with respect to which the polar of P_1 is indeterminate. Dually a line p_1 has the same pole relative to s and s' IF it is the axis of a point-pair of Σ.

8·13. *The general case*

In the general case, which is illustrated in Fig. 22, it was shown in 5·47 that the diagonal triangle LMN of the quadrangle $A_1A_2A_3A_4$ is also the diagonal triangle of the quadrilateral $a_1a_2a_3a_4$, and is a common self-polar triangle of s and s'. There are three line-pairs in the point pencil σ, namely $\|A_2A_3, A_1A_4\|$, $\|A_3A_1, A_2A_4\|$, $\|A_1A_2, A_3A_4\|$ and their vertices are L, M, N. There are three point-pairs, $\|a_2a_3, a_1a_4\|$, $\|a_3a_1, a_2a_4\|$, $\|a_1a_2, a_3a_4\|$ in Σ and their axes are the sides l, m, n of the triangle LMN. Also there can be no other common self-polar triangle of s and s', for no point other than L, M, N has the same polar line relative to s and s'. Thus:

In the general case there is ONE *common self-polar triangle of s and s'.*

It also follows from 4·8 and 4·9 that the only case in which σ contains three distinct line-pairs is the general case. Thus the general case arises IF the equation $\phi(\lambda, \mu) = 0$ has three distinct roots, and this implies that the equation $\Phi(\lambda, \mu) = 0$ also has three distinct roots.

Thus in the general case, taking the common self-polar triangle as triangle of reference, the point equations become

$$\left.\begin{aligned} s &\equiv ax^2+by^2+cz^2 = 0, \\ s' &\equiv a'x^2+b'y^2+c'z^2 = 0, \end{aligned}\right\} \qquad (\cdot 131)$$

and the line equations are accordingly

$$\left.\begin{aligned} S &\equiv bcX^2+caY^2+abZ^2 = 0, \\ S' &\equiv b'c'X^2+c'a'Y^2+a'b'Z^2 = 0.\text{*} \end{aligned}\right\} \qquad (\cdot 132)$$

The equations $\phi(\lambda,\mu) = 0$ and $\Phi(\lambda,\mu) = 0$ become

$$\phi(\lambda,\mu) \equiv (a\lambda+a'\mu)(b\lambda+b'\mu)(c\lambda+c'\mu) = 0,$$

$$\Phi(\lambda,\mu) \equiv (bc\lambda+b'c'\mu)(ca\lambda+c'a'\mu)(ab\lambda+a'b'\mu) = 0.$$

Since s and s' are proper, none of the coefficients a, b, c, a', b', c' can vanish, and since $\phi(\lambda,\mu) = 0$ has distinct roots no two of the ratios $a:a'$, $b:b'$, $c:c'$ can be equal. The common points and common tangents of s and s' are given by

$$x^2:y^2:z^2 = (bc'-b'c):(ca'-c'a):ab'-a'b,$$

$$X^2:Y^2:Z^2 = \left(\frac{1}{bc'}-\frac{1}{b'c}\right):\left(\frac{1}{ca'}-\frac{1}{c'a}\right):\left(\frac{1}{ab'}-\frac{1}{a'b}\right).$$

The polar of P_1 relative to the conic $\lambda s+\mu s' = 0$ is the line

$$\lambda(ax_1x+by_1y+cz_1z)+\mu(a'x_1x+b'y_1y+c'z_1z) = 0,$$

which goes through the point

$$(\{bc'-b'c\}y_1z_1,\quad \{ca'-c'a\}z_1x_1,\quad \{ab'-a'b\}x_1y_1)$$

for all values of $\lambda:\mu$, so this is the point P_1' which is conjugate to P_1 relative to all conics of σ.† The relation between P_1 and P_1' is symmetrical, and, as P_1' varies on a line p_1, P_1 varies on the conic

$$(bc'-b'c)X_1yz+(ca'-c'a)Y_1zx+(ab'-a'b)Z_1xy = 0. \qquad (\cdot 133)$$

This conic is also the locus of the poles of p_1 relative to the conics of σ, and is the eleven-point conic of p_1 and σ.‡ Dually the line p_1' which is conjugate to p_1 relative to all conics of Σ is

$$\left[\left(\frac{1}{bc'}-\frac{1}{b'c}\right)Y_1Z_1,\quad \left(\frac{1}{ca'}-\frac{1}{c'a}\right)Z_1X_1,\quad \left(\frac{1}{ab'}-\frac{1}{a'b}\right)X_1Y_1\right],$$

and, if p_1' contains P_1, then p_1 belongs to the conic scroll

$$\left(\frac{1}{bc'}-\frac{1}{b'c}\right)x_1YZ+\left(\frac{1}{ca'}-\frac{1}{c'a}\right)y_1ZX+\left(\frac{1}{ab'}-\frac{1}{a'b}\right)z_1XY = 0, \qquad (\cdot 134)$$

* Without loss of generality, but with loss of symmetry, the equation of one of the conics may be taken as $x^2+y^2+z^2 = 0$.

† See 5·74, p. 170. ‡ See 5·76 and 5·81, pp. 171, 172.

which is also formed by the polars of P_1 relative to the conics of Σ. This is the eleven-line conic of P_1 and Σ.

The line equation of the conic $\lambda s + \mu s' = 0$ is

$$(\lambda b+\mu b')(\lambda c+\mu c')X^2+(\lambda c+\mu c')(\lambda a+\mu a')Y^2 \\ +(\lambda a+\mu a')(\lambda b+\mu b')Z^2 = 0,$$

or

$$\lambda^2 S+\lambda\mu\{(bc'+b'c)X^2+(ca'+c'a)Y^2+(ab'+a'b)Z^2\} \\ +\mu^2 S' = 0. \quad (\cdot 135)$$

and the point equation of the conic $\lambda S+\mu S' = 0$ is

$$\lambda^2 abcs+\lambda\mu\{(bc'+b'c)aa'x^2+(ca'+c'a)bb'y^2+(ab'+a'b)cc'z^2\} \\ +\mu^2 a'b'c's' = 0. \quad (\cdot 136)$$

Now the conic with line equation ($\cdot$135) is a conic of Σ IF $\lambda\mu = 0$ or if there are numbers ρ and ρ' such that

$$(bc'+b'c)X^2+(ca'+c'a)Y^2+(ab'+a'b)Z^2 \equiv \rho S+\rho' S'.$$

But

$$\begin{vmatrix} bc'+b'c, & ca'+c'a, & ab'+a'b \\ b'c', & c'a', & a'b' \\ bc, & ca, & ab \end{vmatrix} = (bc'-b'c)(ca'-c'a)(ab'-a'b), \quad (\cdot 137)$$

which is not zero, so no numbers ρ and ρ' exist. Thus:

*In the general case no conics of σ belong to Σ except s and s'.**

8·14. *Simple contact*

The case of simple contact is illustrated in Fig. 23. D is the point $a_1 a_2$ and d is the line $A_1 A_2$. B is the point ad and b is the line AD. C is the point bd and c is the line BD. The figure so obtained is self-dual. The only line-pairs of σ are $\| AA_1, AA_2 \|$ and $\| a, d \|$, so that A and B are the only points which have the same polar lines relative to s and s'. Dually the only point-pairs of Σ are $\| aa_1, aa_2 \|$ and $\| A, D \|$, and a and b are the only lines having the same poles relative to s and s'. Since a is the polar of A relative to s and to s' it follows that b is the polar of B relative to s and to s'; hence

$$(BA_1 CA_2) = -1$$

and dually $(ba_1 ca_2) = -1$. Also, since B is conjugate to D relative to s and to s', the polars of D relative to s and to s' pass through B. In fact the contacts of a_1 and a_2 with s lie on a line through B, and

* See also 4·98, p. 145.

similarly for s'. Dually the tangents to s and A_1 and A_2 meet on b and similarly for s'. There is no common self-polar triangle of s and s'.

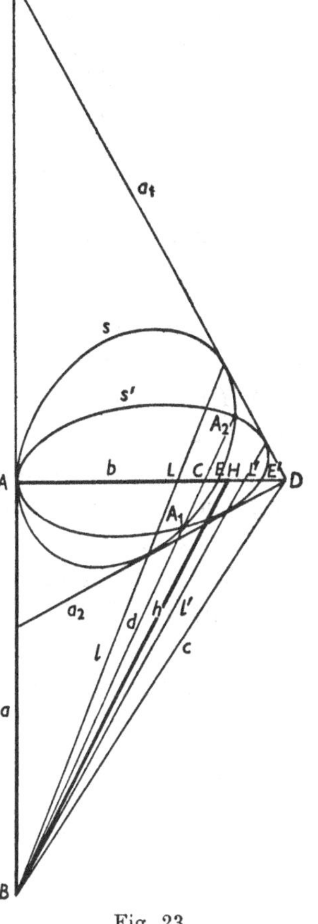

Fig. 23.

Take ABC as triangle of reference and choose co-ordinates so that A_1 is $(0, 1, 1)$. Then A_2 is $(0, 1, -1)$ and the line-pair $\|AA_1, AA_2\|$ is $y^2 - z^2 = 0$. The equations of s and s' may therefore be taken as

$$\left.\begin{aligned} s &\equiv y^2 - z^2 + 2gzx = 0,\\ s' &\equiv y^2 - z^2 + 2g'zx = 0,\end{aligned}\right\} \qquad (\cdot 141)$$

where $gg'(g-g') \neq 0$. Their line equations are

$$\left.\begin{aligned} -S &\equiv X^2 + g^2Y^2 + 2gZX = 0,\\ -S' &\equiv X^2 + g'^2Y^2 + 2g'ZX = 0,\end{aligned}\right\} \qquad (\cdot 142)$$

where S and S' are defined as in 8·12.

The point P_1' which is conjugate to P_1 relative to σ is

$$(-x_1y_1, \quad z_1^2, \quad y_1z_1),$$

and, if P_1' lies on a line p_1, then P_1 lies on the conic

$$Y_1z^2 + Z_1yz - X_1xy = 0, \qquad (\cdot 143)$$

which is also the locus of poles of p_1 relative to the conics of σ. This conic contains the vertices A and B of the line pairs of σ and touches b at A; it also contains the poles of p_1 relative to the point-pairs $\|A_1, A_2\|$, $\|A, A_1\|$, $\|A, A_2\|$, since these are the positions of P_1' when P_1 is at the intersections of p_1 with the sides of the triangle AA_1A_2.

Dually the line p_1' which is conjugate to p_1 relative to Σ is $[gg'X_1Y_1, X_1^2, -gg'Y_1Z_1 - (g+g')X_1Y_1]$, and the conic scroll formed by the polars of a point P_1 relative to the conics of Σ is

$$y_1X^2 - z_1gg'YZ + \{x_1gg' - z_1(g+g')\}XY = 0. \qquad (\cdot 144)$$

This equation is not so simple as (·143) because the choice of triangle of reference was not self-dual. The line-pencil Σ would be better dealt with by taking abc as triangle of reference.

It is actually possible to choose a self-dual triangle of reference. Let E, E' be the intersections of b with s, s' other than A, and denote the lines BE, BE' by e, e'; these are the tangents to s, s' through B other than a. The equations of e and e' are $2gx-z=0$ and $2g'x-z=0$. Also, from ·142, the point-pair $\| A, D \|$ of Σ is given by $(g+g')X^2+2gg'ZX=0$, so that D is the point $(g+g', 0, 2gg')$. Thus the equations of the line-pairs $\| c, d \|$ and $\| e, e' \|$ are

$$2gg'x^2-(g+g')zx=0, \quad 4gg'x^2-2(g+g')zx+z^2=0.$$

The point A has the same polar line relative to these line-pairs, namely the line

$$4gg'x-(g+g')z=0,$$

which passes through B. Call this line h and let H be the point hb. Then H is the pole of a relative to each of the point-pairs $\| C, D \|$ and $\| E, E' \|$, since H and A harmonically separate each of these pairs of points. The whole figure is self-dual, and the dual triangles ABH and abh are identical, apart from the order in which the vertices are taken.

An alternative proof that the harmonic conjugates of A relative to $| C, D |$ and to $| E, E' |$ coincide is as follows. Let l, l' be the polars of D relative to s and s'. Then the conics of σ meet a_1 in pairs of points of an involution with la_1 and $l'a_1$ as double points and $| a_1 d, a_1 a |$ as a pair. Thus $(aldl')=-1$, or $(ALCL')=-1$, where L is bl and L' is bl'. Consider the homography $P \to P'$ on b determined by the relation $(APP'D)=-1$. In this homography $L \to E$, $L' \to E'$, and $C \to H$, where H is the harmonic conjugate of A relative to $| C, D |$; the self-corresponding points are A and D. Hence

$$-1=(ALCL')=(AEHE'),$$

proving that H is also the harmonic conjugate of A relative to $| E, E' |$.

8·15. *Simple contact. Alternative equations*

Take ABH as triangle of reference, and choose co-ordinates so that E is $(1, 0, -2)$, and therefore E' is $(1, 0, 2)$ since $(AEHE')=-1$. Then e and e' are $2x+z=0$ and $2x-z=0$, and the equations of s and s' are

$$s \equiv by^2+z^2+2zx=0, \quad s' \equiv b'y^2-z^2+2zx=0, \qquad (·151)$$

where $bb' \neq 0$. Also $b \neq b'$, for if $b = b'$ the conics have four-point contact at A. The line equations are given by

$$\left.\begin{aligned} -\frac{S}{b} &\equiv -X^2+\frac{Y^2}{b}+2ZX = 0, \\ -\frac{S'}{b'} &\equiv X^2+\frac{Y^2}{b'}+2ZX = 0. \end{aligned}\right\} \qquad (\cdot 152)$$

The line equation of the conic $\lambda s+\mu s' = 0$ is

$$(\lambda-\mu)(\lambda b+\mu b')X^2-(\lambda+\mu)^2Y^2-2(\lambda+\mu)(\lambda b+\mu b')ZX = 0,$$

or $$\lambda^2 S-\lambda\mu\{(b-b')X^2+2Y^2+2(b+b')ZX\}+\mu^2 S' = 0, \qquad (\cdot 153)$$

or $$\lambda(\lambda+\mu)S+\mu(\lambda+\mu)S'-2\lambda\mu(b-b')X^2 = 0, \qquad (\cdot 154)$$

showing that the only conics of σ which belong to Σ are those given by $\lambda\mu = 0$, namely s and s' themselves.

The conic $b's-bs' = 0$ is the line-pair $(b+b')z^2-2(b-b')zx = 0$, so that d is the line $2(b-b')x-(b+b')z = 0$. Dually D is the point $-(b+b')X+2(b-b')Z = 0$. An interesting particular case arises when $b+b' = 0$. Without loss of generality we may take $b = -b' = 1$, so that the conics are

$$s \equiv y^2+z^2+2zx = 0 \quad \text{and} \quad s' \equiv -y^2-z^2+2zx = 0. \qquad (\cdot 155)$$

The common points A_1, A_2 are $(0, 1, \pm i)$ and the common tangents a_1, a_2 are $[1, \pm 1, 0]$. The points C, H, D are coincident and the lines c, h, d are coincident. Harmonic inversion relative to A and h interchanges s and s'. A metrical example of this particular case is two equal circles which touch externally.

8·16. *Three-point contact*

Fig. 24 illustrates the case of three-point contact. B is the point aa_1 and b is the line AA_1. C is the point ba_1 and c is the line BA_1.

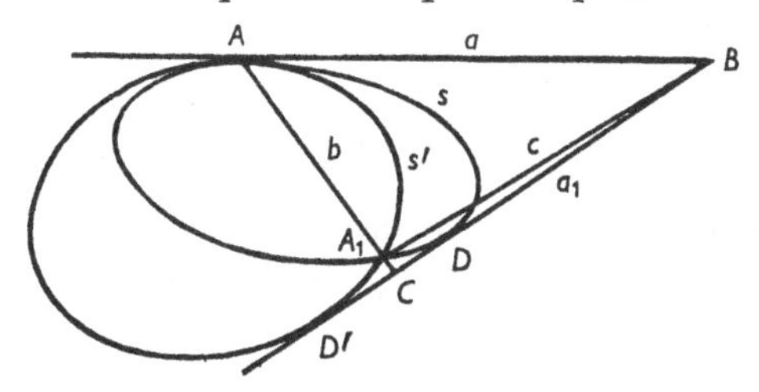

Fig. 24.

D, D' are the contacts of a_1 with s, s' and d, d' are the tangents to s, s' at A_1. The only line-pair of σ is $\|a, b\|$, so A is the only point which has the same polar line relative to s and s'. Dually the only point-pair of Σ is $\| A, B \|$, so a is the only line which has the same pole relative to s and s'. The figure so obtained is self-dual. There is no common self-polar triangle. The conics of σ meet a_1 in pairs of points of an involution of which D, D' are the double points and $| B, C |$ is a pair. Thus $(B, D, C, D') = -1$. Dually $(b, d, c, d') = -1$.

Choose a co-ordinate system so that ABA_1 is triangle of reference and D is the point $(-1, 2, 2)$. Then D' is $(-1, -2, 2)$, a_1 is $2x+z=0$, AD is $y-z=0$, and AD' is $y+z=0$. The equation of s is then of the form

$$(y-z)^2+\lambda(2x+z)z=0,$$

and $\lambda=-1$, since s contains the point $(0, 0, 1)$. Similarly for s'. Thus the equations of s and s' may be taken to be

$$s \equiv y^2-2yz-2zx=0, \quad s' \equiv y^2+2yz-2zx=0. \qquad (\cdot 161)$$

Harmonic inversion relative to B and b interchanges s and s'. The line equations are

$$S \equiv 2ZX-(X-Y)^2=0, \quad S' \equiv 2ZX-(X+Y)^2=0. \qquad (\cdot 162)$$

The point P_1' which is conjugate to P_1 relative to all conics of σ is $(y_1^2+z_1x_1, y_1z_1, -z_1^2)$, and the locus of poles of a line p_1 relative to conics of σ, which is also the locus of P_1 when P_1' lies on p_1, has equation

$$X_1(y^2+zx)+Y_1yz-Z_1z^2=0. \qquad (\cdot 163)$$

This equation may be written in the form

$$X_1s+\{3X_1x+(2X_1+Y_1)y-Z_1z\}z=0,$$

so the locus touches a at A, but does not have three-point contact with s unless p_1 contains A, when $X_1=0$ and the locus degenerates into the lines $z=0$ and $Y_1y-Z_1z=0$. The locus also contains the pole of p_1 relative to $\| A, A_1 \|$ and the double points of the involution cut out on p_1 by σ.

The line equation of the general conic $\lambda s+\mu s'=0$ of the point pencil σ is

$$\lambda^2S+2\lambda\mu(2ZX+X^2-Y^2)+\mu^2S'=0, \qquad (\cdot 164)$$

which may be written in the form

$$\lambda(\lambda+\mu)S+\mu(\lambda+\mu)S'+4\lambda\mu X^2=0.$$

Thus the only conics of the point pencil σ which also belong to the line pencil Σ are those for which $\lambda\mu=0$, namely s and s'.

The triangle formed by the lines b, a, a_1 is not the triangle BAA_1, so the above choice of triangle of reference is not self-dual. But if

H is the point on b such that $(AA_1HC) = -1$, and if h is the line BH, then $(a, a_1, h, c) = -1$, so the figure is still self-dual. The dual of the triangle BAH is the triangle bah, which is the same triangle with the sides taken in a different order. If a co-ordinate system is chosen in which ABH is triangle of reference and D is the point $(-1, 2, 1)$, then D' is $(-1, -2, 1)$ and A_1 is $(1, 0, 1)$. The equations of s and s' are then

$$(y-2z)^2+\lambda(z+x)z = 0 \quad \text{and} \quad (y+2z)^2+\mu(z+x)z = 0,$$

where $\lambda = \mu = -2$ since the conics contain A_1. Thus the equations may be taken in the form

$$s \equiv y^2+2z^2-4yz-2zx = 0, \quad s' \equiv y^2+2z^2+4yz-2zx = 0. \quad (\cdot 165)$$

The line equations are then

$$\left.\begin{aligned} -S &\equiv 2X^2+Y^2-2ZX-4XY = 0,\\ -S' &\equiv 2X^2+Y^2-2ZX+4XY = 0.\end{aligned}\right\} \quad (\cdot 166)$$

8·17. *Four-point contact*

In the case of four-point contact, which is illustrated in Fig. 25, the only line pair in σ is $\|a, a\|$ and the only point-pair in Σ is $\|A, A\|$.

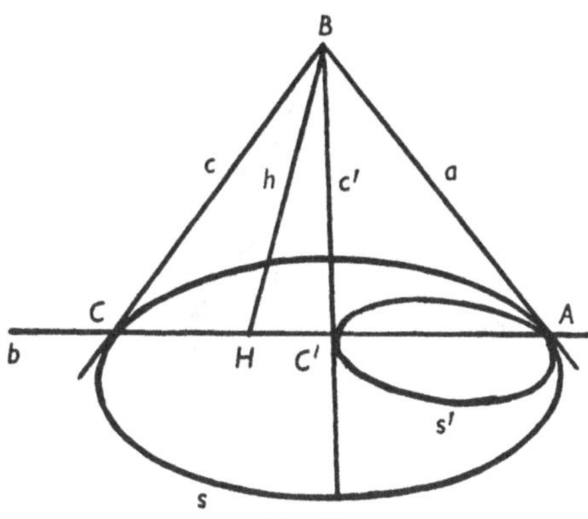

Fig. 25

Thus the only points with the same polar relative to s and s' are the points of a, and dually the only lines which have the same poles relative to s and s' are the lines through A. There is no common self-polar triangle of s and s'. Also any point B of a has the same polar b relative to s and s', and b passes through A. If C, C' are the further intersections of b with s, s', the tangents c, c' at C, C' pass through B. Let H be the point of b such that $(ACHC') = -1$. Then if h is the line BH, $(achc') = -1$. The figure so obtained is self-dual.

Choose co-ordinates so that ABH is triangle of reference and one of the intersections of c' with s is $(1, 2\sqrt{2}, 2)$. Then C, C' are $(-1, 0, 2)$ $(1, 0, 2)$, and the conics are

$$s \equiv y^2 - z^2 - 2zx = 0, \quad s' \equiv y^2 + z^2 - 2zx = 0. \qquad (\cdot 171)$$

The line equations are

$$S \equiv 2ZX - X^2 - Y^2 = 0, \quad S' \equiv 2ZX + X^2 - Y^2 = 0. \quad (\cdot 172)$$

In this case the locus of poles of a line p_1 relative to conics of σ is degenerate, consisting of the lines $X_1 y + Y_1 z = 0$ and $z = 0$. If P_1 is not on a, there is a unique point P_1' which is conjugate to P_1 relative to all conics of σ, namely $(y_1, z_1, 0)$, but if P_1 lies on a, $z_1 = 0$, and P_1 is conjugate to all points of the line $y_1 y - x_1 z = 0$ relative to all conics of σ.

The equations of s, s' are equivalent to $(\cdot 151)$, with $b = b' = -1$, and it follows from $(\cdot 154)$ that in the case of four-point contact the proper conics of σ are also the proper conics of Σ.

8·18. *Proper double contact*

Let B be the point $a_1 a_2$ and b the line $A_1 A_2$ as in Fig. 26. Let H, K be any two points of b such that $(A_1, H, A_2, K) = -1$, and

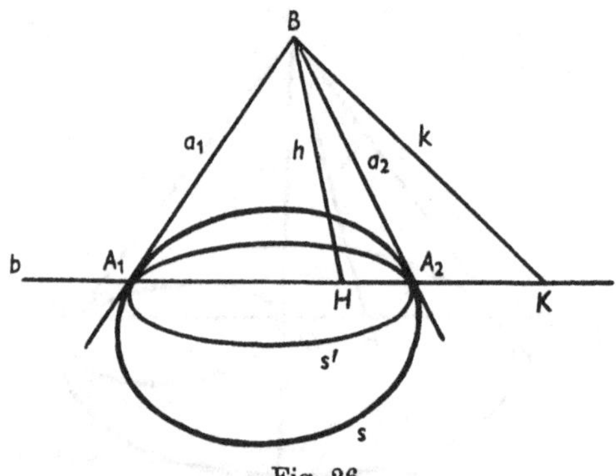

Fig. 26

denote the lines BH, BK by h, k. The line-pairs of σ are $\| a_1, a_2 \|$, $\| b, b \|$ and the point-pairs of Σ are $\| A_1, A_2 \|$ and $\| B, B \|$. The triangle BHK is self-polar relative to s and s', and there are ∞^1 such triangles.

With $A_1 B A_2$ as triangle of reference, the equations of the conics may be taken as

$$s \equiv y^2 + 2gzx = 0, \quad s' \equiv y^2 + 2g'zx = 0.* \qquad (\cdot 181)$$

where $gg' \neq 0$.

* Without loss of generality we could take $g = -\frac{1}{2}$, so that s is the conic $x : y : z = \lambda^2 : \lambda : 1$.

The line equations are

$$-\frac{1}{g}S \equiv gY^2+2ZX = 0, \quad -\frac{1}{g'}S' \equiv g'Y^2+2ZX = 0. \quad (\cdot 182)$$

The point P_1' which is conjugate to P_1 relative to all conics of σ is $(x_1, 0, -z_1)$ unless P_1 lies on $y=0$, when all points of the line $z_1x+x_1z=0$ are conjugate to P_1 relative to all conics of σ. The locus of P_1' as P_1 varies on a general line p_1 degenerates into the lines $X_1x-Z_1z=0$ and $y=0$. The locus of the poles of p_1 relative to the conics of σ degenerates into the same two lines. The line equation of $\lambda s+\mu s'=0$ is

$$\left.\begin{aligned} &-(\lambda g+\mu g')^2Y^2-2(\lambda+\mu)(\lambda g+\mu g')ZX = 0,\\ \text{or} \quad &\lambda^2S-2\lambda\mu\{gg'Y^2+(g+g')ZX\}+\mu^2S' = 0,\\ \text{or} \quad &\lambda^2S+\lambda\mu\left(\frac{g'}{g}S+\frac{g}{g'}S'\right)+\mu^2S' = 0.\end{aligned}\right\} \quad (\cdot 183)$$

So the proper conics of σ are also the proper conics of Σ.

Alternatively, if HBK is taken as triangle of reference, the equations are of the form

$$s\equiv ax^2+by^2+cz^2 = 0, \quad s'\equiv a'x^2+b'y^2+c'z^2 = 0,$$

where $\quad abca'b'c' \neq 0 \quad \text{and} \quad a:a' = c:c'.$

8·19. All the cases have now been discussed, and it should be noticed that the case of double contact, which includes proper double contact and four-point contact, differs from the other cases in that the locus of poles of a general line relative to the conics of σ is a line-pair, the polars of a general point relative to the conics of Σ form a point-pair, and the proper conics of σ are the proper conics of Σ. Thus:

The point pencil σ and the line pencil Σ determined by two proper conics s and s' have no conics in common except s and s' unless s and s' have double contact, when the pencils coincide, apart from their improper members.

It is only in the case of double contact that σ contains a repeated line and Σ a repeated point.

8·2. Reciprocation of one conic relative to another

8·21. We have already introduced in 4·77 the process of reciprocation relative to a proper conic, in which the lines and points of the plane are replaced by their poles and polar lines relative to the conic. A proper conic locus s with its associated conic scroll S reciprocate, with respect to a proper conic γ, into a proper conic scroll S' with its associated conic locus s'. A range of points on a line p reciprocates into a projectively related pencil of lines through P, the pole of p relative to γ. Hence, if two points P_1, P_2 of a line p are conjugate relative to s and so harmonically separate the points (p, s), the reciprocal lines p_1, p_2 harmonically separate the lines $[P, S']$ and are conjugate relative to S'. Thus pole and polar relative to s reciprocate into polar and pole relative to S'.

Let the point and line equations of γ be

$$\gamma \equiv px^2+qy^2+rz^2+2uyz+2vzx+2wxy = 0,$$
$$\Gamma \equiv PX^2+QY^2+RZ^2+2UYZ+2VZX+2WXY = 0,$$

when P, Q, ... are the cofactors of p, q, ... in the determinant of γ, and let s and S have their usual forms

$$s \equiv ax^2+by^2+cz^2+2fyz+2gzx+2hxy = 0,$$
$$S \equiv AX^2+BY^2+CZ^2+2FYZ+2GZX+2HXY = 0.$$

The reciprocal of a point $P(x, y, z)$ relative to γ is, as in 4·77, the line

$$[px+wy+vz, \quad wx+qy+uz, \quad vx+uy+rz].$$

If this line belongs to S the point P lies on s'. Hence the point equation of s' is

$$\begin{aligned} s' \equiv\, & A(px+wy+vz)^2+B(wx+qy+uz)^2+C(vx+uy+rz)^2 \\ & +2F(wx+qy+uz)(vx+uy+rz)+2G(vx+uy+rz)(px+wy+vz) \\ & +2H(px+wy+vz)(wx+qy+uz) = 0; \end{aligned}$$

dually its line equation is

$$S' \equiv a(PX+WY+VZ)^2+\ldots+ = 0,$$

obtained from the preceding by interchanging large and small letters.

For convenience we list the following particular cases, the proof of which is left to the reader:

(i) The reciprocal of the conic $s \equiv ax^2+by^2+cz^2 = 0$ relative to the conic $\gamma \equiv px^2+qy^2+rz^2 = 0$ has point equation

$$\frac{p^2x^2}{a}+\frac{q^2y^2}{b}+\frac{r^2z^2}{c} = 0.$$

(ii) The reciprocal of $s \equiv by^2+cz^2+2gzx = 0$ relative to

$$\gamma \equiv qy^2+rz^2+2vzx = 0$$

is
$$g^2q^2y^2+bv(2gr-cv)\,z^2+2bgv^2zx = 0.$$

(iii) The reciprocal of $s \equiv by^2+2gzx = 0$ relative to

$$\gamma \equiv qy^2+2vzx = 0 \quad \text{is} \quad gq^2y^2+2bv^2zx = 0.$$

8·22. *Reciprocation of a conic into itself*

Suppose that s and s' are two proper conics such that the reciprocal of s relative to s' is the conic s itself. Then, if A is any common point of s and s', the polar line a of A relative to s' must be both a tangent of s and a tangent of s'. Thus the conics must touch wherever they meet, and the only possible cases are proper double contact and four-point contact.

In the case of four-point contact the equations may be taken as

$$s \equiv y^2-z^2-2zx = 0, \quad s' \equiv y^2+z^2-2zx = 0$$

by 8·17, and by 8·21 the reciprocal of s relative to s' is

$$y^2+3z^2-2zx = 0,$$

which does not coincide with s.

In the case of proper double contact the equations may be taken as

$$S \equiv y^2+2gzx = 0, \quad S' \equiv y^2+2g'zx = 0.$$

The reciprocal of s relative to s' is then

$$gy^2+2g'^2zx = 0,$$

which is s IF $g^2 = g'^2$. But if $g = g'$ the conics s and s' are not distinct, so the only possibility is $g+g' = 0$. Since this equation is symmetrical in g and g', it follows that, if s is self-reciprocal relative to s', then s' is self-reciprocal relative to s.

The conics of the point pencil determined by s and s' have equations of the form $y^2+2\rho zx = 0$, and form a $R\infty^1$ with ρ as a parameter. The equation $g+g' = 0$ expresses that the values g, g' of ρ harmonically separate the values $0, \infty$ which correspond to the degenerate conics of the pencil. Thus, if λ is any parameter of the $R\infty^1$ formed by the conics of a point pencil having proper double contact, two conics of the pencil are self-reciprocal relative to each other IF the values of λ which correspond to them harmonically separate the values of λ which correspond to the degenerate conics of the pencil. In particular:

Two distinct conics with point equations $s = 0$ and $s' = 0$ are each self-reciprocal relative to the other IF *they have proper double contact and* IF *the two values of λ for which $s+\lambda s' = 0$ is degenerate are equal and opposite.*

It follows that the following pairs of conics are self-reciprocal relative to each other:

$$y^2+2gzx = 0 \quad \text{and} \quad y^2-2gzx = 0,$$
$$ax^2+by^2+cz^2 = 0 \quad \text{and} \quad -ax^2+by^2+cz^2 = 0,$$
$$by^2+cz^2+2gzx = 0 \quad \text{and} \quad -by^2+cz^2+2gzx = 0.$$

8·23. *Reciprocation of s into s'. General case*

We now consider whether there are conics γ relative to which two given distinct conics s and s' reciprocate into each other or into themselves. Consider first the general case, when the point equations of s and s' are

$$s \equiv ax^2+by^2+cz^2 = 0,$$
$$s' \equiv a'x^2+b'y^2+c'z^2 = 0$$

and the line equations are

$$\frac{1}{abc}S \equiv \frac{X^2}{a}+\frac{Y^2}{b}+\frac{Z^2}{c} = 0,$$
$$\frac{1}{a'b'c'}S' \equiv \frac{X^2}{a'}+\frac{Y^2}{b'}+\frac{Z^2}{c'} = 0.$$

If s' is the reciprocal of s relative to γ, the triangle of reference, being the only common self-polar triangle of s and s', must reciprocate into itself relative to γ, so γ must have an equation of the form

$$\gamma \equiv px^2+qy^2+rz^2 = 0.$$

Then, by 8·21, the reciprocal of s relative to γ is

$$\frac{p^2x^2}{a}+\frac{q^2y^2}{b}+\frac{r^2z^2}{c} = 0,$$

which coincides with s' IF

$$p^2:q^2:r^2 = aa':bb':cc',$$

i.e. IF $\quad p:q:r = \sqrt{aa'}:\pm\sqrt{bb'}:\pm\sqrt{cc'}.$

Thus, using 8·22:

In the general case there are four distinct conics $\gamma_1, \gamma_2, \gamma_3, \gamma_4$ relative to which s and s' are reciprocals, and each of a pair γ_i, γ_j of these four conics is self-reciprocal relative to the other.

The reciprocal of the conic $\lambda s+\mu s' = 0$ relative to any of the conics γ_i is

$$\frac{aa'x^2}{\lambda a+\mu a'}+\frac{bb'y^2}{\lambda b+\mu b'}+\frac{cc'z^2}{\lambda c+\mu c'} = 0,$$

which has line equation

$$\frac{\lambda a+\mu a'}{aa'}X^2+\frac{\lambda b+\mu b'}{bb'}Y^2+\frac{\lambda c+\mu c'}{cc'}Z^2 = 0,$$

or
$$\frac{\mu}{abc}S+\frac{\lambda}{a'b'c'}S' = 0.$$

In fact when s is reciprocated into s' each conic of the point pencil σ reciprocates into a conic of the line pencil Σ, and the same conic of Σ is obtained whichever conic γ_i is used. In particular each line-pair of σ has its vertex at a vertex of the common self-polar triangle and reciprocates into the point-pair of Σ whose axis is the opposite side of the triangle. Each base point A_i of σ reciprocates into one of the four common tangents a_i. The way in which the points A_1, A_2, A_3, A_4 and the lines a_1, a_2, a_3, a_4 are paired depends on which conic γ_i is chosen, but we have the theorem:

The cross-ratio $(A_1A_2A_3A_4)$ *on* s *is equal to the cross-ratio* $(a_ha_ka_la_m)$ *in the scroll of tangents to* s', *where* h, k, l, m *are the numbers* 1, 2, 3, 4 *in some order.*

In the four reciprocations the points A_1, A_2, A_3, A_4 are paired with (i) the lines a_h, a_k, a_l, a_m, (ii) the lines a_k, a_h, a_m, a_l, (iii) the lines a_l, a_m, a_h, a_k, (iv) the lines a_m, a_l, a_k, a_h, these being the permutations of the lines which leaves their cross-ratio unaltered.

In the general case there is no conic γ *relative to which* s *and* s' *are both self-reciprocal.*

For again γ would have to be of the form $px^2+qy^2+rz^2 = 0$, and, for s and s' to be self-reciprocal, p, q, r must satisfy

$$p^2:q^2:r^2 = a^2:b^2:c^2 = a'^2:b'^2:c'^2,$$

and this implies that s and s' have double contact.

8·24. *Simple contact*

If s reciprocates into s' relative to a conic γ, A must reciprocate into a and the line A_1A_2 into the point a_1a_2 (see Fig. 23). Thus γ must touch a at A and contain H. Also since B is ad, the polar of B relative to γ must be b. In fact γ must touch h at H. It was shown in

8·15 that with ABH as triangle of reference the equations may be taken in the form

$$s \equiv by^2+z^2+2zx = 0, \quad s' \equiv b'y^2-z^2+2zx = 0,$$

where $bb'(b-b') \neq 0$. The equation of γ must then be of the form

$$\gamma \equiv \mu y^2+2zx = 0,$$

and by 8·21 the reciprocal of s relative to γ is

$$\mu^2y^2-bz^2+2bzx = 0,$$

which is s' IF $\mu^2 = bb'$. Thus, using 8·22:

In the case of simple contact there are TWO *conics* γ_1 *and* γ_2 *relative to which* s *and* s' *reciprocate into each other, and these conics* γ_1, γ_2 *are self-reciprocal relative to each other.*

Again, it is not possible to find a conic γ relative to which s and s' are both self-reciprocal, for such a conic γ would have to be of the form $\mu y^2+2zx = 0$, and the conic $\mu^2y^2-bz^2+2bzx = 0$ does not coincide with s for any value of μ.

8·25. *Three-point contact*

It was shown in 8·16 that with ABH as triangle of reference the equations may be taken in the form

$$s \equiv y^2+2z^2-4yz-2zx = 0, \quad s' \equiv y^2+2z^2+4yz-2zx = 0.$$

If s' is the reciprocal of s or if s, s' are both self-reciprocal relative to γ, the polars of A and A_1 relative to γ must be a and a_1, and therefore b must be the polar of B and γ must touch a at A and h at H. Hence γ must have an equation of the form

$$\gamma \equiv y^2+2\mu zx = 0,$$

when the reciprocals of s and s' relative to γ are easily seen to be

$$y^2+2\mu^2z^2-4\mu yz-2\mu^2zx = 0$$

and

$$y^2+2\mu^2z^2+4\mu yz-2\mu^2zx = 0.$$

Thus s and s' reciprocate into each other if $\mu = -1$ and into themselves if $\mu = 1$. Thus:

In the case of three-point contact there is ONE *conic* γ *relative to which* s *reciprocates into* s'*, and* ONE *conic* γ' *relative to which* s *and* s' *are both self-reciprocal. The conics* γ *and* γ' *are each self-reciprocal relative to the other.*

The conic γ' has proper double contact with s and s' along the lines

$$y-z = 0 \quad \text{and} \quad y+z = 0.$$

8·26. *Four-point contact*

It was shown in 8·17 that, with ABH as triangle of reference, the equations may be taken as

$$s \equiv y^2 - z^2 - 2zx = 0, \quad s' \equiv y^2 + z^2 - 2zx = 0.$$

The line equations are then

$$S \equiv -X^2 - Y^2 + 2ZX = 0, \quad S' \equiv X^2 - Y^2 + 2ZX = 0.$$

Any conic γ relative to which s and s' reciprocate into each other or are both self-reciprocal must touch a at A, and must therefore be of the form

$$\gamma \equiv by^2 + cz^2 + 2fyz + 2gzx = 0.$$

The polar of P_1 relative to γ is the line $[gz_1, by_1 + fz_1, gx_1 + fy_1 + cz_1]$, which satisfies $S = 0$ IF

$$g^2z_1^2 + (by_1 + fz_1)^2 - 2gz_1(gx_1 + fy_1 + cz_1) = 0,$$

so the reciprocal of s relative to γ has point equation

$$b^2y^2 + (g^2 + f^2 - 2gc)\,z^2 + 2f(b-g)\,yz - 2g^2zx = 0.$$

If this conic is to be s', the coefficient of yz must be zero, so $f(b-g) = 0$. There are thus two possibilities.

If $b = g$ we may take $b = g = 1$,* and then from comparison of the coefficient of z^2 we must have $f^2 - 2c = 0$. Putting $f = 2\mu$ we see that s reciprocates into s' relative to the conic

$$\gamma_\mu \equiv y^2 + 2\mu^2z^2 + 4\mu yz + 2zx = 0$$

for all finite values of μ.†

If $f = 0$ and $b \neq g$, we must have $b^2 = g^2$, from comparison of the coefficients of y^2 and zx, so we may take $b = 1$, $g = -1$. Then from the coefficients of z^2 we must have $c = 0$, giving the conic

$$\gamma' \equiv y^2 - 2zx = 0$$

relative to which s and s' are also reciprocals. Thus:

In the case of four-point contact there is a family of ∞^1 conics γ_μ and one other conic γ' not contained in this family relative to all of which s and s' are reciprocals.

The equation of the conic γ_μ may be written in the form

$$\gamma_\mu \equiv 2(y + \mu z)^2 - (y^2 - 2zx) = 0,$$

so γ_μ has double contact with γ', and the chord of contact is $y + \mu z = 0$, which is not a tangent of γ' since $\mu \neq \infty$. Thus, from 8·22,

* $b = g = 0$ will not provide a solution.

† The value $\mu = \infty$ would make γ degenerate, but γ is not degenerate for finite values of μ.

each conic γ_μ is self-reciprocal relative to γ' and consequently γ' is self-reciprocal relative to each γ_μ.

It follows from the above that there is no conic γ relative to which s and s' are both self-reciprocal.

8·27. *Proper double contact*

It was shown in 8·18 that the equations may be taken as

$$s \equiv y^2 + 2gzx = 0, \quad s' \equiv y^2 + 2g'zx = 0.$$

If γ is any conic relative to which s and s' reciprocate into each other or are both self-reciprocal, then the polars of A_1 and A_2 relative to γ must be a_1 and a_2 or a_2 and a_1. In the first case γ is of the form

$$\gamma \equiv y^2 + 2\mu zx = 0,$$

and by 8·21 the reciprocals of s, s' relative to γ are

$$gy^2 + 2\mu^2 zx = 0, \quad g'y^2 + 2\mu^2 zx = 0.$$

Thus s and s' reciprocate into each other IF $\mu^2 = gg'$, giving two conics

$$\gamma_1 \equiv y^2 + 2\sqrt{(gg')}\,zx = 0, \quad \gamma_2 = y^2 - 2\sqrt{(gg')}\,zx = 0,$$

which are self-reciprocal relative to each other. Also s and s' are both self-reciprocal relative to γ IF $\mu^2 = g^2 = g'^2$, which implies $g + g' = 0$, since s and s' are distinct. Thus s and s' must be self-reciprocal relative to each other and the two conics γ are s and s' themselves.

In the second case, when the polars of A_1 and A_2 relative to γ are a_2 and a_1, the triangle A_1BA_2 is self-polar relative to γ, so γ is of the form

$$\gamma \equiv \lambda x^2 + \mu y^2 + \nu z^2 = 0.$$

The reciprocals of s and s' relative to γ are then

$$g\mu^2 y^2 + 2\nu\lambda zx = 0, \quad g'\mu^2 y^2 + 2\nu\lambda zx = 0,$$

so s reciprocates into s' relative to γ IF $\nu\lambda = gg'\mu^2$. Thus, putting $\nu = 1$, s and s' are reciprocals relative to the conic

$$\gamma_\mu \equiv gg'\mu^2 x^2 + \mu y^2 + z^2 = 0,$$

for all finite values of μ other than $\mu = 0$. The conics γ_1 and γ_2 are self-reciprocal relative to each of these conics γ_μ.

Also s and s' are both self-reciprocal relative to γ IF $\lambda\nu = g^2 = g'^2$. This implies $g = -g'$, and in this case s and s' are self-reciprocal relative to each other and relative to all the conics

$$g^2\mu^2 x^2 + \mu y^2 + z^2 = 0.$$

Thus:

In the case of proper double contact there are ∞^1 *conics* γ_μ *and two other conics* γ_1 *and* γ_2 *relative to all of which* s *and* s' *reciprocate into each other. The conics* γ_1 *and* γ_2 *are self-reciprocal relative to each other and relative to all the conics* γ_μ. *It is not in general possible to find a conic* γ *relative to which* s *and* s' *are both self-reciprocal, but there are* ∞^1 *such conics* γ IF s *and* s' *are self-reciprocal relative to each other.*

8·28. Thus:

If s *and* s' *are two proper conics, it is always possible to find a conic* γ *relative to which* s *and* s' *reciprocate into each other. There are a finite number of such conics* γ *except in the cases of double contact.*

If s *and* s' *have three-point contact there is one conic* γ' *relative to which* s *and* s' *are both self-reciprocal. If* s *and* s' *have proper double contact and are self-reciprocal relative to each other, there are* ∞^1 *other conics* γ' *relative to which* s *and* s' *are both self-reciprocal. Except in these two cases there is no conic* γ' *relative to which* s *and* s' *are both self-reciprocal.*

8·29. *Reciprocation of a line-pair into a point-pair*

It is impossible to reciprocate a line-pair into a line-pair, but there are in general two pencils of conics relative to which a pair of lines l_1 and l_2 are the polars of a pair of points A_1 and A_2. If as in Fig. 27 the line A_1A_2 meets l_1 and l_2 in P_1 and P_2, and if H, K are

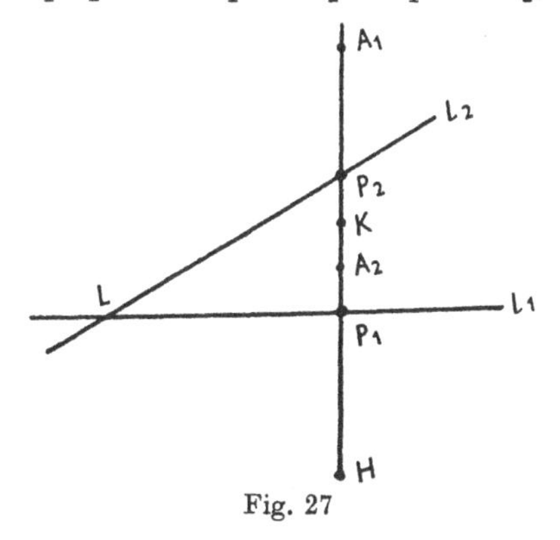

Fig. 27

the double points of the involution determined by the pairs A_1, P_1 and A_2, P_2, then one pencil consists of conics touching LH and LK at H and K, while the other pencil is obtained similarly from the double points of the involution determined by A_1, P_2 and A_2, P_1.

If A_1 coincides with P_1 then A_1 is one of the double points of the first involution and the other is the harmonic conjugate of A_1 relative to A_2, P_2; the second involution is degenerate, with its double points coinciding in A_1. The first pencil of conics exists but there is no proper conic for which l_2 is the polar of A_1 and l_1 the polar of A_2, though the line-pair consisting of any two lines through A_1 which harmonically separate A_1A_2 and l_1 may be regarded as a solution, the polar of A_1 relative to such a line-pair being indeterminate.

If A_1 coincides with P_1 and also A_2 coincides with P_2 the conics of the first pencil touch l_1 at A_1 and l_2 at A_2. The second involution is not completely determined and there is a set of conics, having LA_1A_2 as self-conjugate triangle, relative to which l_2, l_1 are the polars of A_1, A_2 respectively.

We leave it as an exercise to the reader to prove that, if A_1A_2 passes through the point L, the conics relative to which l_1 is the polar of A_1, and l_2 the polar of A_2, form a pencil with four-point contact at L and having A_1A_2 as the common tangent there; a similar pencil is obtained by interchanging the roles of A_1 and A_2. If A_1 coincides with L then any line-pair consisting of two lines through A_1 which harmonically separate A_1A_2 and either of the lines l_1, l_2 may be regarded as a solution.

CHAPTER IX

TWO CONICS. APOLARITY

9·1. Harmonic scroll and locus

9·11. The present chapter is mainly concerned with the geometry of two proper conics. It will be convenient to denote two conic loci by s_1 and s_2 instead of by s and s' as hitherto, in order that we may compare easily the results of this chapter with those of Chapter XI later, in which the use of a dash has a special significance; s_1 as used in this sense must not of course be confused with s_1 as defined in 4·11.

The lines which meet two conic loci s_1 and s_2 in apolar pairs of points form a conic scroll, which is called the harmonic scroll of s_1 and s_2.

Dually the locus of points such that the pairs of tangents through them to two conic scrolls S_1 and S_2 are apolar is a conic locus, called the harmonic locus of S_1 and S_2.

We have already in 5·6 proved the first theorem when s_2 is a line-pair and the second when S_2 is a point-pair. Before proceeding to the general case we need the following:

Lemma: If a homography T is set up between the pairs of two involutions Λ_1 and Λ_2 of pairs of elements of a $R\infty^1$ there are in general TWO *pairs of Λ_1 each of which is apolar in the $R\infty^1$ to the pair of Λ_2 to which it corresponds in T.*

Without loss of generality we can take the $R\infty^1$ to be the points of a proper conic s on which Λ_1, Λ_2 are cut out by lines l_1, l_2 through fixed points L_1, L_2, the two pencils of lines being connected by a relation $\{l_1\}\equiv\{l_2\}$. Let l_2' be the line through L_2 conjugate to l_1. Then we have

$$\{l_2'\}\equiv\{l_1\}\equiv\{l_2\}.$$

The related pencils $\{l_2\}$ and $\{l_2'\}$ have in general two self-corresponding lines. Thus there are in general two pairs of corresponding lines l_1, l_2 which are conjugate relative to s and thus meet it in apolar pairs of points. This proves the lemma.

If now s_1 and s_2 are two proper conic loci, let A be one of their common points and let H be any fixed point in the plane. Let any line l through H meet s_1 in P_1, Q_1 and s_2 in P_2, Q_2. Denote the joins of A to P_1, Q_1, P_2, Q_2 by p_1, q_1, p_2, q_2. Then the pairs of points $|P_1, Q_1|$, $|P_2, Q_2|$ are apolar IF the pairs of lines $|p_1, q_1|$, $|p_2, q_2|$ are apolar.

As l varies the pairs of lines $|p_1, q_1|$, $|p_2, q_2|$ are corresponding pairs of lines in two involution pencils with vertex A whose pairs are homographically related; hence by the lemma applied to the $R\infty^1$ of lines through A there are in general TWO lines l through H such that the pairs $|p_1, q_1|$, $|p_2, q_2|$ are apolar, and the theorem is established.

9·12. A better proof of the preceding theorem is as follows: Let the two conic loci have point equations $s_1 = 0$, $s_2 = 0$ and line equations $S_1 = 0$, $S_2 = 0$. Then by 4·71 the conic with point equation $\lambda s_1 + \mu s_2 = 0$ has line equation

$$\begin{vmatrix} \lambda a_1 + \mu a_2 & \lambda h_1 + \mu h_2 & \lambda g_1 + \mu g_2 & X \\ \lambda h_1 + \mu h_2 & \lambda b_1 + \mu b_2 & \lambda f_1 + \mu f_2 & Y \\ \lambda g_1 + \mu g_2 & \lambda f_1 + \mu f_2 & \lambda c_1 + \mu c_2 & Z \\ X & Y & Z & 0 \end{vmatrix} = 0,$$

or

$$\lambda^2 S_1 + \lambda\mu G + \mu^2 S_2 = 0, \tag{·121}$$

where

$$\begin{aligned} G \equiv (b_1 c_2 + b_2 c_1 - 2f_1 f_2) X^2 + (c_1 a_2 + c_2 a_1 - 2g_1 g_2) Y^2 \\ + (a_1 b_2 + a_2 b_1 - 2h_1 h_2) Z^2 \\ + 2(g_1 h_2 + g_2 h_1 - a_1 f_2 - a_2 f_1) YZ \\ + 2(h_1 f_2 + h_2 f_1 - b_1 g_2 - b_2 g_1) ZX \\ + 2(f_1 g_2 + f_2 g_1 - c_1 h_2 - c_2 h_1) XY. \end{aligned} \tag{·122}$$

The equation (·121), regarded as a quadratic equation in the ratio variable (λ, μ), is a CONDITION for the conic $\lambda s_1 + \mu s_2 = 0$ to touch the line $[X, Y, Z]$. As (λ, μ) varies, the conics $\lambda s_1 + \mu s_2 = 0$ meet the line $[X, Y, Z]$ in pairs of an involution I and (λ, μ) is a ratio parameter of the $R\infty^1$ formed by the pairs of I. The pairs cut out by s_1 and s_2 correspond to the values $(1, 0)$ and $(0, 1)$ of (λ, μ), and the values of (λ, μ) which correspond to the double points of I are the roots of (·121). Hence it follows from 3·63 or 4·65 that the line $[X, Y, Z]$ meets s_1 and s_2 in apolar points IF $G = 0$. This proves the theorem, and the proof is valid when either s_1 or s_2 is a line-pair. The harmonic envelope $G = 0$ will be called the G-conic of s_1 and s_2.*

The roots of (·121), regarded as an equation in (λ, μ), are equal IF the double points of the involution I coincide, that is IF I is an

* It is commonly known as the Φ-conic, but the name introduced here has advantages which will become more apparent in Chapter XI.

involution with a fixed point, which is the case IF the line $[X, Y, Z]$ contains one of the common points of s_1 and s_2. Thus the line equation of the common points of s_1 and s_2 is

$$G^2 = 4S_1 S_2. \qquad (\cdot 123)$$

9·13. Dually if

$$S_1 \equiv A_1X^2 + B_1Y^2 + C_1Z^2 + 2F_1YZ + 2G_1ZX + 2H_1XY = 0,$$
$$S_2 \equiv A_2X^2 + B_2Y^2 + C_2Z^2 + 2F_2YZ + 2G_2ZX + 2H_2XY = 0$$

are any two conic scrolls, the point equation of the conic scroll $\lambda S_1 + \mu S_2 = 0$ is

$$\lambda^2\bar{s}_1 + \lambda\mu f + \mu^2\bar{s}_2 = 0, \qquad (\cdot 131)$$

where the bar sign is used as in 4·72 and

$$\begin{aligned} f \equiv (B_1C_2 + B_2C_1 - 2F_1F_2)\,x^2 + (C_1A_2 + C_2A_1 - 2G_1G_2)\,y^2 \\ + (A_1B_2 + A_2B_1 - 2H_1H_2)\,z^2 \\ + 2(G_1H_2 + G_2H_1 - A_1F_2 - A_2F_1)\,YZ \\ + 2(H_1F_2 + H_2F_1 - B_1G_2 - B_2G_1)\,ZX \\ + 2(F_1G_2 + F_2G_1 - C_1H_2 - C_2H_1)\,XY. \end{aligned} \qquad (\cdot 132)$$

The points such that the pairs of tangents from them to S_1 and S_2 are apolar form the conic locus $f = 0$, the harmonic locus of S_1 and S_2, which we shall also call the f-conic of S_1 and S_2.

The point equation of the common tangents of S_1 and S_2 is

$$f^2 = 4\bar{s}_1\bar{s}_2. \qquad (\cdot 133)$$

9·14. Thus:

The line equation of the harmonic envelope of two conic loci $s_1 = 0$ and $s_2 = 0$ is obtained by equating to zero the coefficient of $\lambda\mu$ in the line equation of the conic locus $\lambda s_1 + \mu s_2 = 0$. Dually the point equation of the harmonic locus of two conic scrolls $S_1 = 0$ and $S_2 = 0$ is obtained by equating to zero the coefficient of $\lambda\mu$ in the point equation of the conic scroll $\lambda S_1 + \mu S_2 = 0$.

9·15. The f-conic is essentially a conic locus determined by two conic scrolls. But if the equations $S_1 = 0$, $S_2 = 0$ of 9·13 are the line equations of two conic loci $s_1 = 0$, $s_2 = 0$, the coefficients $A_1, B_1, \ldots, H_1, A_2, B_2, \ldots, H_2$ being the cofactors of $a_1, b_1, \ldots, h_1, a_2, b_2, \ldots, h_2$ in the determinants

$$\delta_1 = \begin{vmatrix} a_1 & h_1 & g_1 \\ h_1 & b_1 & f_1 \\ g_1 & f_1 & c_1 \end{vmatrix}, \qquad \delta_2 = \begin{vmatrix} a_2 & h_2 & g_2 \\ h_2 & b_2 & f_2 \\ g_2 & f_2 & c_2 \end{vmatrix},$$

then the conic locus $f = 0$ is also said to be the harmonic locus of the conic loci s_1 and s_2. As in 4·73, $\bar{s}_1 \equiv \delta_1 s_1$ and $\bar{s}_2 \equiv \delta_2 s_2$, so the equations (·131) and (·133) become

$$\lambda^2 \delta_1 s_1 + \lambda\mu f + \mu^2 \delta_2 s_2 = 0, \tag{·151}$$

$$f^2 = 4\delta_1 \delta_2 s_1 s_2. \tag{·152}$$

If s_1 is a proper line-pair with vertex P, the equation $S_1 = 0$ represents the repeated point P, and so the conic $f = 0$ is the harmonic locus of the repeated point P and the conic scroll $S_2 = 0$. If s_1 is a repeated line, all the coefficients of S_1 vanish and the conic $f = 0$ is indeterminate.

9·16. The actual equations of the G-conic and f-conic of two proper conics s_1 and s_2 in the five cases that arise may easily be written down. We give the line equations of the G-conic in the different cases, adopting, with an obvious change of notation, standard forms for the equations of s_1 and s_2 which were established in Chapter VIII.

(i) General case:

$$\left.\begin{array}{l} s_1 \equiv a_1 x^2 + b_1 y^2 + c_1 z^2 = 0, \quad s_2 \equiv a_2 x^2 + b_2 y^2 + c_2 z^2 = 0, \\ G \equiv (b_1 c_2 + b_2 c_1) X^2 + (c_1 a_2 + c_2 a_1) Y^2 + (a_1 b_2 + a_2 b_1) Z^2 = 0. \end{array}\right\} \tag{·161}$$

(ii) Simple contact:

$$\left.\begin{array}{r} s_1 \equiv b_1 y^2 + z^2 + 2zx = 0, \quad s_2 \equiv b_2 y^2 - z^2 + 2zx = 0, \\ -G \equiv (b_1 - b_2) X^2 + 2Y^2 + 2(b_1 + b_2) ZX = 0. \end{array}\right\} \tag{·162}$$

(iii) Three-point contact:

$$\left.\begin{array}{c} s_1 \equiv y^2 - 2yz - 2zx = 0, \quad s_2 \equiv y^2 + 2yz - 2zx = 0, \\ \tfrac{1}{2}G \equiv X^2 - Y^2 + ZX = 0. \end{array}\right\} \tag{·163}$$

(iv) Four-point contact:

$$\left.\begin{array}{c} s_1 \equiv y^2 - z^2 - 2zx = 0, \quad s_2 \equiv y^2 + z^2 - 2zx = 0, \\ -\tfrac{1}{2}G \equiv Y^2 - 2ZX = 0. \end{array}\right\} \tag{·164}$$

(v) Proper double contact:

$$\left.\begin{array}{c} s_1 \equiv y^2 + 2g_1 zx = 0, \quad s_2 \equiv y^2 + 2g_2 zx = 0, \\ -\tfrac{1}{2}G \equiv g_1 g_2 Y^2 + (g_1 + g_2) ZX = 0. \end{array}\right\} \tag{·165}$$

9·17. It follows from 9·16 that the only cases in which the G-conic of two proper conics s_1 and s_2 is a proper point-pair are:

(1) The general case, with $b_1c_2+b_2c_1=0$ or $c_1a_2+c_2a_1=0$ or $a_1b_2+a_2b_1=0$.

(2) The case of simple contact, with $b_1+b_2=0$. This is the particular case mentioned in 8·15, when the equations of s_1 and s_2 may be taken in the form 8·155.

In the general case G cannot be a repeated point; for, if

$$c_1a_2+c_2a_1=a_1b_2+a_2b_1=0,$$

then $b_1:b_2=c_1:c_2$, which implies double contact. Thus the only case in which G is a repeated point is the case of proper double contact with $g_1+g_2=0$, and this is the particular case in which s_1 and s_2 are self-reciprocal relative to each other.

The conditions under which the G-conic is degenerate are considered in 9·25.

9·18. It was shown in 8·2 that it is always possible to find a conic γ relative to which two proper conics s_1 and s_2 are reciprocals. If p is the polar line of a point P relative to γ the intersections of p with s_1 and s_2 reciprocate into the tangents to s_2 and s_1 through P; hence, if p is a tangent of G, then P is a point of f. Thus f and G are reciprocals relative to γ. It follows that f is a line-pair or a repeated line IF G is a point-pair or a repeated point.

Also, since the line equation of the conic locus $\lambda s_1+\mu s_2=0$ is

$$\lambda^2S_1+\lambda\mu G+\mu^2S_2=0,$$

it follows that G belongs to the line pencil determined by s_1 and s_2 IF all the conics $\lambda s_1+\mu s_2=0$ belong to this line pencil, that is IF s_1 and s_2 have double contact, by 4·98 or 8·19.

The conics of the point pencil σ determined by s_1 and s_2 reciprocate relative to γ into the conics of the line pencil Σ determined by s_1 and s_2, and f reciprocates into G. Thus f belongs to σ IF G belongs to Σ, that is IF s_1 and s_2 have double contact.

9·19. *If σ is a given point pencil of conics, and if a line p' belongs to the harmonic scroll of the two conics of σ which touch a line p, then p belongs to the harmonic scroll of the two conics of σ which touch p'.*

For, if λ is a parameter of the pencil, let λ_1, λ_2 and λ_1', λ_2' be the values of λ corresponding to the conics of σ which touch p and p'.

Then, by the argument of 9·12, p belong to the G-conic of the two conics of σ which touch p' IF λ_1, λ_2 harmonically separate λ_1', λ_2', and this is a symmetrical CONDITION.

9·2. Properties of the G and f-conics

9·21. We have discussed the G-conic of a proper conic and a line-pair in 5·66. Suppose now that s_1 and s_2 are both proper conics. If P, Q are two points of s_1, the line PQ is a tangent to G IF P and Q are conjugate relative to s_2. Thus:

The tangents to G through any point P of s_1 are the lines joining P to the intersections with s_1 of the polar line of P relative to s_2, and similarly for the tangents to G through a point of s_2.

9·22. The points, or contacts, of G are the points such that the two tangents to G through them coincide. Thus:

The common points of s_1 and G are the points of s_1 whose polar lines relative to s_2 touch s_1, that is the intersections of s_1 with the reciprocal of s_1 relative to s_2; similarly for s_2 and G.

9·23. From 9·21 the tangent to s_1 at P is a tangent of G IF P is self-conjugate relative to s_2, that is IF P lies on s_2. Thus:

The common tangents of s_1 and G are the tangents to s_1 at the intersections of s_1 and s_2; similarly for s_2 and G.

It was shown in 5·49 that in the general case the eight tangents to s_1 and s_2 at their four common points touch a conic, and it now appears that this conic is G.

9·24. If s_1 and s_2 touch at A, it follows from 9·21 that the two tangents to G through A coincide in the common tangent of s_1 and s_2 at A, so that G touches s_1 and s_2 at A. If s_1 and s_2 have three-point contact at A, the only common tangents of s_1 and G are the tangents to s at A and the other intersection of s_1 and s_2. This suggests that G has three-point contact with s_1 and s_2 at A, as is easily verified from (·163). Similarly, if s_1 and s_2 have four-point contact at A, G has four-point contact with s_1 and s_2 at A.

9·25. It was shown in 9·17 that there are only three cases in which G can be degenerate. Consider first the general case in which G can be a proper point-pair but not a repeated point. Let the common points of s_1 and s_2 be A, B, C, D and let the tangents to s_1 and s_2 at these points be $\alpha_1, \beta_1, \gamma_1, \delta_1$ and $\alpha_2, \beta_2, \gamma_2, \delta_2$.

Suppose that G is a point-pair $\| H, K \|$. The eight lines $\alpha_1, \beta_1, \ldots, \delta_2$ being tangents of G, must all contain either H or K. Since no three tangents of s_1 can be concurrent, H and K must be two opposite vertices of the quadrilateral $\alpha_1\beta_1\gamma_1\delta_1$; suppose that H is $\alpha_1\beta_1$ and K is $\gamma_1\delta_1$. Then α_2 and β_2, being distinct from α_1 and β_1, cannot contain H and therefore contain K; similarly γ_2 and δ_2 contain H. The polar line relative to s_1 of the point (α_1, CD) is the line $[A, \gamma_1\delta_1]$, or AK, or α_2. Thus α_1 and α_2 are harmonically separated by the points C, D; similarly β_1 and β_2 are harmonically separated by C, D, while the points A, B harmonically separate γ_1, γ_2 and δ_1, δ_2.

Conversely, if H is $\alpha_1\beta_1$ and K is $\gamma_1\delta_1$, each of the following is a CONDITION for G to be the point-pair $\| H, K \|$:

(1) H lies on γ_2.

(2) Any one line through H other than α_1, β_1 meets s_1 and s_2 in apolar pairs of points.

(3) The points C, D harmonically separate α_1 and α_2.

For, if (1) or (2) is satisfied, G contains three distinct lines through H and is therefore a point-pair $\| H, K' \|$; but γ_1 and δ_1 cannot contain H, so that K' must be K. Also, if (3) is satisfied, the polar line relative to s_1 of the point (α_1, CD) is α_2; therefore α_2 contains the pole of CD relative to s_1, which is K, and the former argument applies.

This particular case admits of a familiar illustration in metrical geometry. If we regard C and D as the circular points at infinity, s_1 and s_2 are two circles through A, B. Let O_1 be the centre of s_1, that is the point $\gamma_1\delta_1$, and O_2 the centre of s_2. Then it follows that each of the following is a CONDITION for the G-conic of s_1 and s_2 to be the point-pair $\| O_1, O_2 \|$:

(*a*) α_2 contains O_1.

(*b*) s_2 contains a pair of distinct inverse points relative to s_1.

(*c*) α_1, α_2 are perpendicular.

If we regard (*c*) as the CONDITION for the circles to be orthogonal, then (*a*) and (*b*) are equivalent conditions, and the G-conic of two orthogonal circles is the point-pair formed by their centres. It should be noted that orthogonality is not a necessary condition for G to be a point-pair, for G may be the point-pair $\| \alpha_1\gamma_1, \beta_1\delta_1 \|$ or the point-pair $\| \alpha_1\delta_1, \beta_1\gamma_1 \|$.

9·26. In the case of simple contact, with the notation of (·162), G is a point-pair IF $b_1+b_2=0$.* G is then the point-pair

$$(b_1-b_2)X^2+2Y^2=0,$$

a pair of points on the tangent at the point of contact. These two points must be $\alpha_1\beta_2$ and $\alpha_2\beta_1$, where α_1, β_1 and α_2, β_2 are the tangents to s_1 and s_2 at their two other intersections.

In the metrical example of two circles which touch, we may take their equations, in homogeneous Cartesian co-ordinates, as

$$x^2+y^2+2g_1zx=0, \quad x^2+y^2+2g_2zx=0,$$

and the G-conic is

$$g_1g_2Y^2-Z^2+(g_1+g_2)ZX=0,$$

which is a point-pair IF $g_1+g_2=0$, that is IF the circles are equal.

9·27. The only remaining case in which G can be degenerate is that of proper double contact, which occurs, in the notation of (·165), IF $g_1+g_2=0$, where G is the repeated point $Y^2=0$, the pole of the chord of contact. Every line through this point Y meets s_1 and s_2 in apolar pairs of points. Conversely, if any line through Y other than the two common tangents meets s_1, s_2 in apolar pairs of points, then G is the repeated point Y.

9·28. The properties of the f-conic of two proper conics s_1 and s_2 are the dual of the properties of G. Thus the points of f on any tangent p of s_1 are the intersections of p with the tangents to s_1 through the pole of p relative to s_2. The common tangents of s_1 and f are the tangents of s_1 whose poles relative to s_2 lie on s_1, that is the common tangents of s_1 and the reciprocal of s_1 relative to s_2. The common points of s_1 and f are the contacts with s_1 of the common tangents of s_1 and s_2. In fact in the general case f contains the eight contacts of the four common tangents of s_1 and s_2. If s_1 and s_2 have two-, three- or four-point contact at A, then f has two-, three- or four-point contact with s_1 and s_2 at A. The cases in which f is degenerate are those in which G is degenerate, which have been discussed. The f-conic of a proper conic and a line-pair was discussed in 5·6.

* The equivalent geometrical condition, in Fig. 23 on p. 217, is that C and D should coincide.

9·3. The v-conics and the W-conics

9·31. *If s_1 is a proper conic locus, and if S_2 is a conic scroll, proper or degenerate, there is* ONE *conic locus v_1 such that S_2 is the harmonic scroll of s_1 and v_1.*

Consider first the case in which S_2 is proper, consisting of the tangents of a proper conic locus s_2. In the general case s_1 and s_2 have four common tangents a, b, c, d. Let A_1, B_1, C_1, D_1 be the contacts of these tangents with s_1, and let a_2, b_2, c_2, d_2 be the other tangents of s_2 through A_1, B_1, C_1, D_1. It follows from 9·23 that the required conic v_1, if it exists, must contain A_1, B_1, C_1, D_1 and touch a_2, b_2, c_2, d_2. But there is ONE conic v_1 which contains A_1, B_1, C_1, D_1 and touches a_2 at A_1, and the harmonic scroll of s_1, and this conic v_1 contains the five lines a, b, c, d, a_2 and therefore coincides with S_2; thus the theorem is established, and the remarkable result follows that the conic v_1 so determined also touches b_1, c_1, d_1 at B_1, C_1, D_1. The reader can easily verify that the theorem remains true if s_1 and s_2 have two-, three- or four-point contact at a point A, and that v_1 has two-, three- or four-point contact with s_1 and s_2 in these cases.

9·32. Now consider the case where S_2 is a proper point-pair $\|H, K\|$. Let a, b and c, d be the tangents to s_1 through H and K, and let A_1, B_1 and C_1, D_1 be their points of contact with s_1. There is ONE conic v_1 which contains A_1, B_1, C_1, D_1 and touches C_1H at C_1, and by 9·25 the harmonic scroll of s_1 and this conic v_1 is $\|H, K\|$; further, the conic v_1 so determined also touches A_1K, B_1K, D_1H at A_1, B_1, D_1. If H lies on s_1, v_1 is the line-pair $\|HC_1, HD_1\|$. If H and K both lie on s_1, v_1 is the repeated line $\|HK, HK\|$.

If S_2 is a repeated point $\|H, H\|$, and if A_1, B_1 are the contacts of the tangents to s_1 through H, the conic v_1, if it exists, must touch s_1 at A_1 and B_1. But, if any line p through H meets s_1 in P, Q, the conics which touch s_1 at A_1 and B_1 cut out on p an involution of which $|P, Q|$ is a pair. Thus ONE such conic meets p in a pair of points harmonically separating P, Q and by 9·27 this is the required conic v_1. If H lies on s_1, and if t_1 is the tangent to s_1 at H, v_1 is the repeated line $\|t_1, t_1\|$. Thus the theorem is completely established.

It is necessary to stipulate that s_1 is proper, for if s_1 were a line-pair $\|l, m\|$, the harmonic scroll of $\|l, m\|$ and any other conic locus contains l and m, so that no conic v_1 can exist unless l and m belong to S_2. If l and m do belong to S_2 there are ∞^2 conics v_1. For let O be the point lm, and let L, M be the contacts of l, m with S_2. Choose any pair of points $| P, Q |$ on l which are apolar to $| O, L |$, and any pair of points $| R, S |$ on m which are apolar to $| O, M |$. Let p be the other tangent to S_2 through P. Then there is ONE conic v_1 through P, Q, R, S which touches p at P, and the polar of O relative to v_1 is LM. By 5·66 the harmonic scroll of v_1 and $\| l, m \|$ touches l, m at L, M and also touches p, and thus coincides with S_2. This conic v_1 also touches the other tangents to S_2 through Q, R, S.

9·33. Dually:

If S_1 is a proper conic scroll and s_2 a conic locus, proper or degenerate, there is ONE *conic scroll W_1 such that s_2 is the harmonic locus of S_1 and W_1.*

In the general case, when s_2 is proper and has four distinct common points A, B, C, D with S_1, if a_1, b_1, c_1, d_1 are the tangents to S_1 at these points and if A_2, B_2, C_2, D_2 are the further intersections of a_1, b_1, c_1, d_1 with s_2, then W_1 touches a_1, b_1, c_1, d_1 at A_2, B_2, C_2, D_2.

9·34. Thus:

Two proper conics s_1 and s_2 uniquely determine two conic loci v_1, v_2 and two conic scrolls W_1, W_2 such that (i) *the G-conic of s_1 and v_1 is s_2*; (ii) *the G-conic of s_2 and v_2 is s_1*; (iii) *the f-conic of s_1 and W_1 is s_2*; (iv) *the f-conic of s_2 and W_2 is s_1.*

9·35. Let γ be one of the conics relative to which s_1 and s_2 are reciprocals. Then, the reciprocation always being relative to γ, the reciprocal of the G-conic of s_1 and v_1 is the f-conic of s_2 and the reciprocal of v_1. Thus the reciprocal of v_1 is W_2 and similarly the reciprocal of v_2 is W_1. It follows that W_1 is a point-pair IF v_2 is a line-pair and that W_2 is a point-pair IF v_1 is a line-pair. If W_1 is a point-pair $\| H, K \|$, then s_2 is the locus of intersections of lines through H and K which are conjugate relative to s_1, so by 5·67 there are ∞^1 ordered quadrangles circumscribed to s_1 and inscribed in s_2. Conversely if there are ∞^1 such quadrangles it follows from 5·68 that s_2 is the f-conic of s_1 and a point-pair. Thus:

If s_1 and s_2 are proper, there are ∞^1 ordered quadrangles circumscribed to s_1 and inscribed in s_2 IF *W_1 is a point-pair or* IF *v_2 is a line-pair. Similarly there are ∞^1 ordered quadrangles inscribed in s_1 and circumscribed to s_2* IF *v_1 is a line-pair or* IF *W_2 is a point-pair.*

9·36. The conic scroll W_1 cannot be a repeated point unless s_2 is a pair of tangents to s_1, for by 5·63 the f-conic of a point-pair $\|A, A\|$ and s_1 is the pair of tangents to s_1 through A.

9·4. Triangles and two conics. Porisms

9·41. Consider two proper conics s_1, s_2 and the conics f, G, v_1, v_2, W_1, W_2 determined by them.

Any triangle ABC which is inscribed in s_1 and self-polar relative to s_2 is circumscribed to G.*

For, since B and C are conjugate relative to s_2, BC meets s_1 and s_2 in apolar pairs of points, and is therefore a tangent of G. Similarly CA and AB touch G. Dually:

Any triangle ABC which is circumscribed to s_1 and self-polar relative to s_2 is inscribed in f.

9·42. *Any triangle ABC which is inscribed in s_1 and circumscribed to s_2 is self-polar relative to v_1 and relative to W_2.*

For s_2 is the G-conic of s_1 and v_1, and therefore BC, being a tangent of s_2, meets s_1 and v_1 in apolar pairs of points. Thus B and C are conjugate relative to v_1; similarly for C and A and for A and B. Also s_1 is the f-conic of s_2 and W_2; A lies on s_1 and therefore the pairs of tangents to s_2 and W_2 through A are harmonically separated. Thus AB is conjugate to AC relative to W_2, and similarly BC is conjugate to BA and CA to CB relative to W_2.

9·43. *If there is one triangle ABC inscribed in s_1 and circumscribed to s_2, then there are ∞^1 such triangles. A general point of s_1 is the vertex of one such triangle, and s_1 is said to be triangularly circumscribed to s_2.*

For, with ABC as triangle of reference, the equations may be taken as

$$s_1 \equiv fyz + gzx + hxy = 0, \quad S_2 \equiv YZ + ZX + XY = 0.$$

The conic v_1 has ABC as self-polar triangle, by 9·42, and its equation is

$$v_1 \equiv ghx^2 + hfy^2 + fgz^2 = 0,$$

* In this chapter we are only thinking of proper triangles with distinct vortices and sides. Degenerate triangles associated with these porisms will be considered in Chapter X.

since substitution in 9·122 shows that the harmonic scroll of this conic and s_1 is S_2. The reciprocal of s_1 relative to v_1 is easily seen to be S_2. If P is any point of s_1, and if the polar of P relative to v_1 meets s_1 in Q and R, it follows that QR is a tangent to S_2 and therefore meets s_1 and v_1 in apolar pairs of points. In fact the triangle PQR is inscribed in s_1 and self-polar relative to v_1, and therefore circumscribed to s_2, by 9·41. Also it is easily verified that the harmonic locus of s_2 and v_1 is s_1, so that v_1 and W_2 coincide. Thus:

9·44. *If s_1 is triangularly circumscribed to s_2, W_2 coincides with v_1, s_1 reciprocates into s_2 relative to v_1, and the ∞^1 triangles inscribed in s_1 and circumscribed to s_2 are all self-polar relative to v_1.*

In general there are three other conics relative to which s_1 and s_2 are reciprocal. If PQR is one of the ∞^1 triangles above, its reciprocal relative to such a conic γ is another triangle $P'Q'R'$ of the same ∞^1 system. The triangles PQR, $P'Q'R'$ are each self-polar relative to v_1. It follows that the reciprocal, γ' say, of γ relative to v_1 is such that the two triangles are reciprocals relative to γ'. Since this is true for all pairs of triangles so derived, γ and γ' must coincide. Thus the other conics (in general three), relative to which s_1 and s_2 are reciprocals, are each self-reciprocal relative to v_1.

The dual of the last two theorems is:

If there is one triangle ABC circumscribed to s_1 and inscribed in s_2, then there are ∞^1 such triangles, and s_1 is said to be triangularly inscribed in s_2. W_1 coincides with v_2, s_1 and s_2 are reciprocals relative to W_1, and the ∞^1 triangles are all self-polar relative to W_1.

9·45. *If there is one triangle ABC inscribed in s_1 and self-polar relative to s_2, then there are ∞^1 such triangles. Each point of s_1 is the vertex of one such triangle and s_1 is said to be outpolar to s_2, or harmonically circumscribed to s_2.*

For ABC is circumscribed to G, by 9·41. It follows from 9·43 that there are ∞^1 triangles inscribed in s_1 and circumscribed to G, and from 9·42 that these triangles are all self-polar relative to s_2. Dually:

9·46. *If there is one triangle ABC circumscribed to s_1 and self-polar relative to s_2, then there are ∞^1 such triangles. s_1 is said to be inpolar to s_2, or harmonically inscribed in s_2.*

9·47. It is possible to choose a conic γ relative to which s_1 and s_2 are reciprocals, and a triangle inscribed in s_1 and self-polar relative to s_2 reciprocates into a triangle circumscribed to s_2 and self-polar relative to s_1. Thus:

If s_1 is outpolar to s_2, then s_2 is inpolar to s_1.

9·48. It also follows that, if s_1 is outpolar to s_2 and s_2 is inpolar to s_1, then the ∞^1 triangles inscribed in s_1 and self-polar relative to s_2 are circumscribed to G, while the ∞^1 triangles circumscribed to s_2 and self-polar relative to s_1 are inscribed in f. The reciprocal of s_1 relative to s_2 is G, and the reciprocal of s_2 relative to s_1 is f. The conic s_1 is the harmonic locus of s_2 and G. The conic s_2 is the harmonic envelope of s_1 and f.

9·49. The theorems of 9·43 and 9·44 are remarkable ones. If s_1 and s_2 are any two proper conics, there are ∞^3 triangles inscribed in s_1 and it is three conditions for a triangle to be circumscribed to s_2. Thus it might be expected that a finite number of triangles can be found which are inscribed in s_1 and circumscribed to s_2. This is not, however, the case, for we have shown that in general there is no such triangle and that if there is one then there are ∞^1. A property of this type is called a *porism*. The theorem of 9·45 is clearly another porism, since three conditions are involved in making a triangle self-polar relative to a conic.

These theorems will be proved by alternative methods in the next section, in 9·8, and in Chapter XI.

9·5. Triangles and two conics. Alternative proofs

9·51. *If the sides of two triangles touch a proper conic, then the vertices of the triangles lie on a conic.*

For, if ABC and $A'B'C'$ are two triangles circumscribed to a proper conic s_1, then by 4·32 the tangents BC, $B'C'$ are met by the tangents AB, AC, $A'B'$, $A'C'$ in subranges with equal cross-ratio. Thus, in Fig. 28, $(BCLM) = (L'M'B'C')$,

or $$A'(BCB'C') = A(BCB'C'),$$

which shows that the six points A, A', B, B', C, C' lie on a conic. Dually:

If the vertices of two triangles lie on a proper conic, then the sides of the triangles touch a conic.

Theorem 9·43 follows immediately. For suppose a triangle ABC is inscribed in s_1 and circumscribed to s_2. Let P be any other point of s_1 and let the tangents to s_2 through P meet s again in Q and R. Then there is a conic touching the sides of the triangles ABC, PQR, and this conic must be s_2, since it has five tangents in common with s_2. Thus QR touches s_2, and each point of s_1 is a vertex of a triangle inscribed in s_1 and circumscribed to s_2.

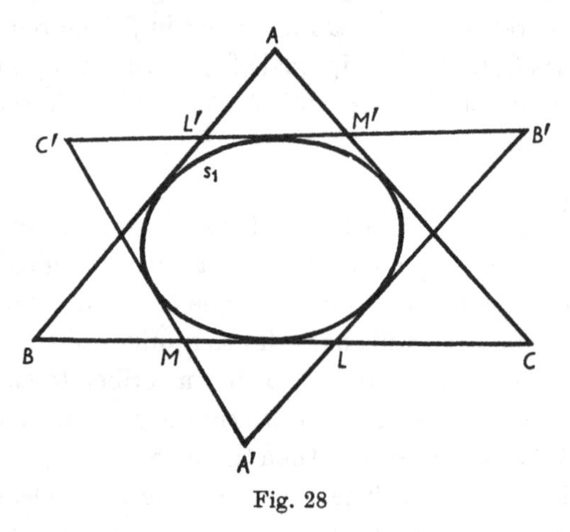

Fig. 28

9·52. *If two triangles are self-polar relative to a proper conic s_1, their vertices lie on a conic.*

For let ABC, $A'B'C'$ be self-polar relative to s_1. Then the lines AB, AC, AB', AC' are conjugate, relative to s_1, to the lines $A'C$, $A'B$, $A'C'$, $A'B'$. Hence

$$A(BCB'C') = A'(CBC'B') = A'(BCB'C').$$

From 4·28 it follows that A, B, C, A', B', C' lie on a conic. Dually:

If two triangles are self-polar relative to a proper conic s_1, their sides touch a conic.

Theorem 9·45 follows immediately. For suppose that ABC is inscribed in s_1 and self-polar relative to s_2. Let P be any point of s_1, let the polar of P relative to s_2 meet s_1 in Q, R, and let the polar of Q relative to s_2 meet QR in R'. Then PQR' is self-polar relative to s_2, and therefore there is a conic through A, B, C, P, Q, R'; since this conic has five points in common with s_1 it must coincide with it. Thus R' is R, and the theorem is proved.

9·53. The theorems that have been proved about outpolar and inpolar conics may be summed up as follows:

If s_1 and s_2 are proper conics, and if it is known that there is one triangle inscribed in s_1 and self-polar relative to s_2, or that there is one triangle circumscribed to s_2 and self-polar relative to s_1, then s_1 is outpolar to s_2, s_2 is inpolar to s_1, and the following corollaries are true:

(1) If P is any point of s_1, the polar of P relative to s_2 meets s_1 in two points Q and R which are conjugate relative to s_2.

(2) If Q and R are any two points of s_1 which are conjugate relative to s_2, the pole of QR relative to s_2 lies on s_1.

(3) If p is any tangent of s_2, and if P is the pole of p relative to s_1, the tangents to s_2 through P are conjugate relative to s_1.

(4) If q and r are any two tangents of s_2 which are conjugate relative to s_1, the polar of qr relative to s_1 is a tangent of s_2.

9·6. Invariants of two conics. Apolarity

9·61. In this section we shall establish the existence of certain functions of the coefficients in the equations of two conic loci, or of two conic scrolls, which are independent of the choice of co-ordinate system.* We consider first the case of two conic loci

$$s_1 \equiv a_1x^2+b_1y^2+c_1z^2+2f_1yz+2g_1zx+2h_1xy = 0,$$
$$s_2 \equiv a_2x^2+b_2y^2+c_2z^2+2f_2yz+2g_2zx+2h_2xy = 0;$$

$A_1, B_1, \ldots, H_1, A_2, B_2, \ldots, H_2$ as usual denote the co-factors of the corresponding small letters in the determinants

$$\delta_1 \equiv \begin{vmatrix} a_1 & h_1 & g_1 \\ h_1 & b_1 & f_1 \\ g_1 & f_1 & c_1 \end{vmatrix} \quad \text{and} \quad \delta_2 \equiv \begin{vmatrix} a_2 & h_2 & g_2 \\ h_2 & b_2 & f_2 \\ g_2 & f_2 & c_2 \end{vmatrix}.$$

The conic $\lambda s_1+\mu s_2 = 0$ is a line-pair IF

$$\begin{vmatrix} a_1\lambda+a_2\mu & h_1\lambda+h_2\mu & g_1\lambda+g_2\mu \\ h_1\lambda+h_2\mu & b_1\lambda+b_2\mu & f_1\lambda+f_2\mu \\ g_1\lambda+g_2\mu & f_1\lambda+f_2\mu & c_1\lambda+c_2\mu \end{vmatrix} = 0.$$

* A fuller treatment will be given later in Chapter XII.

On expanding the determinant this CONDITION becomes

$$\delta_1\lambda^3+\theta_1\lambda^2\mu+\theta_2\lambda\mu^2+\delta_2\mu^3=0, \qquad (\cdot 611)$$

where
$$\left.\begin{aligned}\theta_1 &\equiv A_1a_2+B_1b_2+C_1c_2+2F_1f_2+2G_1g_2+2H_1h_2,\\ \theta_2 &\equiv a_1A_2+b_1B_2+c_1C_2+2f_1F_2+2g_1G_2+2h_1H_2.\end{aligned}\right\} \qquad (\cdot 612)$$

Suppose that with some other system of co-ordinates x', y', z' the conics s_1, s_2 are given by the equations

$$s_1'\equiv a_1'x'^2+b_1'y'^2+c_1'z'^2+2f_1'y'z'+2g_1'z'x'+2h_1'x'y'=0,$$

$$s_2'\equiv a_2'x'^2+b_2'y'^2+c_2'z'^2+2f_2'y'z'+2g_2'z'x'+2h_2'x'y'=0.$$

The co-ordinates x, y, z and x', y', z' are connected by a non-singular linear transformation of the form

$$\left.\begin{aligned}x &= u_1x'+v_1y'+w_1z',\\ y &= u_2x'+v_2y'+w_2z',\\ z &= u_3x'+v_3y'+w_3z',\end{aligned}\right\} \qquad (\cdot 613)$$

and, on making this substitution for x, y, z in the forms s_1, s_2, two quadratic forms s_1'', s_2'' in x', y', z' are obtained. Let them be

$$s_1\equiv s_1''\equiv a_1''x'^2+b_1''y'^2+c_1''z'^2+2f_1''y'z'+2g_1''z'x'+2h_1''x'y',$$
$$s_2\equiv s_2''\equiv a_2''x'^2+b_2''y'^2+c_2''z'^2+2f_2''y'z'+2g_2''z'x'+2h_2''x'y'.$$

With x', y', z' as co-ordinates the conic s_1 is given by $s_1'=0$ and by $s_2''=0$, and therefore

$$s_1'\equiv\rho_1 s_2'',$$

and similarly

$$s_2'\equiv\rho_2 s_2'',$$

where ρ_1, ρ_2 are constants and $\rho_1\rho_2\neq 0$. In fact

$$\left.\begin{aligned}\frac{a_1'}{a_1''}=\frac{b_1'}{b_1''}=\dots=\frac{h_1'}{h_1''}=\rho_1,\\ \frac{a_2'}{a_2''}=\frac{b_2'}{b_2''}=\dots=\frac{h_2'}{h_2''}=\rho_2.\end{aligned}\right\} \qquad (\cdot 614)$$

The expressions δ_1, θ_1, θ_2, δ_2 are homogeneous polynomials in the coefficients of s_1 and s_2, of degree 3, 2, 1, 0 in the former and 0, 1, 2, 3 in the latter. Denote the same functions of the coefficients of s_1', s_2' by δ_1', θ_1', θ_2', δ_2' and define δ_1'', θ_1'', θ_2'', δ_2'' similarly. Then by ($\cdot$614)

$$\delta_1'=\rho_1^3\delta_1'',\quad \theta_1'=\rho_1^2\rho_2\theta_1'',\quad \theta_2'=\rho_1\rho_2^2\theta_2'',\quad \delta_2'=\rho_2^3\delta_2''. \qquad (\cdot 615)$$

9·62. Since on making the substitution (·613) for x, y, z the forms s_1, s_2 become s_1'', s_2'', the form $\lambda s_1 + \mu s_2$ will become $\lambda s_1'' + \mu s_2''$. Thus the equations $\lambda s_1 + \mu s_2 = 0$ and $\lambda s_1'' + \mu s_2'' = 0$ give the same conic. But a CONDITION for $\lambda s_1'' + \mu s_2'' = 0$ to be a line-pair is

$$\delta_1'' \lambda^3 + \theta_1'' \lambda^2 \mu + \theta_2'' \lambda \mu^2 + \delta_2'' \mu^3 = 0.$$

This cubic equation in (λ, μ) must therefore have the same roots as (·611), and so

$$\frac{\delta_1''}{\delta_1} = \frac{\theta_1''}{\theta_1} = \frac{\theta_2''}{\theta_2} = \frac{\delta_2''}{\delta_2} = \sigma, \qquad (\cdot 621)$$

where $\sigma \neq 0$. From (·615) it follows that

$$\delta_1' = \sigma \rho_1^3 \delta_1, \quad \theta_1' = \sigma \rho_1^2 \rho_2 \theta_1, \quad \theta_2' = \sigma \rho_1 \rho_2^2 \theta_2, \quad \delta_2' = \sigma \rho_2^3 \delta_2, \qquad (\cdot 622)$$

whence
$$\frac{\theta_1'^2}{\delta_1' \theta_2'} = \frac{\theta_1^2}{\delta_1 \theta_2} \quad \text{and} \quad \frac{\theta_2'^2}{\theta_1' \delta_2'} = \frac{\theta_2^2}{\theta_1 \delta_2}. \qquad (\cdot 623)$$

Thus:

The values of the expressions $\dfrac{\theta_1^2}{\delta_1 \theta_2}$ *and* $\dfrac{\theta_2^2}{\theta_1 \delta_2}$ *depend only on the two conics and not on the choice of co-ordinates. These expressions are accordingly said to be absolute invariants of the two conics.*

9·63. It also follows from (·622) that $\theta_1 = 0$ implies $\theta_1' = 0$, and that $\theta_2 = 0$ implies $\theta_2' = 0$. The equations $\theta_1 = 0$, $\theta_2 = 0$ are therefore called invariant equations of the two conics.

The fact that the equation $\theta_2 = 0$ is independent of the co-ordinate system leads to another proof of 9·45. For if there is a triangle ABC inscribed in s_1 and self-polar relative to s_2, then with ABC as triangle of reference the equations may be written in the form

$$s_1 \equiv 2uyz + 2vzx + 2wxy = 0, \quad s_2 \equiv x^2 + y^2 + z^2 = 0.$$

Here, in the usual notation,

$$(a_1, b_1, c_1, f_1, g_1, h_1) = (0, 0, 0, u, v, w),$$

$$(A_2, B_2, C_2, F_2, G_2, H_2) = (1, 1, 1, 0, 0, 0),$$

and therefore $\theta_2 = 0$.

It follows that θ_2 will vanish when the equations of s_1 and s_2 are written down in another co-ordinate system. Let P be any point of s_1, and let the polar of P relative to s_2 meet s_1 in Q and R. Let R' be the pole of PQ relative to s_2, so that the triangle PQR' is self-polar relative to s_2. Then, with PQR' as triangle of reference, the equations of s_1 and s_2 may be written in the form

$$s_1 \equiv cz^2+2fyz+2gzx+2hxy = 0, \quad s_2 \equiv x^2+y^2+z^2 = 0.$$

The value of θ_2 is now c. Hence $c = 0$ and R' coincides with R, and the theorem 9·45 is proved.

9·64. The argument of 9·63 is more than a proof of 9·45, for it shows that:

If $s_1 = 0$ and $s_2 = 0$ are proper conics, s_1 is outpolar to s_2 IF $\theta_2 = 0$, *and similarly s_2 is outpolar to s_1* IF $\theta_1 = 0$.

9·65. Dually, if

$$\left.\begin{aligned} S_1 &\equiv A_1X^2+B_1Y^2+C_1Z^2+2F_1YZ+2G_1ZX+2H_1XY = 0,\\ S_2 &\equiv A_2X^2+B_2Y^2+C_2Z^2+2F_2YZ+2G_2ZX+2H_2XY = 0\end{aligned}\right\} \qquad (\cdot 651)$$

are two conic scrolls, the conic scroll $\lambda S_1+\mu S_2 = 0$ is a point-pair IF

$$\begin{vmatrix} A_1\lambda+A_2\mu & H_1\lambda+H_2\mu & G_1\lambda+G_2\mu \\ H_1\lambda+H_2\mu & B_1\lambda+B_2\mu & F_1\lambda+F_2\mu \\ G_1\lambda+G_2\mu & F_1\lambda+F_2\mu & C_1\lambda+C_2\mu \end{vmatrix} = 0,$$

that is IF $$\Delta_1\lambda^3+\Theta_1\lambda^2\mu+\Theta_2\lambda\mu^2+\Delta_2\mu^3 = 0, \qquad (\cdot 652)$$

where $$\Delta_1 \equiv \begin{vmatrix} A_1 & H_1 & G_1 \\ H_1 & B_1 & F_1 \\ G_1 & F_1 & C_1 \end{vmatrix}, \qquad \Delta_2 \equiv \begin{vmatrix} A_2 & H_2 & G_2 \\ H_2 & B_2 & F_2 \\ G_2 & F_2 & C_2 \end{vmatrix},$$

$$\left.\begin{aligned} \Theta_1 &\equiv \bar{a}_1A_2+\bar{b}_1B_2+\bar{c}_1C_2+2\bar{f}_1F_2+2\bar{g}_1G_2+2\bar{h}_1H_2,\\ \Theta_2 &\equiv A_1\bar{a}_2+B_1\bar{b}_2+C_1\bar{c}_2+2F_1\bar{f}_2+2G_1\bar{g}_2+2H_1\bar{h}_2,\end{aligned}\right\} \qquad (\cdot 653)$$

and $\bar{a}_1, \bar{b}_1, \ldots, \bar{h}_1$, $\bar{a}_2, \bar{b}_2, \ldots, \bar{h}_2$ are the co-factors of $A_1, B_1, \ldots, H_1$, $A_2, B_2, \ldots, H_2$ in Δ_1 and Δ_2.

The dual of 9·63 shows that if $\Theta_1 = 0$ for any particular method of writing down the equations of two conic scrolls then $\Theta_1 = 0$ for all methods; similarly for Θ_2. It follows by the dual of 9·64 that:

If $S_1 = 0$ and $S_2 = 0$ are proper, S_1 is inpolar to S_2 IF $\Theta_2 = 0$, *and S_2 is inpolar to S_1* IF $\Theta_1 = 0$.

9·66. If $S_1 = 0$, $S_2 = 0$ are the line equations of $s_1 = 0$, $s_2 = 0$, and $A_1, B_1, \ldots, H_2$ are the co-factors of $a_1, b_1, \ldots, h_2$ in δ_1, δ_2, then, as in 4·73,

$$\bar{a}_1 = \delta_1 a_1, \quad \bar{b}_1 = \delta_1 b_1, \quad \ldots, \quad \bar{h}_1 = \delta_1 h_1,$$
$$\bar{a}_2 = \delta_2 a_2, \quad \bar{b}_2 = \delta_2 b_2, \quad \ldots, \quad \bar{h}_2 = \delta_2 h_2.$$

Hence $$\Delta_1 = \delta_1^2, \quad \Theta_1 = \delta_1 \theta_2, \quad \Theta_2 = \delta_2 \theta_1, \quad \Delta_2 = \delta_2^2 \qquad (\cdot 661)$$

and (·652) becomes

$$\delta_1^2 \lambda^3 + \delta_1 \theta_2 \lambda^2 \mu + \delta_2 \theta_1 \lambda \mu^2 + \delta_2^2 \mu^3 = 0. \qquad (\cdot 662)$$

From (·661) and 9·65 it follows that:

If $s_1 = 0$, $s_2 = 0$ are the point equations of two proper conics, s_1 is inpolar to s_2 IF $\theta_1 = 0$, *and s_2 is inpolar to s_1* IF $\theta_2 = 0$.

This proves 9·47, for $\theta_2 = 0$ is both a CONDITION for s_1 to be outpolar to s_2 and a CONDITION for s_2 to be inpolar to s_1.

9·67. Let S_1, S_2 be the conic scrolls generated by the tangents of two proper conic loci s_1, s_2. When s_1 is outpolar to s_2, and consequently s_2 is inpolar to s_1, there are ∞^1 triangles self-polar relative to s_2 or S_2 whose vertices belong to s_1 and ∞^1 triangles self-polar relative to s_1 whose sides belong to S_2. These two properties are concerned with the points of s_1 and the lines of S_2, so without ambiguity the two equivalent statements 's_1 is outpolar to s_2' and 's_2 is inpolar to s_1' may be replaced by the statement 'the conic locus s_1 and the conic scroll S_2 are *apolar*'. With this convention the statement 'the conic locus s_1 is apolar to the conic scroll S_2' means that s_1 is outpolar to s_2 and s_2 is inpolar to s_1. It is ambiguous to say that two proper conics are apolar without specifying which is to be regarded as a locus and which as a scroll.

9·68. The terms outpolar, inpolar and apolar have so far only been defined when they refer to two proper conics. More generally, if

$$s_1 \equiv a_1 x^2 + b_1 y^2 + c_1 z^2 + 2f_1 yz + 2g_1 zx + 2h_1 xy = 0, \qquad (\cdot 681)$$
$$S_2 \equiv A_2 X^2 + B_2 Y^2 + C_2 Z^2 + 2F_2 YZ + 2G_2 ZX + 2H_2 XY = 0, \qquad (\cdot 682)$$

are a conic locus and scroll, either or both of which may be degenerate, we shall say that s_1 and S_2 are apolar, that s_1 is outpolar to S_2, and that S_2 is inpolar to s_1 IF

$$a_1 A_2 + b_1 B_2 + c_1 C_2 + 2f_1 F_2 + 2g_1 G_2 + 2h_1 H_2 = 0. \qquad (\cdot 683)$$

This extended definition of apolarity only defines a relation between s_1 and S_2 if the truth of the equation (·683) is independent of the choice of co-ordinates. This has already been proved when s_1 and S_2 are both proper. If S_2 is proper, (·682) is the line equation of a proper conic locus $s_2 = 0$, and the argument of 9·61 still applies when s_1 is degenerate. Similarly, if s_1 is proper and S_2 degenerate, the argument of 9·65 shows that (·683) is still an invariant equation. If s_1 is a line-pair $\|p_1, p_1'\|$ and S_2 is a point-pair $\|P_2, P_2'\|$ it will be shown in 9·71 that (·683) is a CONDITION for the points P_2, P_2' to be harmonically separated by the lines p_1, p_2, and is therefore independent of the co-ordinate system.

9·69. *Mutually apolar conics*

Let $s_1 = 0$, $s_2 = 0$ be the point equations and $S_1 = 0$, $S_2 = 0$ the associated line equations of two proper conics. If both θ_1 and θ_2 vanish, then s_1 and S_2 are apolar and S_1 and s_2 are apolar, and we can say without ambiguity that the conics are mutually apolar, or that each is both inpolar and outpolar to the other. In this case the equation (·611) reduces to

$$\delta_1 \lambda^3 + \delta_2 \mu^3 = 0,$$

which necessarily has distinct roots in (λ, μ), since neither of δ_1, δ_2 is zero. Thus there are three line-pairs in the pencil $\lambda s_1 + \mu s_2 = 0$, and the two conics have four distinct common points and, dually, four distinct common tangents.

It follows from 9·48 that the G and f conics coincide in a conic s_3 which is the reciprocal of either of s_1, s_2 relative to the other. It is in fact the case that the three conics s_1, s_2, s_3 form a perfectly symmetrical set; the configuration determined by them will be discussed more fully in 11·48.

The two conics are also triangularly inscribed and circumscribed to each other. For let BC be a chord of s_2 which touches s_1 and let the other tangents from B and C to s_1 meet in A. Referred to ABC as triangle of reference we can take the equations as

$$s_1 \equiv x^2 + y^2 + z^2 - 2yz - 2zx - 2xy = 0,$$
$$S_1 \equiv 2YZ + 2ZX + 2XY = 0,$$
$$s_2 \equiv ax^2 + 2fyz + 2gzx + 2hxy = 0,$$
$$S_2 \equiv -f^2X^2 - g^2Y^2 - h^2Z^2 + 2(gh - af)\,YZ + 2hf\,ZX + 2fg\,ZX = 0,$$

where $fgh \neq 0$. The CONDITIONS for mutual apolarity are

$$f+g+h = 0, \quad f^2+g^2+h^2+2(gh-af)+2hf+2fg = 0,$$

from which it follows that $a = 0$ and A lies on s_2. Thus, by 9·43, s_2 is triangularly circumscribed to s_1, and similarly s_1 is triangularly circumscribed to s_2. Another proof of this will be given in 10·28.

9·7. Apolarity in degenerate cases

9·71. Suppose that the conic locus s_1 of 9·68 is a line-pair $\|p_1, p_1'\|$, so that its equation may be taken as

$$s_1 \equiv (X_1x+Y_1y+Z_1z)(X_1'x+Y_1'y+Z_1'z) = 0.$$

The equation (·683) becomes

$$A_2X_1X_1'+B_2Y_1Y_1'+C_2Z_1Z_1'+F_2(Y_1Z_1'+Y_1'Z_1)$$
$$+G_2(Z_1X_1'+Z_1'X_1)+H_2(X_1Y_1'+X_1'Y_1) = 0,$$

which is a CONDITION for p_1 and p_1' to be conjugate relative to S_2. Thus:

A line-pair $\|p_1, p_1'\|$ is apolar to a conic scroll S_2 IF *p_1 and p_1' are conjugate relative to S_2. In particular a repeated line $\|p_1, p_1\|$ is apolar to S_2* IF *p_1 belongs to S_2.*

This holds when S_2 is degenerate, so a line-pair $\|p_1, p_1'\|$ and a point-pair $\|P_2, P_2'\|$ are apolar IF P_2, P_2' harmonically separate p_1, p_1'. Dually:

A point-pair $\|P_2, P_2'\|$ is apolar to a conic locus s_1 IF *P_2 and P_2' are conjugate relative to s_1. In particular a repeated point $\|P_2, P_2\|$ is apolar to s_2* IF *P_2 lies on s_1.*

9·72. The properties of 9·53 still hold with slight modifications when one of the conics is a proper line-pair or a proper point-pair. Suppose for example that s_1 is a proper line-pair $\|a, b\|$ and that S_2 is either a proper conic scroll or a proper point-pair. Let A, B be the poles of a, b relative to S_2 and denote the point ab by C. If there is a triangle inscribed in s_1 and self-polar relative to s_2, either a or b must contain two vertices of the triangle, so the third vertex must be A or B, the lines a and b must be conjugate relative to S_2, and s_1 is therefore apolar to s_2 in accordance with the definition of 9·68. Also, if P is any point of s_1 other than A and B, the polar of P relative to S_2 meets s_1 in two points Q and R which are conjugate relative to S_2, since one of them is A or B. Thus there are ∞^1

triangles inscribed in s_1 and self-polar relative to S_2. There are also ∞^1 triangles circumscribed to S_2 and self-polar relative to s_1, for the tangents to S_2 through C together with any other tangent to S_2 form such a triangle.

When s_1 is a repeated line, or when S_2 is a repeated point there are no triangles inscribed in s_1 or circumscribed to S_2. (For this reason the definition of 9·68 is preferable to that of 9·45, to which it is equivalent in all other cases.) But when a repeated line $\|a, a\|$ is apolar to a conic scroll S_2, a belongs to S_2. If A is the contact of a with S_2, the polar relative to S_2 of a general point of a meets a in A, which is self-conjugate relative to S_2.

9·73. When a conic γ_1 is said to be outpolar to a conic γ_2 it is implied that γ_1 is a conic locus and that γ_2 is regarded as a conic scroll. Thus a point-pair cannot be said to be outpolar to a conic, and a conic cannot be said to be outpolar to a line-pair. In fact if $s_1 = 0$, $s_2 = 0$ are the point equations of two conic loci, $\theta_2 = 0$ is only a CONDITION for s_1 to be outpolar to s_2 if s_2 is proper. If $s_2 = 0$ is a line-pair $\|p_2, q_2\|$, $\theta_2 = 0$ is a CONDITION for s_1 to be outpolar to the repeated point $\|p_2 q_2, p_2 q_2\|$, since the line equation $S_2 = 0$ represents this repeated point. If $s_2 = 0$ is a repeated line, all the coefficients of S_2 vanish and θ_2 is identically zero.

9·8. Linear systems of conic loci and conic scrolls. Apolar linear systems. Apolarity in general

9·81. When a system of co-ordinates x, y, z is chosen, the equation of any conic locus

$$s \equiv ax^2 + by^2 + cz^2 + 2fyz + 2gzx + 2hxy = 0$$

is determined apart from a constant factor when the conic locus is given. There is thus a (1, 1) correspondence between the conic loci and the ∞^5 ratio sets (a, b, c, f, g, h) formed by their coefficients. Particular conic loci will be denoted by

$$s_i \equiv a_i x^2 + b_i y^2 + c_i z^2 + 2f_i yz + 2g_i zx + 2h_i xy = 0.$$

The m conic loci $s_1, s_2, \ldots, s_m$ are said to be dependent or independent according as the corresponding ratio sets are dependent or independent. Thus if $m > 6$ the conic loci are necessarily dependent, but if $m \leqslant 6$ they are in general independent.

If s_1, s_2, ..., s_{r+1} are independent, so that necessarily $r < 6$, the corresponding ratio sets $(a_i, b_i, c_i, f_i, g_i, h_i)$ determine a linear ∞^r

system of ratio sets $(\kappa_i a_i, \kappa_i b_i, \kappa_i c_i, \kappa_i f_i, \kappa_i g_i, \kappa_i h_i)$, where the summation convention is used for $i = 1, 2, \ldots, r+1$. These are the ratio sets corresponding to the conic loci

$$\kappa_1 s_1 + \kappa_2 s_2 + \ldots + \kappa_{r+1} s_{r+1} = 0;$$

the system of conic loci obtained by giving all values to the ratio set $(\kappa_1, \kappa_2, \ldots, \kappa_{r+1})$ is therefore called a linear ∞^r system of conic loci, and will be denoted by σ^r. The same system σ^r is determined by any $r+1$ independent conic loci of σ^r. In particular a σ^1 has been called a pencil of conic loci, or a point pencil of conics. A σ^2 will be called a *net* of conic loci, or a point net of conics. A σ^0 is a unique conic locus. A σ^5 contains all conic loci.

Dually there is a (1, 1) correspondence between the conic scrolls

$$S \equiv AX^2 + BY^2 + CZ^2 + 2FYZ + 2GZX + 2HXY = 0,$$

and the ratio sets (A, B, C, F, G, H), and conic scrolls are said to be dependent or independent according as the corresponding ratio sets are dependent or independent. If $r < 6$, $r+1$ independent conic scrolls $S_1, S_2, \ldots, S_{r+1}$ determine a linear ∞^r system Σ^r of conic scrolls with equations

$$\kappa_1 S_1 + \kappa_2 S_2 + \ldots + \kappa_{r+1} S_{r+1} = 0.$$

A Σ^0 is a unique conic scroll, a Σ^1 a line pencil of conics, and a Σ^2 a line net of conics.

9·82. A condition on the ratio set (a, b, c, f, g, h) is a homogeneous polynomial equation

$$\phi(a, b, c, f, g, h) = 0,$$

and is said to be a condition on the conic locus s. The degree of ϕ is called the degree of the condition. In particular a linear condition is an equation of the form

$$Pa + Qb + Rc + 2Uf + 2Vg + 2Wh = 0,$$

and will be denoted by $\mathscr{C}(P, Q, R, U, V, W)$; the coefficient 2 is merely introduced for convenience later. There is thus a (1, 1) correspondence between the linear conditions on a conic locus s and the ratio sets (P, Q, R, U, V, W), and linear conditions are said to be dependent or independent according as the corresponding ratio sets are dependent or independent. Dually a linear condition on the conic scroll S is an equation of the form

$$pA + qB + rC + 2uF + 2vG + 2wH = 0,$$

and will be denoted by $\mathscr{C}(p, q, r, u, v, w)$.

It follows from 9·68 and 9·71 that any linear condition

$$\mathscr{C}(P, Q, R, U, V, W)$$

on s is a CONDITION for s to be apolar to the conic scroll

$$\Gamma \equiv PX^2 + QY^2 + RZ^2 + 2UYZ + 2VZX + 2WXY = 0.$$

In particular if Γ is a point pair $\| H, K \|$ the condition expresses that H and K are conjugate relative to s, and, if Γ is a repeated point $\| H, H \|$, that H lies on s. Dually any linear condition $\mathscr{C}(p, q, r, u, v, w)$ on S is a CONDITION for S to be apolar to the conic locus

$$\gamma \equiv px^2 + qy^2 + rz^2 + 2uyz + 2vzx + 2wxy = 0.$$

If γ is a line-pair $\| h, k \|$ or a repeated line $\| h, h \|$, the condition expresses that h and k are conjugate relative to S or that h belongs to S.

The theory of linear conditions on conic loci or conic scrolls is simply the theory of homogeneous linear equations in six variables. Thus from 1·3 and 1·4:

If $s_1, s_2, \ldots, s_{r+1}$ $(r < 5)$ are independent conic loci determining a linear system σ^r, then the conic scrolls which are apolar to $s_1, s_2, \ldots, s_{r+1}$ form a linear system Σ^{4-r} of freedom $4-r$, and every conic locus of σ^r is apolar to every conic scroll of Σ^{4-r}; σ^r is said to be apolar to Σ^{4-r}.

Dually:

If $S_1, S_2, \ldots, S_{r+1}$ $(r < 5)$ are independent conic scrolls determining a linear system Σ^r, then the conic loci which are apolar to $S_1, S_2, \ldots, S_{r+1}$ form a linear system σ^{4-r} of freedom $4-r$, and every conic scroll of Σ^r is apolar to every conic locus of σ^{4-r}; Σ^r is said to be apolar to σ^{4-r}.

Also:

If each of m conic loci $s_1, s_2, \ldots, s_m$ is apolar to each of n conic scrolls $S_1, S_2, \ldots, S_n$, then every conic locus of the linear point system determined by $s_1, s_2, \ldots, s_m$ is apolar to every conic scroll of the linear line system determined by $S_1, S_2, \ldots, S_n$.

Five independent conic scrolls $S_1, S_2, \ldots, S_5$ determine ONE conic locus which is apolar to them all. If $S_1, S_2, \ldots, S_5$ are not independent but belong to a $\Sigma^{4-\lambda}$ with $\lambda > 0$, then the conic loci which are apolar to $S_1, S_2, \ldots, S_5$ form a σ^λ. There is no danger of a porism where only linear conditions are concerned. Thus in general, if $p+q+r = 5$, there is ONE conic locus which contains p given points, has q given pairs of points as conjugate points, and is apolar to r given proper conic scrolls, but these will be an infinity of such conic loci if the conditions are dependent. This will happen if $p \geqslant 4$ and four of the given points lie on a line l, for any conic locus through three of these points contains l and therefore contains the fourth point. Similarly if $q \geqslant 3$ and if three of the given pairs of points are pairs of an involution I on a line l, then a conic locus relative to which two of these pairs are conjugate must contain the double points of I, and therefore has the third pair as a pair of conjugate points, and so again the conditions are dependent.

9·83. *If* $P_i \equiv x_i X + y_i Y + z_i Z = 0$ $(i = 1, 2, \ldots, 6)$ *are the line equations of six points, then the* CONDITION *that the six points should lie on a conic locus is that there should exist numbers* $\lambda_1, \lambda_2, \ldots, \lambda_6$, *not all zero, such that* $\lambda_1 P_1^2 + \lambda_2 P_2^2 + \ldots + \lambda_6 P_6^2 \equiv 0$.

For, if the six points lie on a conic locus s, each of the repeated points $P_i^2 = 0$ is apolar to s. The conic scrolls apolar to s form a linear ∞^5 system, any six members of which are necessarily dependent; thus, in particular, $P_1^2 = 0$, $P_2^2 = 0$, ..., $P_6^2 = 0$ are dependent. Conversely, if $P_1^2, P_2^2, \ldots, P_6^2$ satisfy the identity above, suppose that $\lambda_1 \neq 0$. Then there exists a conic locus s through the five points $P_2, P_3, \ldots, P_6$, and s is therefore apolar to each of

$$P_2^2 = 0, P_3^2 = 0, \ldots, P_6^2 = 0$$

and hence to every conic scroll of the system

$$\mu_2 P_2^2 + \mu_3 P_3^2 + \ldots + \mu_6 P_6^2 = 0;$$

since $$-\lambda_1 P_1^2 \equiv \lambda_2 P_2^2 + \lambda_3 P_3^2 + \ldots + \lambda_6 P_6^2,$$

the degenerate scroll $P_1^2 = 0$ belongs to this system, and thus s contains the point P_1 also.

Dually:

If $p_i \equiv X_i x + Y_i y + Z_i z = 0$ $(i = 1, 2, \ldots, 6)$ *are the point equations of six lines, the* CONDITION *that the six lines should belong to a conic scroll is that there should exist numbers* $\lambda_1, \lambda_2, \ldots, \lambda_6$, *not all zero, such that* $\lambda_1 p_1^2 + \lambda_2 p_2^2 + \ldots + \lambda_6 p_6^2 \equiv 0$.

9·84. Let $P_1 = 0$, $P_2 = 0$, $P_3 = 0$ be the line equations of the vertices of a triangle, and $p_1 = 0, p_2 = 0, p_3 = 0$ the point equations of the opposite sides. The conic scroll $P_1^2 = 0$ is apolar to every conic locus through the point P_1 and in particular to the conic loci $p_2p_3 = 0$, $p_3p_1 = 0$ and $p_1p_2 = 0$. In fact the ∞^2 system

$$\lambda_1 P_1^2 + \lambda_2 P_2^2 + \lambda_3 P_3^2 = 0,$$

of conic scrolls for which the triangle is self-polar, is apolar to the ∞^2 system

$$\mu_1 p_2 p_3 + \mu_2 p_3 p_1 + \mu_3 p_1 p_2 = 0$$

of conic loci through P_1, P_2, P_3.

Dually the ∞^2 system of conic loci

$$\lambda_1 p_1^2 + \lambda_2 p_2^2 + \lambda_3 p_3^2 = 0,$$

relative to which the triangle is self-polar, is apolar to the ∞^2 system

$$\mu_1 P_2 P_3 + \mu_2 P_3 P_1 + \mu_3 P_1 P_2 = 0$$

of conic scrolls containing p_1, p_2, p_3.

Suppose now that $P_1, P_2, \ldots, P_6$ are six points of a conic locus, so that, by 9·83, there is an identity

$$\lambda_1 P_1^2 + \lambda_2 P_2^2 + \ldots + \lambda_6 P_6^2 \equiv 0,$$

with not all of the λ_i zero. This can be written in the form

$$-\lambda_1 P_1^2 - \lambda_2 P_2^2 - \lambda_3 P_3^2 \equiv \lambda_4 P_4^2 + \lambda_5 P_5^2 + \lambda_6 P_6^2,$$

where neither side vanishes identically. It follows that there exists a conic scroll Γ, whose equation can be written either as

$$\lambda_1 P_1^2 + \lambda_2 P_2^2 + \lambda_3 P_3^2 = 0$$

or as

$$\lambda_4 P_4^2 + \lambda_5 P_5^2 + \lambda_6 P_6^2 = 0.$$

The two forms of writing down the equation show that each of the triangles $P_1P_2P_3$ and $P_4P_5P_6$ is self-polar to Γ. Thus:

If the vertices of two triangles lie on a conic locus there exists a conic scroll relative to which the triangles are self-polar. Dually, if the sides of two triangles belong to a conic scroll there exists a conic locus relative to which the triangles are self-polar.

Conversely, if two triangles $P_1P_2P_3$ and $P_4P_5P_6$ are each self-polar relative to a conic scroll Σ, the equation of Σ can be written down in either of the forms

$$\lambda_1 P_1^2 + \lambda_2 P_2^2 + \lambda_3 P_3^2 = 0 \quad \text{or} \quad \lambda_4 P_4^2 + \lambda_5 P_5^2 + \lambda_6 P_6^2 = 0,$$

where not all of λ_1, λ_2, λ_3 and not all of λ_4, λ_5, λ_6 are zero. Hence there exists a number ρ ($\neq 0$) such that

$$\lambda_1 P_1^2 + \lambda_2 P_2^2 + \lambda_3 P_3^2 + \rho\lambda_4 P_4^2 + \rho\lambda_5 P_5^2 + \rho\lambda_6 P_6^2 \equiv 0,$$

and therefore, by 9·83, the six vertices lie on a conic locus. We thus have another proof of the theorems of 9·52, including the degenerate cases discussed in 9·72:

If two triangles are self-polar relative to a conic scroll their vertices lie on a conic locus; and, dually, if two triangles are self-polar relative to a conic locus their sides belong to a conic scroll.

9·85. We saw in 9·83 that if six points P_i lie on a conic locus s they satisfy an identity

$$\lambda_1 P_1^2 + \lambda_2 P_2^2 + \lambda_3 P_3^2 + \lambda_4 P_4^2 + \lambda_5 P_5^2 + \lambda_6 P_6^2 \equiv 0,$$

where not every λ_i is zero. We now prove that, if s is a proper conic locus, none of the λ_i is zero. Suppose that λ_1 vanishes. Then, if $P_0 \equiv x_0 X + y_0 Y + z_0 Z = 0$ is a general point, we have the identity

$$0 \,.\, P_0^2 + \lambda_2 P_2^2 + \lambda_3 P_3^2 + \lambda_4 P_4^2 + \lambda_5 P_5^2 + \lambda_6 P_6^2 \equiv 0,$$

which shows that $P_0, P_2, P_3, \ldots, P_6$ lie on a conic locus. There is thus a conic locus through the five points $P_2, P_3, \ldots, P_6$ and any other point P_0, which can only be the case if at least four of the points $P_2, P_3, \ldots, P_6$ are collinear, on a line which is necessarily part of s. Incidentally it appears that none of the λ_i is zero in the case when s is a line-pair, each line containing three of the six points.

If then $P_1, P_2, \ldots, P_6$ lie on a proper conic locus s, no λ_i is zero and therefore the conic scroll Γ of 9·84 is proper. We can thus reciprocate relative to Γ and prove the theorem of 9·51: *If the vertices of two triangles $P_1P_2P_3$ and $P_4P_5P_6$ lie on a proper conic locus s their sides belong to a proper conic scroll*, namely the reciprocal of s relative to Γ. If s is a line-pair $\| h, k \|$, and if h contains P_1, P_5, P_6 and k contains P_2, P_3, P_4 and none of these points is hk, then Γ is still proper; the reciprocal of s relative to Γ is the point-pair $\| P_1, P_4 \|$, and the sides of the two triangles belong to this degenerate conic scroll. If h contains P_1, P_4 and k contains P_2, P_3, P_5, P_6, the two triangles have a common side and the theorem is trivial, for there is certainly a conic scroll containing this and the other four sides of the triangles.

9·86. If two conic loci s_1, s_2 are apolar to a conic scroll S then, by 9·82, every conic locus of the pencil $\lambda s_1 + \mu s_2 = 0$ is apolar to S; in particular, if $\| h, k \|$ is a line-pair of the pencil, the lines h and k are conjugate relative to S. Thus we have:

If two pairs of opposite sides of a quadrangle $ABCD$ are conjugate relative to a conic scroll S, then the third pair are conjugate relative to S, and all the conic loci through A, B, C, D are apolar to S. The quadrangle is then said to be outpolar to S.

Dually:

If two pairs of opposite vertices of a quadrilateral abcd are conjugate relative to a conic locus s, then the third pair are conjugate relative to s, and all the conic scrolls containing a, b, c, d are apolar to s. The quadrangle is then said to be inpolar to s.

If $P_i \equiv x_i X + y_i Y + z_i Z = 0$ $(i = 1, 2, 3, 4)$ are the (distinct) base points of a pencil σ of conic loci, each degenerate conic scroll $P_i^2 = 0$ is apolar to every member of σ, and the ∞^3 system of conic scrolls apolar to σ is

$$\lambda_1 P_1^2 + \lambda_2 P_2^2 + \lambda_3 P_3^2 + \lambda_4 P_4^2 = 0.$$

That this is an ∞^3 system follows from the fact that, if there existed an identity

$$\mu_1 P_1^2 + \mu_2 P_2^2 + \mu_3 P_3^2 + \mu_4 P_4^2 \equiv 0,$$

where not every μ_i is zero, then, by an argument similar to that of 9·85, there would be a conic locus through P_1, P_2, P_3, P_4 and any two other points; this would mean that P_1, P_2, P_3, P_4 were collinear. Thus:

Any conic scroll relative to which the quadrangle $P_1P_2P_3P_4$ is outpolar has an equation of the form $\lambda_1 P_1^2 + \lambda_2 P_2^2 + \lambda_3 P_3^2 + \lambda_4 P_4^2 = 0$.

Dually:

Any conic locus relative to which the quadrilateral $p_1p_2p_3p_4$ is inpolar has an equation of the form $\lambda_1 p_1^2 + \lambda_2 p_2^2 + \lambda_3 p_3^2 + \lambda_4 p_4^2 = 0$.

The converse theorems are left to the reader.

Two triangles which are reciprocals relative to a proper conic are in perspective.

For let $A'B'C'$ be the reciprocal of ABC relative to s, and let BB', CC' meet in O. The pair of lines $|AC, BO|$ are conjugate relative to s, since BO contains the pole B' of AC. Similarly $|AB, CO|$ are conjugate relative to s, and therefore the third pair of opposite sides $|BC, AO|$ of the quadrangle $ABCO$ are conjugate relative to s; thus AO contains A'. Conversely

If two triangles ABC, $A'B'C'$ are in perspective from O, there is a conic relative to which they are reciprocals.

For it is possible to choose a conic s relative to which $B'C'$ is the polar of A, $C'A'$ is the polar of B and C is conjugate to A'. If the polar of C meets $B'C'$ in B'', the triangles ABC, $A'B''C'$ are reciprocals relative to s and are therefore in perspective. Thus BB'' contains O, and B'' coincides with B'. This theorem has also been proved in 5·29.

9·87. *If* $P_i \equiv x_i X + y_i Y + z_i Z = 0$ $(i = 1, 2, 3, 4, 5)$ *are the equations of any five points of a proper conic locus s, the equation of any conic scroll apolar to s can be written in the form*

$$\lambda_1 P_1^2 + \lambda_2 P_2^2 + \lambda_3 P_3^2 + \lambda_4 P_4^2 + \lambda_5 P_5^2 = 0.$$

Dually:

If $p_i \equiv X_i x + Y_i y + Z_i z = 0$ $(i = 1, 2, 3, 4, 5)$ *are the equations of any five lines of a proper conic scroll S, the equation of any conic locus apolar to S can be written in the form*

$$\lambda_1 p_1^2 + \lambda_2 p_2^2 + \lambda_3 p_3^2 + \lambda_4 p_4^2 + \lambda_5 p_5^2 = 0.$$

The proof of these and the converse theorems are left to the reader.

9·88. *Apolarity in general. The Clebsch-Aronhold notation**

The notions of polarity and apolarity may be extended to plane curves and scrolls of any order and class, and we give here a brief account of the general theory.

It is convenient to use point co-ordinates (x_1, x_2, x_3) and line co-ordinates $[X_1, X_2, X_3]$ instead of the usual (x, y, z) and $[X, Y, Z]$. The equation of a general curve of order n can be written in the form

$$\sum_{r+s+t=n} \frac{n!}{r!\,s!\,t!} a_{rst} x_1^r x_2^s x_3^t = 0, \qquad (\cdot 881)$$

and a general scroll of class n is

$$\sum_{r+s+t=n} \frac{n!}{r!\,s!\,t!} A_{rst} X_1^r X_2^s X_3^t = 0. \qquad (\cdot 882)$$

The curve and the scroll are said to be apolar if

$$\sum_{r+s+t=n} \frac{n!}{r!\,s!\,t!} a_{rst} A_{rst} = 0. \qquad (\cdot 883)$$

The reader will verify that this is in agreement with 9·68 in the case $n = 2$.

* 9·88 and 9·89 may be omitted on a first reading.

The above expressions may be greatly simplified by the use of a notation due to Clebsch and Aronhold. Compare the left-hand side of the equation ·881 with the expression $(a_1x_1+a_2x_2+a_3x_3)^n$. The coefficient of $x_1^r x_2^s x_3^t (r+s+t=n)$ in the latter is $\frac{n!}{r!s!t!} a_1^r a_2^s a_3^t$, which differs from the corresponding coefficient in (·881) only in that $a_1^r a_2^s a_3^t$ replaces a_{rst}. If therefore we regard a_1, a_2, a_3 as symbols which can be multiplied together in the ordinary way, and such that the expression $a_1^r a_2^s a_3^t$ is to be replaced by a_{rst} if $r+s+t=n$ and is meaningless if $r+s+t \neq n$, then we can write (·881) symbolically as

$$(a_1x_1+a_2x_2+a_3x_3)^n = 0,$$

and this can further be contracted to

$$(ax)^n = 0,$$

where (ax) stands for $(a_1x_1+a_2x_2+a_3x_3)$.

In the same way we can write (·882) symbolically as

$$(X_1A_1+X_2A_2+X_3A_3)^n = 0,$$

or as

$$(XA)^n = 0.$$

In this notation the condition (·883) that the curve $(ax)^n = 0$ and the scroll $(XA)^n = 0$ should be apolar takes the simple form

$$(aA)^n \equiv (a_1A_1+a_2A_2+a_3A_3)^n = 0. \qquad (·884)$$

If $\gamma \equiv (ax)^m = 0$ is a curve of order m and $\Gamma \equiv (XA)^n = 0$ a scroll of class $n(>m)$, the scroll of class $n-m$ whose equation is

$$(aA)^m (XA)^{n-m} = 0, \qquad (·885)$$

is called the *polar scroll* of γ relative to Γ. If $(aA)^m (XA)^{n-m} \equiv 0$ identically, then γ is said to be apolar to Γ. In the case $m=1$, $n=2$ the polar scroll is a point, in fact the pole of the line γ relative to the conic scroll Γ; if Γ is a point-pair $\| H, K \|$ (or a repeated point $\| H, H \|$), and γ the axis HK of the point-pair (or a line through H), then Γ and γ are apolar in the above extended sense.

Dually, if in the general case $n<m$, Γ has a *polar curve*

$$(aA)^n (ax)^{m-n} = 0, \qquad (·886)$$

of order $m-n$, relative to γ, and, if this equation is identically true, Γ and γ are apolar.

Consider the composite curve of order n made up of the curve $(ax)^m = 0$ and the straight line $X_1^{(1)}x_1 + X_2^{(1)}x_2 + X_3^{(1)}x_3 = 0$, or $(X^{(1)}x) = 0$, counted $n-m$ times. Its symbolic equation is

$$(ax)^m (X^{(1)}x)^{n-m} = 0,$$

and it is apolar to the scroll $(XA)^n = 0$ if

$$(aA)^m (X^{(1)}A)^{n-m} = 0,$$

(by ·884).

By comparison with (·885) we see that the polar scroll of the curve $\gamma \equiv (ax)^m = 0$ relative to the scroll $\Gamma \equiv (XA)^n = 0$ consists of just those lines p such that the curve of order n consisting of γ and the line p counted $n-m$ times is apolar to Γ.

Dually the polar curve of a scroll Γ of class n relative to a curve γ of order $m(>n)$ consists of those points which, counted $m-n$ times, make up with Γ a scroll of class m apolar to γ.

A point $P^{(1)}$ with co-ordinates $(x_1^{(1)}, x_2^{(1)}, x_3^{(1)})$ and line equation $X_1 x_1^{(1)} + X_2 x_2^{(1)} + X_3 x_3^{(1)} = 0$, or $(Xx^{(1)}) = 0$, has a polar curve of order $n-1$ relative to a curve $\gamma \equiv (ax)^n = 0$. Its equation is, by (·886),

$$(ax^{(1)})(ax)^{n-1} = 0.$$

This is called the *first polar curve* of the point $P^{(1)}$ relative to γ. The *second polar curve* of $P^{(1)}$ is the polar curve relative to γ of the scroll consisting of the point $P^{(1)}$ taken twice. It is of order $n-2$ and its equation is

$$(ax^{(1)})^2 (ax)^{n-2} = 0.$$

Generally the *rth polar curve* of $P^{(1)}$ is the polar curve of $P^{(1)}$ counted r times. It is of order $n-r$ and its equation is

$$(ax^{(1)})^r (ax)^{n-r} = 0.$$

If the equation of γ is expressed in the usual way by

$$f(x_1, x_2, x_3) = 0,$$

where f is a homogeneous polynomial of order n, then it may be verified that the expression just given for the rth polar curve of $P^{(1)}$ relative to γ is precisely the same as

$$\left(x_1^{(1)} \frac{\partial}{\partial x_1} + x_2^{(1)} \frac{\partial}{\partial x_2} + x_3^{(1)} \frac{\partial}{\partial x_3}\right)^r f(x_1, x_2, x_3) = 0.$$

This is in fact easily proved, in the case where γ actually consists of a straight line $a_1 x_1 + a_2 x_2 + a_3 x_3 = 0$ counted n times, by repeated application of the *operator*

$$\left(x_1^{(1)} \frac{\partial}{\partial x_1} + x_2^{(1)} \frac{\partial}{\partial x_2} + x_3^{(1)} \frac{\partial}{\partial x_3}\right)$$

to the expression $(a_1x_1+a_2x_2+a_3x_3)^n$; and we may proceed as though this were the case if it is understood at every stage that $a_1^r a_2^s a_3^t$ is merely a symbolic expression for a_{rst}.

Dually, a line $p^{(1)}$ with point equation $(X^{(1)}x)=0$ has an *rth polar scroll*, of class $n-r$,

$$(X^{(1)}A)^r(XA)^{n-r}=0$$

relative to a scroll $\Gamma \equiv (XA)^n = 0$ of class n.

It is important to realize that the property of apolarity, the polar curve of a curve relative to a scroll, and so on, depend only on the original curve and scroll for which they are defined and not on the co-ordinate system employed. Suppose in fact that we have a new system of point co-ordinates (x_1', x_2', x_3') defined by the non-singular transformation

$$x_i = \rho_{ij}x_j', \tag{$\cdot$887}$$

where $i, j = 1, 2, 3$, and the presence of a repeated suffix implies summation. The symbolic expression (ax) or (a_ix_i) becomes $(a_i\rho_{ij}x_j')$.

If we write $$a_j' = a_i\rho_{ij},$$

or, equivalently, $$a_i' = a_j\rho_{ji}, \tag{$\cdot$888}$$

then (a_ix_i) has become $(a_i'x_i')$. In fact, comparing (·887) and (·888) with 1·651 and 1·653, we see that the symbols (a_1, a_2, a_3) are transformed into the symbols (a_1', a_2', a_3') precisely as though they were line co-ordinates. In a similar way the symbols (A_1, A_2, A_3) are transformed into symbols (A_1', A_2', A_3') as though they were point co-ordinates. We saw in 1·65 that, if (x, y, z), $[X, Y, Z]$ are the original co-ordinates of a point and a line and (x', y', z'), $[X', Y', Z']$ their new co-ordinates, then

$$xX+yY+zZ \equiv x'X'+y'Y'+z'Z'.$$

In the same way here the symbolic expressions (ax), (XA), (aA) are transformed into $(a'x')$, $(X'A')$, $(a'A')$. If the equation of the curve (·881), referred to the new co-ordinates, is

$$\sum_{r+s+t=n} \frac{n!}{r!\,s!\,t!} a_{rst}' x_1'^r x_2'^s x_3'^t = 0,$$

we can find the values of the coefficients in this equation in terms of those of (·881) in the following way. We write a_{rst}' symbolically as $a_1'^r a_2'^s a_3'^t$, then replace a_1', a_2', a_3' by their expressions (·888) in terms of a_1, a_2, a_3, and then, in the final expression, replace any term $a_1^k a_2^l a_3^m$, $(k+l+m=n)$, by a_{klm}. Since all the processes we have discussed involve simple operations on symbols of the form (ax), (XA), (aA) it follows that the results obtained do not depend on the choice of co-ordinate system.

9·89. The theory of 9·88 can be readily extended to space of more than two dimensions. It is also applicable to the theory of sets of points on a line. In the geometry of points of a line, however, the dual of a point is a point. It is thus convenient to use simultaneously two different systems of point co-ordinates (x_1, x_2) and $[X_1, X_2]$, where, if these two ratio sets represent the same point,

$$x_1 X_1 + x_2 X_2 = 0, \quad \text{or} \quad x_1 : x_2 = X_2 : -X_1.$$

Thus, if we have a set of n points given, in the first co-ordinate system, by

$$a_{n,0} x_1^n + n a_{n-1,1} x_1^{n-1} x_2 + \frac{n(n-1)}{2!} a_{n-2,2} x_1^{n-2} x_2^2 + \ldots + a_{0,n} x_2^n = 0, \tag{·891}$$

their equation, in the second co-ordinate system, is

$$a_{n,0} X_2^n - n a_{n-1,1} X_2^{n-1} X_1 + \ldots + (-1)^n a_{0,n} X_1^n = 0. \tag{·892}$$

The set of points (·891) can be written symbolically in the usual way as $(ax)^n = 0$, and the set is said to be apolar to the set $(XA)^n = 0$ if $(aA)^n = 0$.

Suppose we have a set of n points given by

$$b_{n,0} x_1^n + n b_{n-1,1} x_1^{n-1} x_2 + \frac{n(n-1)}{2!} b_{n-2,2} x_1^{n-2} x_2^2 + \ldots + b_{0,n} x_2^n = 0. \tag{·893}$$

To find the condition that the sets (·891) and (·893) should be apolar it is convenient first to express one of the sets in the co-ordinates $[X_1, X_2]$ and then apply the general theory. The condition thus is that the sets (·892) and (·893) should be apolar, namely

$$b_{n,0} a_{0,n} - n b_{n-1,1} a_{1,n-1} + \frac{n(n-1)}{2!} b_{n-2,2} a_{2,n-2} - \ldots + (-1)^n b_{0,n} a_{n,0} = 0. \tag{·894}$$

This is a symmetrical condition, showing that it is immaterial which of the sets (·891) and (·893) is transformed into the $[X_1, X_2]$ system.

If we have a set of n points on the line determined by either of the equations

$$(ax)^n = 0, \quad (XA)^n = 0,$$

and a set of $m(<n)$ points determined by either of the equations

$$(bx)^m = 0, \quad (XB)^m = 0,$$

then the latter set has a *polar set* of $n-m$ points relative to the former, and its equation can be taken as either of

$$(aB)^m(ax)^{n-m} = 0, \quad (bA)^m(XA)^{n-m} = 0.$$

The *rth polar set* of a point P_1 relative to a set S_n of n points consists of $n-r$ points, and it is the polar set relative to S_n of the point P_1 counted r times. If P_1 is $(x_1^{(1)}, x_2^{(1)})$ and S_n is given by $(ax)^n = 0$, the rth polar set is given by

$$(ax^{(1)})^r (ax)^{n-r} = 0.$$

9·9. Examples from metrical geometry

9·91. We conclude this chapter with a few metrical examples of the preceding theory. As usual we denote the circular points at infinity by I and J.

If s is a parabola with focus O, the triangle OIJ is circumscribed to s, and therefore, by 9·51, if ABC is any other triangle circumscribed to s the six points A, B, C, O, I, J lie on a conic. Hence:

*The circumcircle of a triangle circumscribed to a parabola passes through the focus.**

This can be expressed in the form:

The locus of the foci of parabolas inscribed in a triangle is the circumcircle of the triangle.

Since there is ONE parabola touching four lines in general position, we deduce *Wallace's Theorem*:

The circumcircles of the four triangles formed by sets of three out of four given lines in general position are concurrent.

9·92. If s is a rectangular hyperbola with centre C, the triangle CIJ is self-polar relative to s. If PQR is any other triangle self-polar relative to s then, by 9·84, there is a conic touching the sides and a conic through the vertices of the two triangles CIJ, PQR. Hence:

If a conic is inpolar to a rectangular hyperbola s and has a focus at the centre of s, it is a parabola.

Circles outpolar to a rectangular hyperbola pass through its centre.

* Cf. 6·28, p. 191.

The locus of centres of rectangular hyperbolas, relative to which a given triangle is self-polar, is the circumcircle of the triangle.

9·93. Suppose γ_1, γ_2 are two of the ∞^2 circles outpolar to a proper conic s, and let their intersections, apart from the circular points, be A and B. By 9·86 all circles of the coaxal system, or point pencil, determined by γ_1 and γ_2 are outpolar to s. One member of this system is the line-pair $\|AB, IJ\|$; thus AB and IJ are conjugate relative to s and therefore AB contains the centre C of s. Hence C has the same power k relative to γ_1, γ_2, where $CA \,.\, CB = k^2$, and γ_1 and γ_2, and similarly all other circles outpolar to s, are cut orthogonally by the circle Γ whose centre is C and radius k. This circle Γ is in particular the locus of the centres of point-circles outpolar to s, that is the locus of points P such that the lines PI, PJ are conjugate relative to s, that is the director circle of s. *Hence circles outpolar to a conic are orthogonal to the director circle of the conic, and, in particular, circles outpolar to a parabola have their centres on the directrix.* We deduce *Gaskin's Theorem:**

The circumcircle of a triangle self-polar relative to a conic is orthogonal to the director circle of the conic. In particular the circumcentre of a triangle self-polar relative to a parabola lies on the directrix.

The second theorem of 9·92 is also a particular case of Gaskin's Theorem, for the director circle of a rectangular hyperbola with centre C is the line-pair $\|CI, CJ\|$ or, in real geometry, the point-circle C.

9·94. There is in general ONE conic locus γ satisfying the five linear conditions of passing through I and J and having a given triangle ABC as a self-polar triangle. We call γ the circle conjugate to the triangle. If D, E, F are the feet of the perpendiculars from A, B, C to the opposite sides, meeting in the orthocentre H, then γ has centre H and radius ρ, where

$$HA \,.\, HD = HB \,.\, HE = HC \,.\, HF = \rho^2.$$

The circle γ is outpolar to any conic s inscribed in the triangle ABC. Thus, if s is a rectangular hyperbola, we deduce from 9·92:

The locus of the centres of rectangular hyperbolas inscribed in a triangle is the circle conjugate to the triangle.

* Cf. $R(2)$, p. 48.

There are TWO rectangular hyperbolas touching four lines in general position. Hence:

The four circles conjugate to the triangles formed by sets of three out of four given lines in general position have two common points.

If s is a parabola we deduce from 9·93:

The orthocentre of a triangle circumscribed to a parabola lies on the directrix.

Since there is ONE parabola touching four lines in general position, we deduce:

The orthocentres of the four triangles formed by sets of three out of four given lines in general position are collinear.

9·95. Consider the line pencil Σ of conics touching four lines a, b, c, d. There are ∞^1 circles outpolar to two, and therefore to all, conics of Σ; they form a coaxal system σ_1. Let p be the line of centres and q the radical axis of σ_1. The director circles of conics of Σ are, by (9·93), orthogonal to all circles of σ_1, and therefore form the orthogonal system σ_2 of coaxal circles having q as line of centres and p as radical axis. The centres of the director circles are also the centres of the conics of Σ, and so q is the locus of centres of conics of Σ, containing, in particular, the centres of the point-pairs of Σ, that is the mid-points of the diagonals of the quadrilateral $abcd$; q is in fact the conjugate of the line at infinity relative to Σ. The line p is, by 9·93, the directrix of the unique parabola of Σ, and, by 9·94, it contains the orthocentres of the triangles bcd, acd, abd, abc.

CHAPTER X

TWO-TWO CORRESPONDENCES

10·1. (2, 1) and (2, 2) correspondences

10·11. Considerable use has been made of the theory of (1, 1) correspondences developed in Chapters II and III. The general theory of (m, n) correspondences, with m and n greater than one, is beyond the scope of this book, but symmetric (2, 2) correspondences must be discussed here, as they are intimately connected with the properties of two conics. As an introduction, a few remarks will be made about (2, 1) correspondences and general (2, 2) correspondences.

By 2·56 any algebraic (2, 1) correspondence $T(2, 1)$ between two ratio variables (λ, μ) and (λ', μ') is given by an equation of the form

$$(a_1\lambda^2 + 2a_2\lambda\mu + a_3\mu^2)\lambda' + (b_1\lambda^2 + 2b_2\lambda\mu + b_3\mu^2)\mu' = 0. \quad (\cdot 111)$$

Regarding (λ, μ) and (λ', μ') as ratio parameters of the points of two lines p and p', each point P of p corresponds to ONE point P' of p', and a general point P' of p' corresponds to TWO points P_1, P_2 of p. The points B'_1, B'_2 of p' which satisfy

$$(a_2\lambda' + b_2\mu')^2 = (a_1\lambda' + b_1\mu')(a_3\lambda' + b_3\mu') \quad (\cdot 112)$$

are such that the corresponding points of p coincide, and they are called *branch points*. From (·111) it appears that, as P' varies on p', the pairs $|P_1, P_2|$ form an involution I on p; the pairs of I are homographically related to the points of p', the double points of I corresponding to the branch points B'_1, B'_2.

The theorem of 3·63 is equivalent to the statement that the pairs of points of p corresponding to two points P'_1 and P'_2 of p' are apolar IF $(B'_1P'_1B'_2P'_2) = -1$. In fact the representation of the pairs of an involution on any $R\infty^1$ by the values of a ratio parameter (κ_1, κ_2) may be regarded as a (2, 1) correspondence between the $R\infty^1$ and (κ_1, κ_2).

The branch points B'_1, B'_2 given by (·112) coincide IF

$$(2a_2b_2 - a_1b_3 - a_3b_1)^2 = 4(a_2^2 - a_1a_3)(b_2^2 - b_1b_3),$$

and this is a CONDITION for

$$a_1\lambda^2 + 2a_2\lambda\mu + a_3\mu^2 \quad \text{and} \quad b_1\lambda^2 + 2b_2\lambda\mu + b_3\mu^2$$

to have a common factor, since the double points of I coincide IF I has a fixed point.

If in a particular $T(2, 1)$ the line p' contains more than two branch points, then (·112) must be identically satisfied by all (λ', μ'), so every point of p' is a branch point. This happens IF

$$a_1 : b_1 = a_2 : b_2 = a_3 : b_3 \quad \text{and} \quad a_2^2 = a_1 a_3,$$

that is IF (·111) is of the form

$$(a\lambda + b\mu)^2 (p\lambda' + q\mu') = 0.$$

Since a $T(2, 1)$ between (λ, μ) and (λ', μ') depends on the ratios of the six coefficients $a_1, a_2, \ldots, b_3$ in (·111), it follows that a $T(2, 1)$ between two $R\infty^1$ is in general uniquely determined when five pairs of corresponding elements are given.

10·12. Consider a $T(2, 1)$ between two pencils of lines. Taking the vertices of the pencils as the vertices Y, Z of the triangle of reference, the lines of the pencils are

$$\mu x - \lambda z = 0, \quad \mu' x - \lambda' y = 0. \tag{·121}$$

Since (λ, μ), (λ', μ') are ratio parameters, the $T(2, 1)$ is given by an equation of the form (·111), and on substituting $\lambda : \mu = x : z$, $\lambda' : \mu' = x : y$ in this equation it appears that the locus of meets of corresponding lines of the pencils is the cubic curve γ given by

$$(a_1 x^2 + 2a_2 xz + a_3 z^2)\, x + (b_1 x^2 + 2b_2 xz + b_3 z^2)\, y = 0, \tag{·122}$$

which contains Z and has a double point* at Y.

The line YZ, or $x = 0$, belongs to both pencils and corresponds to the value $(0, 1)$ of both (λ, μ) and (λ', μ'). If YZ is regarded as a line of the first pencil, the corresponding line of the second pencil is the tangent to γ at Z. If YZ is regarded as a line of the second pencil, the two corresponding lines of the first pencil are the tangents to γ at Y.

The line YZ is self-corresponding IF (·111) is satisfied when both (λ, μ) and (λ', μ') are $(0, 1)$, that is IF $b_3 = 0$. In this case x is a factor of (·122), and γ reduces to the line YZ and the conic

$$a_1 x^2 + 2a_2 xz + a_3 z^2 + b_1 xy + 2b_2 yz = 0,$$

which contains Y but not Z.

Conversely, if P is a variable point of a proper conic s, if Y is a fixed point of s, and if Z is a fixed point not on s, then the correspondence between the lines YP and ZP is a $T(2, 1)$ with YZ as a self-corresponding line.

* See 2·44, p. 53.

10·13. By 2·56 an algebraic (2, 2) correspondence $T(2,2)$ between (λ,μ) and (λ',μ') is given by an equation of the form

$$\begin{aligned}\phi(\lambda,\mu;\,\lambda',\mu') \equiv (a_1\lambda^2+2a_2\lambda\mu+a_3\mu^2)\,\lambda'^2 \\ +2(b_1\lambda^2+2b_2\lambda\mu+b_3\mu^2)\,\lambda'\mu' \\ +(c_1\lambda^2+2c_2\lambda\mu+c_3\mu^2)\,\mu'^2 = 0, \quad (\cdot 131)\end{aligned}$$

which may also be written in the form

$$\begin{aligned}(a_1\lambda'^2+2b_1\lambda'\mu'+c_1\mu'^2)\,\lambda^2 \\ +2(a_2\lambda'^2+2b_2\lambda'\mu'+c_2\mu'^2)\,\lambda\mu \\ +(a_3\lambda'^2+2b_3\lambda'\mu'+c_3\mu'^2)\,\mu^2 = 0. \quad (\cdot 132)\end{aligned}$$

Regarding (λ,μ), (λ',μ') as ratio parameters of two lines p, p', a general point P of p corresponds to TWO points P_1', P_2' of p'. P is said to be a *branch point* IF P_1' and P_2' coincide, so there are in general four branch points B_1, B_2, B_3, B_4 on p given by

$$(b_1\lambda^2+2b_2\lambda\mu+b_3\mu^2)^2 = (a_1\lambda^2+2a_2\lambda\mu+a_3\mu^2)\,(c_1\lambda^2+2c_2\lambda\mu+c_3\mu^2). \quad (\cdot 133)$$

Similarly there are in general four branch points B_1', B_2', B_3', B_4' on p' given by

$$\begin{aligned}(a_2\lambda'^2+2b_2\lambda'\mu'+c_2\mu'^2) = (a_1\lambda'^2+2b_1\lambda'\mu'+c_1\mu'^2) \\ \times(a_2\lambda'^2+2b_2\lambda'\mu'+c_2\mu'^2). \quad (\cdot 134)\end{aligned}$$

The points B_1, B_2, B_3, B_4 on p correspond to repeated pairs of points $|D_1',D_1'|$, $|D_2',D_2'|$, $|D_3',D_3'|$, $|D_4',D_4'|$ on p', and the points D_1', D_2', D_3', D_4' are called the *double points* on p'. Similarly there are four double points D_1, D_2, D_3, D_4 on p, corresponding to the branch points on p'.

If in a particular $T(2,2)$ there are more than four branch points on p, (·133) is an identity and every point of p is a branch point. This happens IF (·131) is of the form

$$(a\lambda^2+2b\lambda\mu+c\mu^2)\,(p'\lambda'+q'\mu')^2 = 0,$$

or of the form

$$\{(a_1\lambda+a_2\mu)\,\lambda'+(b_1\lambda+b_2\mu)\,\mu'\}^2 = 0.$$

10·14. When (λ,μ), (λ',μ') are regarded as values of a ratio parameter (ξ,η) of a line p (or of any $R\infty^1$) the equation ·131 gives the general (2, 2) correspondence between the points of p (or between the elements of the $R\infty^1$). As in 2·81 we must distinguish between the forward correspondence $(\lambda,\mu)\to(\lambda',\mu')$ and the reverse correspondence $(\lambda',\mu')\to(\lambda,\mu)$. Thus in the forward correspondence the point (ξ_1, η_1) corresponds to the points given by $\phi(\xi_1, \eta_1; \xi, \eta) = 0$, while in the reverse correspondence (ξ_1, η_1) corresponds to the points given by

$$\phi(\xi, \eta; \xi_1, \eta_1) = 0.$$

There are in general four *united points*, or self-corresponding points, given by the quartic equation $\phi(\lambda, \mu; \lambda, \mu) = 0$. Thus, if in a particular $T(2, 2)$ there are more than four united points, then every point of p is a united point and $\mu'\lambda - \lambda'\mu$ is a factor of $\phi(\lambda, \mu; \lambda', \mu')$.

The branch points in the forward and reverse correspondences are given by (·133) and (·134). There are also two sets of double points. The branch points and double points must not be confused with the united points.

The correspondence is said to be *symmetric* IF the forward and reverse correspondences are identical, that is IF the equation (·131) is unaltered by interchanging (λ, μ) and (λ', μ'). Now this equation may be written

$$a_1\lambda^2\lambda'^2 + 4b_2\lambda\mu\lambda'\mu' + c_3\mu^2\mu'^2 + 2(a_2\lambda\mu\lambda'^2 + b_1\lambda'\mu'\lambda^2) + (c_1\lambda^2\mu'^2 + a_3\lambda'^2\mu^2) + 2(c_2\lambda\mu\mu'^2 + b_3\lambda'\mu'\mu^2) = 0.$$

On interchanging (λ, μ) and (λ', μ') the equation becomes

$$a_1\lambda^2\lambda'^2 + 4b_2\lambda\mu\lambda'\mu' + c_3\mu^2\mu'^2 + 2(b_1\lambda\mu\lambda'^2 + a_2\lambda'\mu'\lambda^2) + (a_3\lambda^2\mu'^2 + c_1\lambda'^2\mu^2) + 2(b_3\lambda\mu\mu'^2 + c_2\lambda'\mu'\mu^2) = 0.$$

If a_1, b_2, c_3 are not all zero, these equations are the same IF

$$a_2 = b_1, \quad a_3 = c_1, \quad b_3 = c_2.$$

In this case $\phi(\lambda, \mu; \lambda', \mu')$ and $\phi(\lambda', \mu'; \lambda, \mu)$ are identical. If $a_1 = b_2 = c_3 = 0$, the equations are the same IF

$$\frac{a_2}{b_1} = \frac{b_1}{a_2} = \frac{a_3}{c_1} = \frac{c_1}{a_3} = \frac{b_3}{c_2} = \frac{c_2}{b_3},$$

that is IF either

$$a_2 - b_1 = a_3 - c_1 = b_3 - c_2 = 0,$$

or

$$a_2 + b_1 = a_3 + c_1 = b_3 + c_2 = 0.$$

In the first case $\phi(\lambda, \mu; \lambda', \mu')$ and $\phi(\lambda', \mu'; \lambda, \mu)$ are again identical; in the second case they differ in sign, and the equation (·131) is of the form

$$(\lambda\mu' - \lambda'\mu)\{p\lambda\lambda' + q(\lambda\mu' + \lambda'\mu) + r\mu\mu'\} = 0. \qquad (\cdot 141)$$

The symmetric $T(2, 2)$ is then the trivial one consisting of the sum* of the identical correspondence and an involutory homography.

* See 2·52, p. 59.

10·15. A symmetric $T(2, 2)$ *which is not given by an equation of the form* (·141) will be denoted by $S(2, 2)$.* It is conveniently represented by an equation of the form

$$(a\lambda^2+h\lambda\mu+g\mu^2)\,\lambda'^2+(h\lambda^2+b\lambda\mu+f\mu^2)\,\lambda'\mu' \\ +(g\lambda^2+f\lambda\mu+c\mu^2)\,\mu'^2=0, \quad (\cdot 151)$$

which is clearly equivalent to (·131) when $a_2=b_1$, $a_3=c_1$, $b_3=c_2$. In an $S(2,2)$ on a line the forward and reverse correspondences coincide, so there is only one set of branch points and one set of double points. The united points are given by

$$(a\xi^2+h\xi\eta+g\eta^2)\,\xi^2+(h\xi^2+b\xi\eta+f\eta^2)\,\xi\eta+(g\xi^2+f\xi\eta+c\eta^2)\,\eta^2=0. \quad (\cdot 152)$$

In general, if P is a united point, only one of the points corresponding to P coincides with P. If a point U is such that both the points corresponding to U coincide in U, then U is called a *doubly united point*; it is also a branch point and a double point. A doubly united point U counts twice among the set of united points. For if the parameter (ξ,η) is so chosen that U is the point $(0, 1)$, then λ^2 must be a factor of the equation obtained by putting $\lambda'=0$ in (·151), so that $f=c=0$, and consequently ξ^2 is a factor of (·152).

The degenerate symmetric $T(2, 2)$ given by the equation (·141), in which every point is a united point, has two doubly united points, given by

$$p\xi^2+2q\xi\eta+r\eta^2=0, \quad (\cdot 153)$$

and can have no more. It follows that if a symmetric $T(2, 2)$ possesses three distinct doubly united points, counting as six united points, then it cannot be given by an equation of the form (·141), and must therefore be an $S(2, 2)$ given by an equation of the form (·151). The equation (·152) which gives the united points must be

* In the matrix notation which will be introduced in Chapter XI the equation (·131) may be written in the form

$$(\lambda'^2 \quad 2\lambda'\mu' \quad \mu'^2)\begin{pmatrix} a_1 & a_2 & a_3 \\ b_1 & b_2 & b_3 \\ c_1 & c_2 & c_3 \end{pmatrix}\begin{pmatrix} \lambda^2 \\ 2\lambda\mu \\ \mu^2 \end{pmatrix}=0,$$

and 10·14 shows that this gives a symmetric $T(2, 2)$ IF the matrix

$$\begin{pmatrix} a_1 & a_2 & a_3 \\ b_1 & b_2 & b_3 \\ c_1 & c_2 & c_3 \end{pmatrix}$$

is symmetric or skew-symmetric. In the former case the correspondence is an $S(2, 2)$, and in the latter it is a degenerate correspondence of type (·141).

identically satisfied by all (ξ, η), so that $a = h = b+2g = f = c = 0$, and ($\cdot$151) reduces to

$$g(\lambda\mu' - \lambda'\mu)^2 = 0. \qquad (\cdot 154)$$

Thus every point is a doubly united point, and $S(2, 2)$ is said to be *doubly identical.*

If an $S(2, 2)$ is known to possess two distinct doubly united points and one other united point, counting as five united points, then again $\cdot$152 must be an identity, and $S(2, 2)$ is doubly identical.

10·16. The $S(2, 2)$ which is given by ($\cdot$151) is reducible* if the polynomial on the left can be expressed as the product of two or more polynomial factors $\phi_i(\lambda, \mu;\ \lambda', \mu')$. If (λ, μ) and (λ', μ') are interchanged the product is unaltered and therefore the factors ϕ_i are either unaltered or are interchanged in pairs. There are essentially only two main types of reducible $S(2, 2)$, given respectively by equations of the form

$$\{p_1\lambda\lambda' + q_1(\lambda\mu' + \lambda'\mu) + r_1\mu\mu'\}\{p_2\lambda\lambda' + q_2(\lambda\mu' + \lambda'\mu) + r_2\mu\mu'\} = 0 \qquad (\cdot 161)$$

and

$$\{p\lambda\lambda' + q\lambda\mu' + r\lambda'\mu + s\mu\mu'\}\{p\lambda\lambda' + r\lambda\mu' + q\lambda'\mu + s\mu\mu'\} = 0, \qquad (\cdot 162)$$

where $q \neq r$.

The $S(2, 2)$ given by the equation ($\cdot$161) consists of the sum of two involutory homographies I_1 and I_2, which may coincide. In particular either or both of I_1, I_2 may be improper and have a fixed element†; if, for example, I_1 has a fixed element (α, β) the equation ($\cdot$161) takes the form

$$(\beta\lambda - \alpha\mu)(\beta\lambda' - \alpha\mu')\{p_2\lambda\lambda' + q_2(\lambda\mu' + \lambda'\mu) + r_2\mu\mu'\} = 0, \qquad (\cdot 163)$$

and, in $S(2, 2)$, a general (λ, μ) corresponds to the fixed element (α, β) and to the mate of (λ, μ) in I_2.

The $S(2, 2)$ represented by ($\cdot$162) consists of the sum of a non-involutory homography H and its reverse H^{-1}. If H (and therefore also H^{-1}) is improper, ($\cdot$162) is of the form

$$\{(\beta\lambda - \alpha\mu)(\delta\lambda' - \gamma\mu')\}\{(\delta\lambda - \gamma\mu)(\beta\lambda' - \alpha\mu')\} = 0,$$

which can be re-written as

$$\{(\beta\lambda - \alpha\mu)(\beta\lambda' - \alpha\mu')\}\{(\delta\lambda - \gamma\mu)(\delta\lambda' - \gamma\mu')\} = 0, \qquad (\cdot 164)$$

which shows that the trivial $S(2, 2)$ in which a general element corresponds to two fixed elements can be regarded as a special case of either ($\cdot$161) or ($\cdot$162).

The particular case where H is the identical homography leads to the doubly identical S (2, 2).

* See 2·52, p. 59. † See 3·58, p. 94.

10·2. Symmetric (2, 2) correspondence on a proper conic. Poncelet's Theorem

10·21. If a single variable is used to represent a ratio variable, it follows from 10·15 that an $S(2,2)$ between the values u, v of a variable λ is given by an equation of the form

$$S(2,2) \equiv (au^2+hu+g)\,v^2+(hu^2+bu+f)\,v+(gu^2+fu+c) = 0. \quad (\cdot 211)$$

In this paragraph λ will be regarded as a parameter of the proper conic γ whose equation is $x:y:z = \lambda^2:\lambda:1$, so that $S(2,2)$ is a correspondence between the points P, Q of γ whose co-ordinates are $(u^2, u, 1)$, $(v^2, v, 1)$.

Now ($\cdot$211) is a CONDITION for P and Q to be conjugate relative to the conic locus s given by

$$s \equiv ax^2+by^2+cz^2+2fyz+2gzx+2hxy = 0.$$

Thus:

The pairs of corresponding points of an $S(2,2)$ on a proper conic γ are conjugate relative to a conic locus s.

Conversely, if γ is any proper conic and s is a conic locus, the correspondence between points P and Q of γ which are conjugate relative to s is an $S(2,2)$.

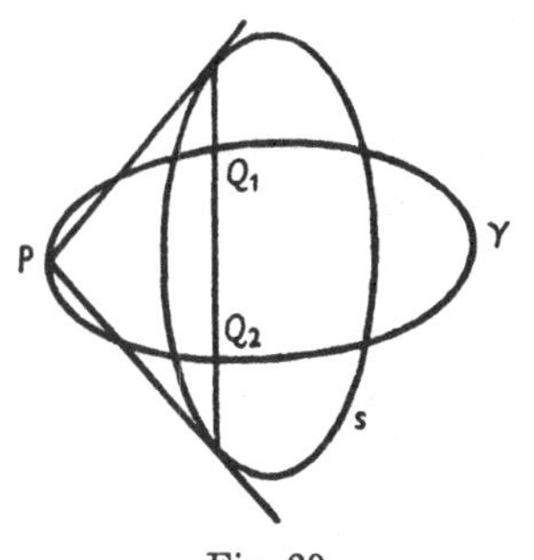

Fig. 29

The two points Q_1, Q_2 which correspond to a point P of γ are the intersections of γ with the polar of P relative to s, as in Fig. 29. The united points of $S(2,2)$ are the intersections of γ with s. If s is a proper conic locus the branch points are the intersections of γ with the reciprocal of γ relative to s, and the double points are the contacts with γ of the common tangents of γ and the reciprocal of γ relative to s.

If s is a line-pair $\|h, k\|$ with vertex O, which is the case IF the determinant

$$\delta \equiv \begin{vmatrix} a & h & g \\ h & b & f \\ g & f & c \end{vmatrix}$$

is of rank two, the branch points are the intersections of γ with the harmonic conjugates, relative to $\|h, k\|$, of the two tangents to γ from O. The double points, which coincide in pairs, are the contacts with γ of the tangents through O. In particular, if O lies on γ, $S(2, 2)$ is reducible of the type ($\cdot$163). The particular type of $S(2, 2)$ for which δ is of rank two will be discussed more fully in 10·4.

If s is a repeated line, which is the case IF δ is of rank one, the $S(2, 2)$ is reducible of the type ($\cdot$164).

If s and γ coincide the $S(2, 2)$ is doubly identical.

10·22. The equation ($\cdot$211) may be written in the form

$$c+g(u+v)^2+au^2v^2+huv(u+v)+(b-2g)\,uv+f(u+v) = 0, \quad (\cdot 221)$$

so that it is a CONDITION for the chord

$$x-(u+v)\,y+uvz = 0$$

joining the points P, Q of γ to belong to the conic scroll S' given by

$$S' \equiv cX^2+gY^2+aZ^2-hYZ-(2g-b)\,ZX-fXY = 0.$$

Thus:

The joins of corresponding points of an $S(2, 2)$ on a proper conic γ are the lines of a conic scroll S'.

Conversely, if γ is a proper conic and S' is a conic scroll, the correspondence between points P and Q of γ such that PQ belongs to S' is an $S(2, 2)$.

The two points Q_1, Q_2 which correspond to a point P of γ are the further intersections of γ with the two lines of S' through P, as in Fig. 30. The united points are the contacts with γ of the tangents of γ which belong to S'. If S' is proper the branch points are the common points of γ and S', and the double points are the further intersections with γ of the tangents to S' at these common points.

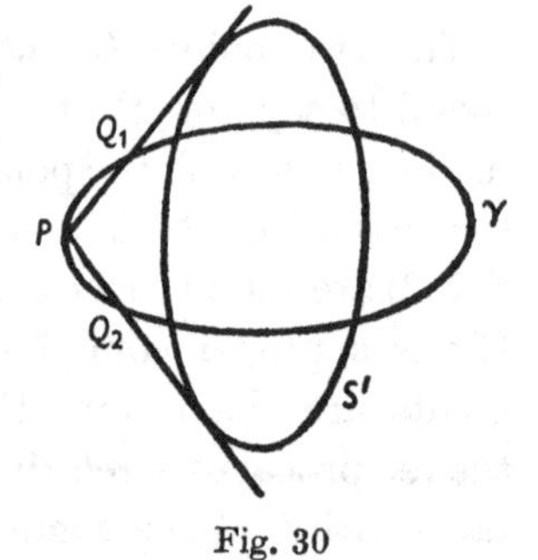

Fig. 30

If S' is a point-pair $\|H, K\|$, $S(2, 2)$ is the sum of the two involutions on γ whose centres are H and K. The branch and double points coincide in the two points (HK, γ). If S' is a repeated point $\|H, H\|$, $S(2, 2)$ is merely the involution with centre H counted twice. The CONDITIONS for these two cases are that the determinant

$$\begin{vmatrix} 2c & -f & b-2g \\ -f & 2g & -h \\ b-2g & -h & 2a \end{vmatrix}$$

should be of rank two and one respectively.

The case where S' is a point-pair $\| H, K \|$ corresponds to the reducible $S(2, 2)$ of the type (·161). We can obtain the sub-types of this by supposing either or both of H, K to lie on γ; again, if K coincides with H and S' is the repeated point $\| H, H \|$, the point H may in particular lie on γ.

The other main type of reducible $S(2, 2)$ given by (·162) arises when (·211) is of the form

$$(auv+bu+cv+d)(auv+cu+bv+d) = 0.$$

The equations

$$auv+bu+cv+d = 0, \quad auv+cu+bv+d = 0,$$

give a homography H on γ and its reverse H^{-1}, and $S(2, 2)$ is the sum of these. The conic scroll S' is formed by the joins of corresponding points in H (or in H^{-1}), and has double contact with γ, as was shown in 7·1.

Conversely, the $S(2, 2)$ on γ determined by the tangents of a conic scroll S' which has double contact with γ is the sum of a homography and its reverse. For the equation of S' is of the form

$$S' \equiv (pX+qY+vZ)^2-\rho^2(Y^2-4ZX) = 0,$$

and the equation of $S(2,2)$ is

$$\{ruv+(q+\rho)u+(q-\rho)v+p\}\{ruv+(q-\rho)u+(q+\rho)v+p\} = 0^*.$$

10·23. If γ and s are given, the $S(2,2)$ on γ in which P, Q correspond when they are conjugate relative to s is determined. It then follows from 10·22 that the joins PQ are the tangents of a conic scroll S'. But these lines PQ are just the lines which meet γ and s in apolar pairs of points; thus the theorem of 9·11 is established, the harmonic scroll of γ and s being S'. It should be noticed that this proof only applies when γ is proper.

Conversely, if γ and S' are given, the $S(2, 2)$ and the conic locus s are uniquely determined; thus the theorem of 9·31 is established, s being the v-conic of γ and S'.

The dual of 2·66 showed that an algebraic system of lines is a conic scroll when it is known that TWO lines of the system pass through a general point of the plane. It now appears that it is only necessary to know that TWO lines of the system pass through a general point of a proper conic locus γ. For it then follows that the pair of points

* Cf. 7·14, p. 195.

P, Q is which γ is met by the lines of the system are corresponding points in a $S(2, 2)$ on γ, so that, by 10·22, the system forms a conic scroll.

10·24. If there is one set of distinct points P, Q, R on γ such that each of the three points corresponds to the remaining two in $S(2, 2)$, then the triangle PQR is self-polar relative to s and circumscribed to S'. Thus the theorems of 9·43 and 9·45 are both equivalent to the following theorem.

If in a symmetric (2, 2) *correspondence* $S(2, 2)$ *among the elements of an* $R\infty^1$ *there is one set of three distinct elements such that each pair of the set are corresponding elements in the* $S(2, 2)$, *then there are* ∞^1 *such sets of three elements, and each element of the* $R\infty^1$ *belongs to one set. When this happens the* $S(2, 2)$ *is said to be compounded of triads.*

The conic γ, or $y^2 - zx = 0$, is outpolar to the conic s of 10·21 IF $ac - g^2 = hf - bg$. Thus:

The $S(2, 2)$ *given by*

$$(au^2 + hu + g)\,v^2 + (hu^2 + bu + f)\,v + (gu^2 + fu + c) = 0$$

is compounded of triads IF $ac - g^2 = hf - bg$.

10·25. It has been shown that the correspondence between points of a proper conic γ whose joins touch a conic scroll S_1' is a symmetric (2, 2) correspondence, which will now be denoted by $S_1(2, 2)$. A sequence of points $P_0, P_1, P_2, \ldots, P_n$ of γ, as shown in Fig. 31, which is such that $P_{i-1}P_i$ and P_iP_{i+1} are the two lines of S_1' through P_i for $i = 1, 2, \ldots, n-1$, will be called a *tangential sequence*.

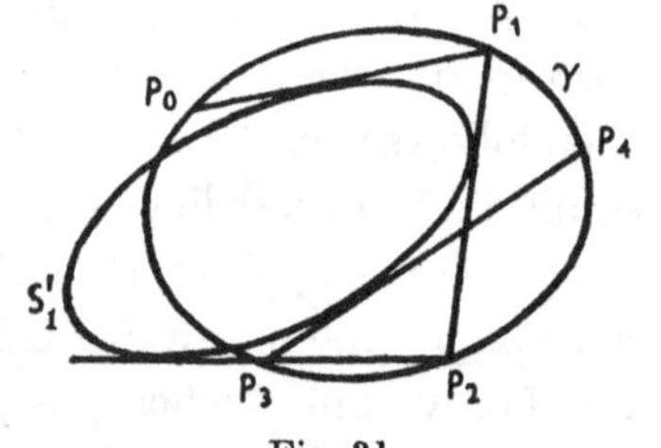

Fig. 31

Clearly $P_n, P_{n-1}, \ldots, P_0$ is also a tangential sequence. For convenience we shall in the following suppose that S_1' is a proper conic scroll, but the argument is applicable equally well to the case when S_1' is a point-pair $\| H, K \|$, provided neither H nor K lies on γ; if either or both of H, K lie on γ, a tangential sequence, with the possible exception of the first two points, consists merely of the two points (HK, γ) taken alternately.

A general point P_0 determines two tangential sequences, since P_1

may have one of two positions. Thus there is a symmetric (2, 2) correspondence between P_0 and P_n, which will be denoted by $S_n(2, 2)$. We must first show that $S_n(2, 2)$ is of the general type (·151) and is not a degenerate symmetric correspondence of the type (·141), which is the sum of an involutory (1, 1) correspondence and the identical correspondence. If $S_n(2, 2)$ is of this latter type, every point P_0 is a united point and, for every point P_0, we have a tangential sequence $P_0, P_1, P_2, \ldots, P_{n-1}, P_0$. But in this case $P_0, P_{n-1}, P_{n-2}, \ldots P_1, P_0$ is also a tangential sequence starting from P_0, and therefore, provided P_1 and P_{n-1} are distinct, P_0 is a doubly united point. If P_{n-1} coincides with P_1 then, since there cannot be three tangents of S'_1 through P_1, P_2 must coincide with P_{n-2}, and similarly P_3 with P_{n-3}, P_4 with P_{n-4} and so on. If $n = 2m$ we come to a point P_m such that the two tangents to S'_1 from it coincide, that is to a point of intersection of γ and S'_1. If $n = 2m+1$ we come to a point $P_{m-1}(\equiv P_{m+1})$ at which the tangent to γ is also a tangent to S'_1. Clearly neither of these cases can occur if P_0 is a general point of γ. Thus if the general point P_0 is a united point of $S_n(2, 2)$ it is a doubly united point; $S_n(2, 2)$ cannot be of the type (·141) but is the special case (·154) of the general type (·151).

If the general P_0 is not a united point of $S_n(2, 2)$ it follows from 10·21 and 10·22 that as P_0 varies on γ the pairs of points P_0, P_n are conjugate relative to a conic locus s_n and the lines P_0P_n belong to a conic scroll S'_n. In particular $S_2(2, 2)$ is the correspondence between points of γ which correspond to the same point in $S_1(2, 2)$. Thus, in Figs. 29 and 30, as P varies on γ the points Q_1, Q_2 are conjugate relative to a conic locus s_2 and the lines Q_1Q_2 belong to a conic scroll S'_2, which, if s is proper, is of course the reciprocal of γ relative to s.

Suppose there is an ordered polygon whose n vertices

$$Q_0, Q_1, \ldots, Q_{n-1}$$

lie on γ and whose n sides Q_0Q_1, Q_1Q_2, ..., $Q_{n-1}Q_0$ touch S'. We assume that the vertices are all distinct. Each vertex is then a doubly united point of $S_n(2, 2)$. By 10·15 an $S(2, 2)$ with three or more doubly united points is doubly identical, every point being a doubly united point. Thus the following theorem, usually known as Poncelet's Theorem, which has already been proved for $n = 3$ and $n = 4$, is now established for $n \geqslant 3$:

If there is one ordered polygon of n sides inscribed in γ and circumscribed to S_1', then there are ∞^1 such polygons, and a general point of γ is a vertex of ONE *of the polygons.*

This theorem is essentially a theorem about an $S(2, 2)$, and may therefore be applied to any $R\infty^1$. It should be noticed that when $n = 2m$ each point P_0 determines ONE point P_m, since both

$$P_0, P_1, \ldots, P_m \quad \text{and} \quad P_0, P_{n-1}, P_{n-2}, \ldots, P_m$$

are tangential sequences. Thus the pairs of points $| P_0, P_m |$ belong to an involution on γ, and consequently the joins of opposite vertices of the ∞^1 polygons pass through a fixed point. This was proved for the case $n = 4$ in 5·68.

It is possible to distinguish between the two tangential sequences determined by a general point P_0 of γ in the case when S_1' has double contact with γ at the points A_1, A_2. For then, as we have seen in 10·22, $S_1(2, 2)$ is the sum of a homography H and its reverse H^{-1}; it follows that $S_n(2, 2)$ is the sum of H^n and its reverse H^{-n}, and S_n' has double contact with γ at A_1, A_2. In particular, if S_1' has double contact with γ and if there is one polygon P_0, $P_1 \ldots P_{n-1}$ inscribed in γ and circumscribed to S_1', then $S_1(2, 2)$ is the sum of a cyclic homography of period n and its reverse.

10·26. In the proof of Poncelet's Theorem which has just been given, and in the various proofs for the cases of the triangle and ordered quadrangle, it has been understood that the vertices of the given ordered polygon are all distinct, and nothing has been said about the degenerate triangles, ordered quadrangles, and ordered polygons of the ∞^1 systems. The reader may have noticed that there are some points of γ which give rise to degenerate polygons, and the matter must now be investigated.

An $S_n(2, 2)$ in general has four united points and they arise as follows:

(*a*) If $n = 2m$, let A_0 be one of the common points of γ and S_1', that is one of the branch points of $S_1(2, 2)$. Then, as is shown in Fig. 32, A_0 determines only one tangential sequence $A_0, A_1, \ldots, A_m$. Thus $A_m, A_{m-1}, \ldots, A_1, A_0, A_1, \ldots, A_m$ is a tangential sequence of $n+1$ points and A_m is a united point of $S_n(2, 2)$. The tangent to γ at A_m is a tangent of S_n', and the four united points of $S_n(2, 2)$ arise in this manner from the four intersections of γ and S_1'.

(*b*) If $n = 2m+1$, let B_0 be the point of contact with γ of a common tangent of γ and S_1', that is one of the united points of $S_1(2,2)$. Let B_1 be the further intersection with γ of the other tangent to S_1' through B_0. Then, as in Fig. 33, B_0 determines two tangential sequences $B_0, B_1, B_2, \ldots, B_m$ and $B_0, B_0, B_1, \ldots, B_m$. Thus $B_m, B_{m-1}, \ldots, B_1, B_0, B_0, B_1, \ldots, B_m$ is a tangential sequence of $n+1$ points, and so B_m is a united point of $S_n(2,2)$; in general four united points arise in this way.

It must now be decided whether these degenerate polygons $A_m, A_{m-1}, \ldots, A_{m-1}, A_m$ and $B_m, B_{m-1}, \ldots, B_{m-1}, B_m$ are to be considered as circumscribed to S_1'. If a polygon $Q_0, Q_1, \ldots, Q_{n-1}$ with distinct vertices is inscribed in γ and circumscribed to S_1', then the two points corresponding to Q_0 in $S_1(2,2)$ are Q_1 and Q_{n-1}, and the

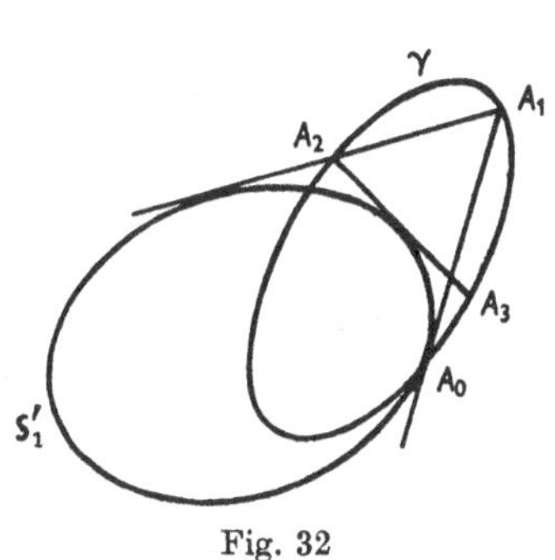

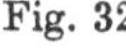
Fig. 32

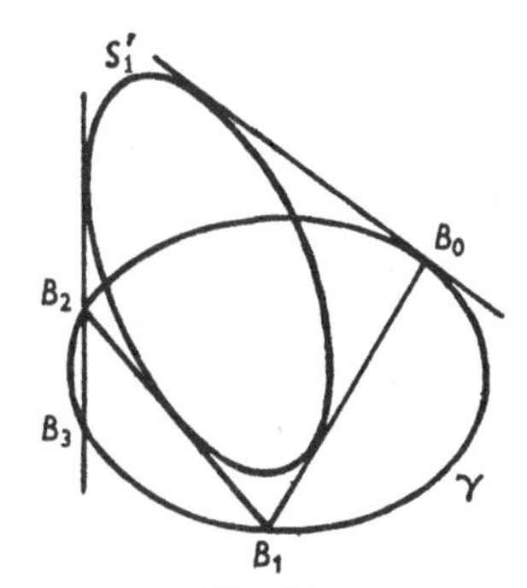

Fig. 33

next point in the tangential sequence $Q_0, Q_1, \ldots, Q_{n-1}, Q_0$ is Q_1; in fact both tangential sequences determined by Q_0 repeat themselves after n steps, and, as we have seen already, Q_0 and similarly the other vertices are doubly united points of $S_n(2,2)$. The degenerate polygon $A_m, A_{m-1}, \ldots, A_{m-1}, A_m$ will thus be said to be circumscribed to S_1' IF the sequence $A_m, A_{m-1}, \ldots, A_{m-1}, A_m$ repeats itself. In general this is not the case, the next point in the sequence being A_{m+1}, where A_{m-1} and A_{m+1} are the two points, not in general the same, which correspond to A_m in $S_1(2,2)$. These two points are the same IF A_m is a branch point of $S_1(2,2)$. Thus the degenerate polygon $A_m, A_{m-1}, \ldots, A_{m-1}, A_m$ is said to be circumscribed to S_1' IF A_m is a branch point of $S_1(2,2)$. Similarly the degenerate polygon $B_m, B_{m-1}, \ldots, B_{m-1}, B_m$ is said to be circumscribed to S_1' IF B_m is a branch point of $S_1(2,2)$.

10·27. The following definition of an inscribed and circumscribed polygon covers the degenerate cases:

A polygon whose n vertices $P_0, P_1, \ldots, P_{n-1}$ *lie on* γ *is said to be circumscribed to* S_1' IF $P_0, P_1, \ldots, P_{n-1}, P_0, P_1$ *is a tangential sequence.*

With this definition it is possible to prove a more complete form of Poncelet's Theorem, namely:

If there is one polygon $Q_0, Q_1, \ldots, Q_{n-1}$, *degenerate or not, which is inscribed in* γ *and circumscribed to* S_1', *and if at least three of the vertices are distinct, then each point of* γ *is a vertex of one such polygon.*

For the two tangential sequences determined by Q_0 (which may coincide) are

$$Q_0, Q_1, \ldots, Q_{n-1}, Q_0, Q_1, \ldots$$

and

$$Q_0, Q_{n-1}, \ldots, Q_1, Q_0, Q_{n-1} \ldots,$$

so Q_0 is a doubly united point of $S_n(2, 2)$. Similarly $Q_1, Q_2, \ldots, Q_{n-1}$ are doubly united points of $S_n(2, 2)$ and by 10·15 every point of γ is a doubly united point of $S_n(2, 2)$. Consequently, if $P_0, P_1, \ldots, P_n$ is any tangential sequence, P_n coincides with P_0. So one of the tangential sequences determined by P_1 is $P_1, P_2, \ldots, P_{n-1}, P_0$, and the next point in this sequence must be P_1, since P_1 is a doubly united point of $S_n(2, 2)$. Thus $P_0, P_1, \ldots, P_{n-1}, P_0, P_1$ is a tangential sequence, and the polygon $P_0 P_1 \ldots P_{n-1}$ is circumscribed to S_1'.

There is a subtle point in this proof that should be mentioned. Having pointed out that P_n coincides with P_0 the proof is completed by considering P_1. This is necessary because the fact that P_n coincides with P_0 is not sufficient to show that the polygon $P_0 P_1 \ldots P_{n-1}$ is circumscribed to S'. For it is conceivable that P_{n-1} might coincide with P_1, and that the other point P_1' corresponding to P_0 in $S_1(2, 2)$ gives rise to another tangential sequence $P_0, P_1', P_2', \ldots$ in which P_n' must of course coincide with P_0 since P_0 is a doubly united point of $S_n(2, 2)$. The sequence which begins with P_0, P_1, is then

$$P_0, P_1, \ldots, P_{n-1}, P_0, P_1', P_2', \ldots, P_{n-1}', P_0, P_1, \ldots$$

so the polygon $P_0 P_1 \ldots P_{n-1}$ is not circumscribed to S'. This possibility will occur if two of the points A_m or two of the points B_m obtained in 10·26 coincide, and if P_0 is the point of coincidence. For example, consider the case in which γ and S_1' are as in Fig. 34, and $n = 3$. When P_0 is A, the two sequences P_0, P_1, P_2, P_0 and P_0, P_1', P_2', P_0 are A, B, B, A and A, C, C, A. Thus A is a doubly united point of

$S_3(2,2)$. But $A, B, B, A, C, C, A, B \ldots$ is a tangential sequence, so in $S_3(2,2)$ B is not a doubly united point but a branch point, corresponding to C counted twice.

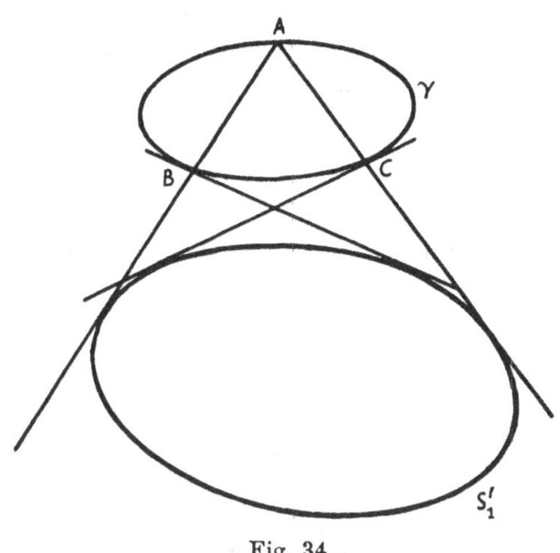

Fig. 34

10·28. Consider the case $n = 4$. In Fig. 32, although A_2, A_1, A_0, A_1, A_2 is a tangential sequence, the quadrangle $A_2 A_1 A_0 A_1$ is only said to be circumscribed to S' if A_2 is a branch point of $S_1(2,2)$, that is if A_2 lies on S', when A_3 will coincide with A_1. It follows from 10·27 that if there is one quadrangle inscribed in γ and circumscribed to S', then A_2 must lie on S'. Conversely it also follows that if A and B are two distinct intersections of γ and S', and if the tangents to S' at A and B meet in a point C of γ as in Fig. 35, then there are ∞^1 quadrangles inscribed in γ and circumscribed to S'. For A, C, B, C, A, C is a tangential sequence, so A, C, B are doubly united points of $S_4(2,2)$.

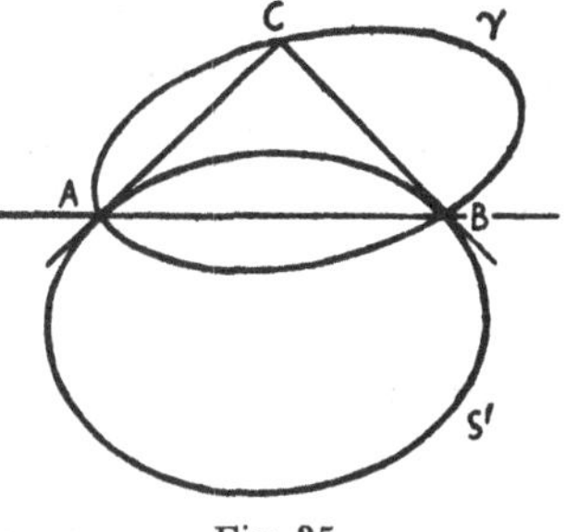

Fig. 35

If γ and S' touch at A, the point A determines only one tangential sequence, namely A, A, A, ..., so A is a doubly united point of $S_n(2,2)$ for all values of n. But of course it does not follow that there are ∞^1 polygons of n sides inscribed in γ and circumscribed to S', since it is only known that $S_n(2,2)$ has one doubly united point.

Consider now the case $n = 3$. In Fig. 33, although $B_1 B_0 B_0 B_1$ is a tangential sequence, the degenerate triangle $B_1 B_0 B_0$ is only said to be circumscribed to S' if B_1 is a branch point of $S_1(2, 2)$, that is if B_1 lies on S', when B_2 coincides with B_0. It follows from 10·27 that, if there is one triangle with distinct vertices inscribed in γ and circumscribed to S', then there are ∞^1 such triangles, and B_1 lies on S'.

Conversely, if the tangent to S' at a simple intersection A with γ meets γ again in a point B other than A, and if B is a point of contact with γ of a common tangent of γ and S', as in Fig. 35, then γ is triangularly circumscribed to S'.

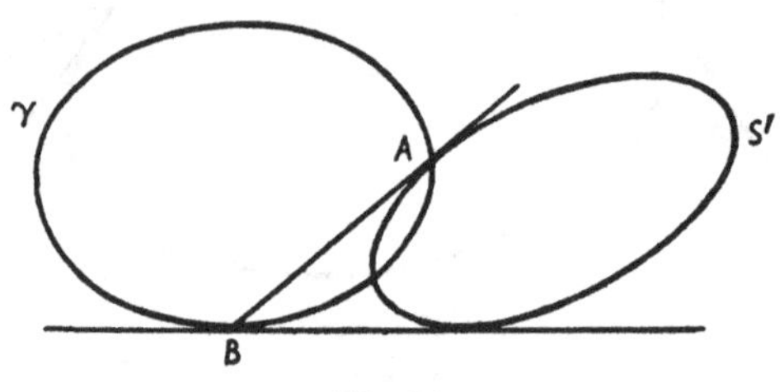

Fig. 36

This, however, does not follow from 10·27, because the fact that A and B are doubly united points of $S_3(2, 2)$ is not enough to show that every point of γ is a doubly united point. But it is assumed that A is a simple intersection of γ and S', so the two conics have at least two common points and at least two common tangents. Thus there is a common tangent whose contact B_0 with γ is distinct from B. Then the point B_1 of Fig. 33 is a united point of $S_3(2, 2)$ distinct from A and B, so by 10·15 every point of γ is a doubly united point of $S_3(2, 2)$, and the converse is established.

Of course if γ and S' touch at A the degenerate triangle AAA is inscribed in γ and circumscribed to S', but no further information can be deduced. Thus, with the exact definition of an inscribed and circumscribed triangle, the theorem of 9·43 may be extended as follows:

If there is one triangle with two or three distinct vertices which is inscribed in γ and circumscribed to S', then each point of γ is a vertex of one triangle which is inscribed in γ and circumscribed to S'.

The theorem of 9·45 may be extended in a similar manner.

Consider two proper conics s and s' with a simple intersection A.

Let the tangents to s and s' at A meet s' and s again in B' and B as in Fig. 37. Then it has been shown that s is triangularly circumscribed to s' IF the tangent to s at B touches s'. Similarly s' is triangularly circumscribed to s IF the tangent to s' at B' touches s.

Similarly s is outpolar to s' IF AB' is the polar of B relative to s', that is IF s' touches BB' at B'. Also s is inpolar to s' IF s touches BB' at B. Thus s and s' are mutually apolar IF BB' is a common tangent of s and s'. It follows that if s and s' are mutually apolar, then s is both triangularly inscribed and circumscribed to s'.* The converse does not hold, for if s is triangularly inscribed and triangularly circumscribed to s', then B, B' are both points of contact of common tangents, but they need not be points of contact of the same common tangent.

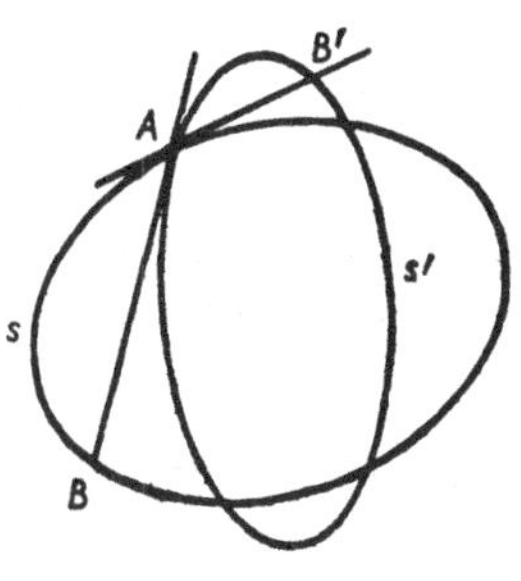

Fig. 37

10·29. Poncelet's Theorem is a porism in the sense in which the term was defined in 9·49. But it should be noted that the poristic nature of a theorem may depend on the way in which the theorem is stated. Thus, for $n = 3$, we should expect to find a finite number of points P of γ such that the two points Q, R which correspond to P in $S_1(2,2)$ are themselves a pair of corresponding points in $S_1(2,2)$. Indeed these points are actually the united points of $S_1(2,2)$. Analytically if $S_1(2,2)$ is given by (·211), and if P is the point $\lambda = u$, then Q, R are given by $\lambda = v_1, v_2$, where

$$1 : -(v_1+v_2) : v_1 v_2 = (au^2+hu+g) : (hu^2+bu+f) : (gu^2+fu+c), \qquad (\cdot 291)$$

so, writing (·211) in the form (·221), it follows that Q and R are corresponding points in $S_1(2,2)$ IF

$$\begin{aligned} c(au^2+hu+g)^2 + g(hu^2+bu+f)^2 + a(gu^2+fu+c)^2 \\ -h(hu^2+bu+f)(gu^2+fu+c) \\ -(2g-b)(gu^2+fu+c)(au^2+hu+g) \\ -f(au^2+hu+g)(hu^2+bu+f) = 0. \end{aligned} \qquad (\cdot 292)$$

This equation reduces to

$$(ac+bg-hf-g^2)\,\phi(u) = 0, \qquad (\cdot 293)$$

where $$\phi(u) \equiv au^4 + 2hu^3 + (b+2g)\,u^2 + 2fu + c. \qquad (\cdot 294)$$

* Cf. 9·69, p. 252.

But the quartic equation $\phi(u) = 0$ gives the united points of $S_1(2,2)$, so if the equation (·293) is satisfied by any point $\lambda = u$ other than a united point it must be identically satisfied for all values of u, and consequently either

$$\phi(u) \equiv 0 \quad \text{or} \quad ac+bg = hf+g^2.$$

Also $\phi(u) \equiv 0$ IF S' coincides with γ.

Thus, if S' does not coincide with γ, there are always positions of P for which Q and R correspond in $S_1(2,2)$, namely the united points of $S_1(2,2)$. But if there is one position of P which is not a united point of $S_1(2,2)$ and for which Q, R correspond in $S_1(2,2)$, then every point P of γ has this property. A CONDITION *for this to happen is*

$$ac - g^2 = hf - bg.*$$

This theorem is equivalent to the theorem of 10·24, but when stated in this form it depends on the fact that a quartic polynomial which vanishes for more than four values of the variable is identically zero, and is not a porism.

Stated in terms of a parameter λ of any $R\infty^1$ the theorem becomes:

If there is one set of values $\lambda_1, \lambda_2, \lambda_3$ of λ_1 not all equal, such that three corresponding pairs of values in an $S(2,2)$ are $|\lambda_2, \lambda_3|$, $|\lambda_3, \lambda_1|$, $|\lambda_1, \lambda_2|$ then the $S(2,2)$ is compounded of triads.

Also for $n \geqslant 3$

If there is one set of values $\lambda_1, \lambda_2, \ldots, \lambda_n$ of λ, not all equal, such that n corresponding pairs of values in an $S(2,2)$ are

$$|\lambda_n, \lambda_2|, \ |\lambda_1, \lambda_3|, \ |\lambda_2, \lambda_4|, \ \ldots, \ |\lambda_{n-2}, \lambda_n|, \ |\lambda_{n-1}, \lambda_1|,$$

then there are ∞^1 such sets.

10·3. $T(2,2)$ compounded of involutions

10·31. If I and I' are involutions on the lines p and p', any homography H between the pairs of I and the pairs of I' determines a $T(2,2)$ between p and p'. For each point of p belongs to ONE pair of I which corresponds to ONE pair of I', and similarly each point of p' determines ONE pair of points of p. This particular type of $T(2,2)$ is said to be *compounded of involutions*. In this $T(2,2)$ the points of any pair of I correspond to the same pair of points of p', and the points of any pair of I' correspond to the same pair of points of p.

* Cf. 10·24, p. 278.

The branch points on p are the two pairs of points corresponding to the double points of I', and similarly for the branch points on p'. The double points of $T(2,2)$ on p and p' are the double points of I and I' each counted twice. It will now be proved that:

In a general $T(2,2)$ between two lines p and p' no two distinct points of p correspond to the same pair of points on p'. If however two particular distinct points of p correspond to the same pair of points of p', then the $T(2,2)$ is compounded of involutions. A CONDITION *for the $T(2,2)$ given by (·131) to be compounded of involutions is*

$$\delta \equiv \begin{vmatrix} a_1 & a_2 & a_3 \\ b_1 & b_2 & b_3 \\ c_1 & c_2 & c_3 \end{vmatrix} = 0.$$

A corollary is:

If two of the double points on p' coincide, then the two corresponding branch points on p coincide, unless $T(2,2)$ is compounded of involutions.

The pair of points $|P_1', P_2'|$ of p' which correspond to the point (λ, μ) of p are given by

$$\xi\lambda'^2 + 2\eta\lambda'\mu' + \zeta\mu'^2 = 0, \qquad (\cdot 311)$$

where

$$\xi : \eta : \zeta = a_1\lambda^2 + 2a_2\lambda\mu + a_3\mu^2 : b_1\lambda^2 + 2b_2\lambda\mu + b_3\mu^2 : c_1\lambda^2 + 2c_2\lambda\mu + c_3\mu^2. \qquad (\cdot 312)$$

If $\delta \neq 0$, the equations (·312) give

$$\lambda^2 : 2\lambda\mu : \mu^2 = A_1\xi + B_1\eta + C_1\zeta : A_2\xi + B_2\eta + C_2\zeta : A_3\xi + B_3\eta + C_3\zeta, \qquad (\cdot 313)$$

where $A_1, B_1, \ldots, C_3$ are the co-factors of $a_1, b_1, \ldots, c_3$ in δ. On varying (λ, μ), an ∞^1 system of pairs $|P_1', P_2'|$ is obtained, but each of these pairs determines a unique set of ratios $\xi : \eta : \zeta$, and by (·313) no such set of ratios $\xi : \eta : \zeta$ can arise from two distinct values of $\lambda : \mu$. Thus, if two distinct points of p correspond to the same pair of points of p', then $\delta = 0$.

Conversely, suppose that $\delta = 0$. Then there are numbers p, q, r not all zero such that

$$p(a_1\lambda^2 + 2a_2\lambda\mu + a_3\mu^2) + q(b_1\lambda^2 + 2b_2\lambda\mu + b_3\mu^2) + r(c_1\lambda^2 + 2c_2\lambda\mu + c_3\mu^2) \equiv 0 \qquad (\cdot 314)$$

for all (λ, μ). There are also numbers p', q', r' not all zero such that

$$p'(a_1\lambda'^2 + 2b_1\lambda'\mu' + c_1\mu'^2) + q'(a_2\lambda'^2 + 2b_2\lambda'\mu' + c_2\mu'^2) + r'(a_3\lambda'^2 + 2b_3\lambda'\mu' + c_3\mu'^2) \equiv 0 \qquad (\cdot 315)$$

for all (λ', μ'). It follows from (·314) that all the pairs (P'_1, P'_2) are given by an equation of the form (·311) in which

$$p\xi + q\eta + r\zeta = 0,$$

so these ∞^1 pairs belong to an involution I' on p'. Similarly it follows from (·315) that the ∞^1 pairs of points of p which correspond to points of p' belong to an involution I on p. Now any point P_i of p corresponds to a pair of points $|P'_i, P'_j|$ of p' belonging to I'. If P_j is the mate of P_i in I, both P'_i and P'_j must correspond to the pair $|P_i, P_j|$, and consequently P_j corresponds to $|P'_i, P'_j|$. Thus the $T(2,2)$ is compounded of the involutions I and I', and the theorem is proved.

In fact in any $T(2,2)$ between p and p', if a point P of p corresponds to the points P'_1, P'_2 of p', then P'_1 corresponds to P and another point P_1, while P'_2 corresponds to P and another point P_2. The theorem shows that in general P_1 and P_2 do not coincide for any position of P other than the branch points on p, but that, if P_1 and P_2 coincide for any one position of P other than the branch points, then P_1 and P_2 always coincide.

As an example the correspondence between points of two lines which lie on the same conic of a given point pencil of conics is a $T(2,2)$ which is compounded of involutions.

10·32. A general $T(2,2)$ between two distinct lines p and p' determines an $S(2,2)$ between points of p' which correspond to the same point of p. The branch points of $T(2,2)$ on p' are the branch points of $S(2,2)$, and the double points of $T(2,2)$ on p' are the united points of $S(2,2)$. An alternative proof of the first part of the theorem of 10·31 can be obtained by remarking that if two distinct points of p correspond in $T(2,2)$ to the same pair of points $|P'_1, P'_2|$ on p', then P'_1 and P'_2 are each branch points of $S(2,2)$ which are not branch points of $T(2,2)$. Thus $S(2,2)$ has more than its complement of branch points and every point of p' is a branch point of $S(2,2)$. Hence it follows that, if any point P' of p' corresponds in $T(2,2)$ to points P_1, P_2 of p, then the points of p', other than P', to which P_1 and P_2 correspond in $T(2,2)$, must coincide. Consequently the pairs $|P_1, P_2|$ on p satisfy the conditions of 3·52 and therefore form an involution.

10·4. $S(2,2)$ compounded of an involution

10·41. Suppose that on a conic γ there is an involution I, and that J is an involution among the pairs of I. Then each point P_1 of γ belongs to one pair $|P_1, P_2|$ of I, and this pair $|P_1, P_2|$ belongs to one pair of J, consisting of $|P_1, P_2|$ and another pair $|Q_1, Q_2|$ of I. Thus P_1 determines the two points Q_1, Q_2, and the correspondence so established is clearly an $S(2,2)$. This particular type of $S(2,2)$ is said to be compounded of the involution I. In this $S(2,2)$ the points of any pair of I correspond to the same pair of points.

The argument of 10·31 holds, but an $S(2,2)$ on γ is given by an equation of the form (·211), so the determinant δ of 10·31 becomes the symmetric determinant

$$\delta = \begin{vmatrix} a & h & g \\ h & b & f \\ g & f & c \end{vmatrix},$$

and, if $\delta = 0$, the ratios $p:q:r$ and $p':q':r'$ determined by (·314) and (·315) are equal, so the involutions I and I' coincide. Also the $(1,1)$ correspondence H between the pairs of I and I' becomes an involutory $(1,1)$ correspondence between the pairs of I. Thus:

In a general $S(2,2)$ on a proper conic γ, no two distinct points of γ correspond to the same pair of points. If, however, two particular distinct points of γ correspond to the same pair of points, then the $S(2,2)$ is compounded of an involution. A CONDITION *for the $S(2,2)$ given by (·211) to be compounded of an involution is*

$$\delta \equiv \begin{vmatrix} a & h & g \\ h & b & f \\ g & f & c \end{vmatrix} = 0.$$

10·42. An alternative proof follows at once from 10·21, for, IF $\delta = 0$, the conic s is a line-pair; hence, if P corresponds to $|Q_1, Q_2|$ in $S(2,2)$, the line Q_1Q_2 passes through the vertex O of s, and the other intersection of OP with γ also corresponds to $|Q_1, Q_2|$.

10·43. The preceding theorem gives yet another proof of Poncelet's Theorem in the case $n = 4$. For suppose there is a quadrangle whose vertices A, B, C, D lie on γ and whose sides AB, BC, CD, DA touch S', and let AC, BD meet in O. Then in the $S(2,2)$ on γ, in which two points correspond when their join touches S', the points A and C correspond to the same pair of points $|B, D|$. The $S(2,2)$ is therefore compounded of the involution I cut out on γ by lines through O, and there are ∞^1 pairs of lines through O, meeting γ in pairs of points $|A', C'|$, $|B', D'|$, such that $A'B'$, $B'C'$, $C'D'$, $D'A'$ are tangents of S'.

If S' is given by

$$A'X^2 + B'Y^2 + C'Z^2 + 2F'YZ + 2G'ZX + 2H'XY = 0,$$

the $S(2,2)$ is given by ($\cdot$211), where

$$a:b:c:f:g:h = C':2(B'+G'):A':-2H':B':-2F'.$$

Thus:

There are ∞^1 *ordered quadrangles inscribed in* γ *and circumscribed to* S' IF

$$\begin{vmatrix} C' & -2F' & B' \\ -2F' & 2(B'+G') & -2H' \\ B' & -2H' & A' \end{vmatrix} = 0.$$

10·5. Geometrical representation of a general $T(2,2)$

10·51. A geometrical representation of the general $T(2,2)$ given by ($\cdot$131) is obtained by regarding (λ, μ) and (λ', μ') as ratio parameters of the points of the two conic loci s and s' with equations

$$x:y:z = (a_1\lambda^2 + 2a_2\lambda\mu + a_3\mu^2):$$
$$(b_1\lambda^2 + 2b_2\lambda\mu + b_3\mu^2):(c_1\lambda^2 + 2c_2\lambda\mu + c_3\mu^2)$$

and

$$x:y:z = \mu'^2 : -\lambda'\mu' : \lambda'^2.$$

For the tangent to s' at (λ', μ') is

$$\lambda'^2 x + 2\lambda'\mu' y + \mu'^2 z = 0,$$

and ($\cdot$131) is a CONDITION for this line to contain the point (λ, μ) of s. In fact the points of s' which correspond to a point P of s are the contacts of the tangents to s' through P, while the points of s which correspond to a point P' are the intersections of s with the tangent to s' at P'.

10·52. Provided $\delta \neq 0$ s and s' are both proper and have in general four common points B_1, B_2, B_3, B_4 and four common tangents b_1, b_2, b_3, b_4. Then the branch points on s are B_1, B_2, B_3, B_4 and the branch points on s' are the contacts B_1', B_2', B_3', B_4' of b_1, b_2, b_3, b_4 with s'. It is possible to choose a conic relative to which the points of s reciprocate into the tangents of s', and when this is done B_1, B_2, B_3, B_4 reciprocate into b_i, b_j, b_k, b_l, where (i,j,k,l) is a permutation of $(1, 2, 3, 4)$. Then the cross-ratio (B_1, B_2, B_3, B_4) on s is equal to the cross-ratio (b_i, b_j, b_k, b_l) among the tangents of s', which is equal to the cross-ratio (B'_i, B'_j, B'_k, B'_l) on s'. Thus:

In a general $T(2,2)$ between two $R\infty^1$ the cross-ratio of the four branch elements on one $R\infty^1$ is equal to the cross-ratio of the four branch elements on the other $R\infty^1$ arranged in some order.

The double points of the $T(2,2)$ on s are the contacts of b_1, b_2, b_3, b_4 with s, while the double points on s' are B_1, B_2, B_3, B_4, and by a similar argument the cross-ratio of the two sets of points are equal, when the points are suitably arranged.

This theorem is still true for the branch points when the $T(2,2)$ is compounded of involutions, but the proof just given breaks down because $\delta = 0$ and s is therefore a repeated line $\|p, p\|$.* A general point P of p is given by two distinct values of (λ, μ), which coincide for two positions P_1, P_2, and the branch points on s' are the contacts of tangents to s' through P_1, P_2. The branch values of (λ, μ) are the four values which give the two meets of p and s'. The proof breaks down because it is impossible to reciprocate s into s', but the theorem is easily verified analytically. For suppose that a $T(2,2)$ between two lines p, p' is compounded of two involutions I, I', and choose ratio parameters (λ, μ), (λ', μ') so that the double points of both involutions are $(1, 0)$, $(0, 1)$. Then the pairs of the involutions are given by

$$u\lambda^2 + v\mu^2 = 0, \quad u'\lambda'^2 + v'\mu'^2 = 0,$$

and the homography H between the pairs of I and I' is given by an equation of the form

$$a^2uu' + b^2uv' + c^2u'v + d^2vv' = 0.$$

The double points of I' are given by the values $(1, 0)$, $(0, 1)$ of (u', v'), which correspond in H to the values $(c^2, -a^2)$, $(d^2, -b^2)$ of (u, v), so the branch points B_1, B_2, B_3, B_4 on p are given by the values (a, c), $(-a, c)$, (b, d), $(-b, d)$ of (λ, μ). Similarly the branch points B'_1, B'_2, B'_3, B'_4 on p' are given by the values (a, b), $(-a, b)$, (c, d), $(-c, d)$ of (λ', μ'), and

$$(B_1, B_2, B_3, B_4) = \frac{4abcd}{(ad + bc)^2} = (B'_1, B'_2, B'_3, B'_4).$$

Thus in a homography between the pairs of two involutions I and I' on two lines p and p' the four points of p which correspond to the double

* See 4·42, p. 113.

points of I' have the same cross-ratios as the four points of p' which correspond to the double points of I. Also, if the two pairs of points of p which correspond to the double points of I' are apolar, then the two pairs of points of p' which correspond to the double points of I are apolar.

In particular, given a point pencil of conics σ and two lines p and p', the four points in which p is met by the two conics of σ which touch p' have the same cross-ratio as the four points in which p' is met by the two conics of σ which touch p. Also, if p belongs to the harmonic scroll of the two conics of σ which touch p', then p' belongs to the harmonic scroll of the two conics of σ which touch p, as was shown in 9·19.

10·53. Consider the general $T(2, 2)$ between the lines of two pencils

$$\mu x - \lambda z = 0 \quad \text{and} \quad \mu' x - \lambda' y = 0,$$

given by (·131). The locus of meets of corresponding lines is the quartic curve

$$(a_1 x^2 + 2a_2 xz + a_3 z^2)\, x^2 + 2(b_1 x^2 + 2b_2 xz + b_3 z^2)\, xy + (c_1 x^2 + 2c_2 xz + c_3 z^2)\, y^2 = 0, \quad (\cdot 531)$$

which has double points* at the vertices Y, Z of the pencils. The branch lines of the first pencil are the lines through Y which touch the curve elsewhere, and similarly the branch lines of the other pencil are the tangents through Z.

Conversely, if P is a variable point of a quartic curve C^4 which has two double points D_1 and D_2, then the correspondence between $D_1 P$ and $D_2 P$ is a $T(2, 2)$, since a general line through either double point meets C^4 in two further points. Thus by 10·52:

If a quartic curve C^4 has two double points, there are in general four lines through each double point which touch the curve elsewhere, and these two sub-pencils have equal cross-ratio.

The line YZ, or $x = 0$, is self-corresponding IF $c_3 = 0$, when the quartic curve (·531) reduces to the line $x = 0$ and a cubic curve. Conversely, if P is a variable point and A, B two fixed points on a general cubic curve C^3, the correspondence between AP and BP is a $T(2, 2)$ in which AB is self-corresponding, so by 10·52:

* See 2·44, p. 53.

Through a point A of a general cubic curve pass four tangents of the curve, and the cross-ratio of this subpencil remains constant as A varies on the curve.

The line YZ regarded as a member of either pencil corresponds to itself counted twice IF $c_3 = b_3 = c_2 = 0$, when the quartic curve (·531) reduces to the repeated line $x^2 = 0$ and a conic.

Conversely, if P is a variable point on a conic and A, B are two fixed points not lying on the conic, the correspondence between AP and BP is a $T(2,2)$ of this type. Exceptionally the conic may contain Y or Z, when the $T(2,2)$ is degenerate.

As a metrical example the locus of the foci of the conics of a general line pencil is a cubic curve, but, if the axis of one of the point pairs of the pencil is the line at infinity, the locus is a conic. In fact the foci of conics which touch the sides of a parallelogram lie on a conic.

10·54. The $T(2,2)$ between (λ,μ) and (λ',μ') given by (·131) depends on the ratios of the nine coefficients $a_1, a_2, \ldots, c_3$. A pair of values (λ_1,μ_1) and (λ_1',μ_1') will correspond IF

$$(a_1\lambda_1^2+2a_2\lambda_1\mu_1+a_3\mu_1^2)\lambda_1'^2 \\ +2(b_1\lambda_1^2+2b_2\lambda_1\mu_1+b_3\mu_1^2)\mu_1'^2+(c_1\lambda_1^2+2c_2\lambda_1\mu_1+c_3\mu_1^2)=0,$$

a linear equation in $a_1, a_2, \ldots, c_3$. Thus eight given corresponding pairs of values determine ONE $T(2,2)$ unless the eight linear equations are dependent. If the eight linear equations are dependent the eight pairs are said to be associated.

But if (λ_i,μ_i), (λ_i',μ_i'), $(i = 1,2,\ldots,7)$, are seven given pairs, there is in general ONE other pair (λ_8,μ_8), (λ_8',μ_8') which is a corresponding pair in all the ∞^1 $T(2,2)$ in which the given pairs are corresponding. This pair (λ_8,μ_8), (λ_8',μ_8') is called the eighth associated pair of the seven given pairs. To prove this regard (λ,μ) and (λ',μ') as ratio parameters of two pencils of lines with vertices A and B, the representation being such that the line AB corresponds to the values (λ_7,μ_7) and (λ_7',μ_7'). Let P_i be the meet of the lines (λ_i,μ_i), (λ_i',μ_i'), for $i = 1,2,\ldots,6$. Then for any $T(2,2)$ in which the seven given pairs are corresponding, the locus of meets of corresponding lines is a cubic curve through A, B, P_1, P_2, $\ldots$, P_6. But in general all cubic curves through these eight points contain ONE other point P_8* so the lines AP_8, BP_8 give the required eighth associated pair.

* See for example H. J. Hilton, *Plane Algebraic Curves* (Oxford, 1920), p. 186.

An exceptional case in which the associated point is not unique arises when the six points $P_1, P_2, \ldots, P_6$ lie on a conic through A or through B, let us say the former. In this case all the cubic curves contain seven points of the conic and therefore contain the whole conic, the residual part being a variable line through B. Thus, if P is any point of the conic, the lines AP, BP give a pair of values of (λ, μ), (λ', μ') which correspond in a $T(2, 1)$ which is *contained* in all the $T(2, 2)$ in which the seven given pairs correspond. The equation of any such $T(2, 2)$ is of the form

$$(p\lambda' + q\mu')\{(a_1\lambda^2 + 2a_2\lambda\mu + a_3\mu^2)\lambda' + (b_1\lambda^2 + 2b_2\lambda\mu + b_3\mu^2)\mu'\} = 0.$$

In this degenerate $T(2, 2)$ a general value of (λ, μ) corresponds to two values of (λ', μ'), of which one is the fixed value $(-q, p)$; to a general value of (λ', μ') there corresponds two values of (λ, μ), but to the particular value $(-q, p)$ of (λ', μ') there correspond all values of (λ, μ).

CHAPTER XI

APPLICATION OF MATRIX ALGEBRA

11·1. Introduction. The conic, polarity and reciprocation

11·11. A little knowledge of matrix algebra will be required in this chapter. A good introduction, sufficient for the present purpose, will be found in *Advanced Algebra*, vol. III, by Durell and Robson.* The following is a summary of the notation that will be used, and of the knowledge that will be required.

Clarendon type will be used to denote matrices and a dash will denote transposition. A matrix is said to be zero or to vanish IF all its coefficients are zero. Let $\mathbf{M}$ denote the square matrix (m_{ij}), where $i, j = 1, 2, \ldots, n$; that is the matrix whose coefficient in the ith row and jth column is m_{ij}. The determinant of the matrix will be denoted by $|m_{ij}|$ or by $|\mathbf{M}|$. The rth power of $|\mathbf{M}|$ will be written $|\mathbf{M}|^r$. The matrix (p_{ij}) in which p_{ij} is the co-factor of m_{ji} in (m_{ij}) is called the adjoint of $\mathbf{M}$ and will be denoted by $\mathbf{M}_a$. If $|\mathbf{M}| \neq 0$ the matrix (q_{ij}) in which $q_{ij} = \dfrac{p_{ij}}{|\mathbf{M}|}$ is called the inverse of $\mathbf{M}$ and is denoted by $\mathbf{M}^{-1}$; then $\mathbf{M}_a = |\mathbf{M}| \times \mathbf{M}^{-1}$. If $|\mathbf{M}| = 0$, the matrix $\mathbf{M}$ is said to be singular, and the inverse matrix is undefined.

If λ is a scalar, $\lambda\mathbf{M}$ is the matrix obtained by multiplying every coefficient of $\mathbf{M}$ by λ. Thus:

$$|\lambda\mathbf{M}| = \lambda^n |\mathbf{M}|, \quad (\lambda\mathbf{M})_a = \lambda^{n-1}\mathbf{M}_a, \quad (\lambda\mathbf{M})^{-1} = \lambda^{-1}\mathbf{M}^{-1}.$$

If $\mathbf{N} = (n_{ij})$ is another square matrix of order n, the product $\mathbf{MN}$ is the matrix $(m_{ik}n_{kj})$, where the dummy suffix summation convention is used and $k = 1, 2, \ldots, n$. Then

$$|\mathbf{MN}| = |\mathbf{NM}| = |\mathbf{M}| \times |\mathbf{N}|,$$
$$(\mathbf{MN})' = \mathbf{N}'\mathbf{M}',$$
$$(\mathbf{MN})_a = \mathbf{N}_a\mathbf{M}_a$$

and, if $|\mathbf{M}| \neq 0$ and $|\mathbf{N}| \neq 0$,

$$(\mathbf{MN})^{-1} = \mathbf{N}^{-1}\mathbf{M}^{-1}.$$

Similar equations may be deduced for the product $\mathbf{MN}\ldots\mathbf{P}$ of any number of square matrices of the same order.

* London: G. Bell and Sons, Ltd. 1937.

In general the product $\mathbf{NM}$ is not equal to $\mathbf{MN}$, though the equality may hold in certain particular cases. For instance, if $\mathbf{I}_n$ is the unit matrix of order n,

$$\mathbf{MI}_n = \mathbf{I}_n\mathbf{M} = \mathbf{M}, \quad \mathbf{MM}_a = \mathbf{M}_a\mathbf{M} = |\mathbf{M}| \times \mathbf{I}_n,$$

and, if $|\mathbf{M}| \neq 0$, $\qquad \mathbf{MM}^{-1} = \mathbf{M}^{-1}\mathbf{M} = \mathbf{I}_n.$

It follows that $\qquad |\mathbf{M}_a| = |\mathbf{M}|^{n-1}$

and, provided $|\mathbf{M}| \neq 0$,

$$|\mathbf{M}^{-1}| = \frac{1}{|\mathbf{M}|}.$$

Since $(\mathbf{M}_a)_a\mathbf{M}_a = (\mathbf{MM}_a)_a = (|\mathbf{M}| \times \mathbf{I}_n)_a = |\mathbf{M}|^{n-1}\mathbf{I}_n$ it follows on multiplying after by $\mathbf{M}$ that

$$|\mathbf{M}|(\mathbf{M}_a)_a = |\mathbf{M}|^{n-1}\mathbf{M}.$$

Thus, if $|\mathbf{M}| \neq 0$, $\qquad (\mathbf{M}_a)_a = |\mathbf{M}|^{n-2}\mathbf{M}.$

Also, if $|\mathbf{M}| \neq 0$, then $\qquad (\mathbf{M}^{-1})' = (\mathbf{M}')^{-1}.$

A matrix $\mathbf{M}$ is said to be symmetric IF it is a square matrix and is equal to its transposed matrix $\mathbf{M}'$; it is skew-symmetric if $\mathbf{M} = -\mathbf{M}'$.

If $\mathbf{M}$ and $\mathbf{N}$ are not both square matrices, the product $\mathbf{MN}$ is still defined if the number of columns in $\mathbf{M}$ is equal to the number of rows of $\mathbf{N}$, and it is still true that $(\mathbf{MN})' = \mathbf{N}'\mathbf{M}'$.

11·12. In this chapter the co-ordinates of the points and lines of a plane will be (x_1, x_2, x_3) and $[X_1, X_2, X_3]$ instead of (x, y, z) and $[X, Y, Z]$. A particular point P_i will be (x_{1i}, x_{2i}, x_{3i}) and a particular line p_i will be $[X_{1i}, X_{2i}, X_{3i}]$. The points will be represented by column matrices

$$\mathbf{x} \equiv \begin{pmatrix} x_1 \\ x_2 \\ x_3 \end{pmatrix} \quad \text{and} \quad \mathbf{x}_i \equiv \begin{pmatrix} x_{1i} \\ x_{2i} \\ x_{3i} \end{pmatrix},$$

and the lines by column matrices

$$\mathbf{X} \equiv \begin{pmatrix} X_1 \\ X_2 \\ X_3 \end{pmatrix} \quad \text{and} \quad \mathbf{X}_i \equiv \begin{pmatrix} X_{1i} \\ X_{2i} \\ X_{3i} \end{pmatrix}.$$

If λ is a non-zero scalar, $\lambda\mathbf{x}$ represents the same point as $\mathbf{x}$ and $\lambda\mathbf{X}$ represents the same line as $\mathbf{X}$. The point equation of the line p_i is $\mathbf{X}_i'\mathbf{x} = 0$ or $\mathbf{x}'\mathbf{X}_i = 0$ and the line equation of the point P_i is

$\mathbf{X}'\mathbf{x}_i = 0$ or $\mathbf{x}_i'\mathbf{X} = 0$. Any column matrix $\mathbf{r} \equiv (r_1, r_2, r_3)'$, whose three coefficients are not all zero, may be regarded as representing a point, which will be called the point $\mathbf{r}$; similarly any non-vanishing column matrix $\mathbf{R} \equiv (R_1, R_2, R_3)'$ may be regarded as representing a line, which will be called the line $\mathbf{R}$.

11·13. The point equation of a conic locus is

$$s \equiv a_{11}x_1^2 + a_{22}x_2^2 + a_{33}x_3^2 + 2a_{23}x_2x_3 + 2a_{31}x_3x_1 + 2a_{12}x_1x_2 = 0.$$

By taking $\quad a_{32} = a_{23}, \quad a_{13} = a_{31}, \quad a_{21} = a_{12},$

this equation may be written in the form

$$s \equiv \mathbf{x}'\mathbf{s}\mathbf{x} = 0,$$

where $\mathbf{s}$ is the symmetrical matrix

$$\begin{pmatrix} a_{11} & a_{12} & a_{13} \\ a_{21} & a_{22} & a_{23} \\ a_{31} & a_{32} & a_{33} \end{pmatrix},$$

which is called the matrix of the quadratic form s. If the conic locus is given, the matrix $\mathbf{s}$ is not uniquely determined, but the ratios of the coefficients of $\mathbf{s}$ are uniquely determined. The matrix $\mathbf{s}$ is called a point matrix of the conic locus, and, if λ is a non-zero scalar, $\lambda\mathbf{s}$ is a point matrix of the same conic locus. We shall in the future occasionally speak of *the* point matrix of a conic locus; although this is not strictly accurate, it is unlikely to lead to confusion.

Any non-vanishing symmetrical square matrix $\mathbf{m}$ of order three determines a unique conic locus, $\mathbf{x}'\mathbf{m}\mathbf{x} = 0$, of which it is a point matrix, and this will be called the conic locus $\mathbf{m}$. The conic locus $\mathbf{m}$ is a line-pair IF $|\mathbf{m}| = 0$, and the equation of the line-pair formed by p_1 and p_2 is

$$\mathbf{x}'\mathbf{X}_1\mathbf{X}_2'\mathbf{x} = 0.$$

Dually, any conic scroll is given by an equation of the form

$$\mathbf{X}'\mathbf{S}\mathbf{X} = 0,$$

where $\mathbf{S}$ is a symmetrical matrix of order three, which is called a line matrix of the conic scroll. The conic scroll $\mathbf{S}$ is a point-pair IF $|\mathbf{S}| = 0$, and the line equation of the point-pair $\| P_1, P_2 \|$ is

$$\mathbf{X}'\mathbf{x}_1\mathbf{x}_2'\mathbf{X} = 0.$$

It should be noticed that the point matrix of a conic locus was defined to be a symmetric matrix. Thus a non-symmetric three-row square matrix $\mathbf{a}$ is not regarded as a point matrix of the conic locus whose equation is $\mathbf{x}'\mathbf{a}\mathbf{x} = 0$, but the symmetric matrix $\mathbf{b}$ determined by $2\mathbf{b} = \mathbf{a}+\mathbf{a}'$ is a point matrix of this conic locus.* In fact when such a phrase as 'the conic locus $\mathbf{m}$' occurs, it implies that $\mathbf{m}$ is a symmetric non-zero matrix of order three, and similarly for scrolls. As far as possible small Clarendon letters will denote point matrices and capital Clarendon letters will denote line matrices.

11·14. Expressed in the new notation the results of 4·71 and 4·72 are as follows. Consider first a conic locus $\mathbf{s}$ with point equation $\mathbf{x}'\mathbf{s}\mathbf{x} = 0$. The tangents of $\mathbf{s}$ are the lines which satisfy the equation $\mathbf{X}'\mathbf{s}_a\mathbf{X} = 0$. Two points P_1 and P_2 are conjugate relative to $\mathbf{s}$ IF $\mathbf{x}_1'\mathbf{s}\mathbf{x}_2 = 0$ or IF $\mathbf{x}_2'\mathbf{s}\mathbf{x}_1 = 0$.† Two lines p_1 and p_2 are conjugate relative to $\mathbf{s}$ IF $\mathbf{X}_1'\mathbf{s}_a\mathbf{X}_2 = 0$ or IF $\mathbf{X}_2'\mathbf{s}_a\mathbf{X}_1 = 0$. The polar of P_1 relative to $\mathbf{s}$ is the line $\mathbf{s}\mathbf{x}_1$ and the pole of p_1 relative to $\mathbf{s}$ is the point $\mathbf{s}_a\mathbf{X}_1$.

Dually consider a conic scroll $\mathbf{S}$ with line equation $\mathbf{X}'\mathbf{S}\mathbf{X} = 0$. The contacts of $\mathbf{S}$ are the points which satisfy the equation $\mathbf{x}'\mathbf{S}_a\mathbf{x} = 0$. Two lines p_1 and p_2 are conjugate relative to $\mathbf{S}$ IF $\mathbf{X}_1'\mathbf{S}\mathbf{X}_2 = 0$ or IF $\mathbf{X}_2'\mathbf{S}\mathbf{X}_1 = 0$. Two points P_1 and P_2 are conjugate relative to $\mathbf{S}$ IF $\mathbf{x}_1'\mathbf{S}_a\mathbf{x}_2 = 0$ or IF $\mathbf{x}_2'\mathbf{S}_a\mathbf{x}_1 = 0$. The pole of p_1 relative to $\mathbf{S}$ is the point $\mathbf{S}\mathbf{X}_1$ and the polar of P_1 relative to $\mathbf{S}$ is the line $\mathbf{S}_a\mathbf{x}_1$.

If $|\mathbf{S}| \neq 0$, $(\mathbf{s}_a)_a = |\mathbf{s}|\,\mathbf{s}$, so the locus of contacts of the conic scroll formed by the tangents of the conic locus $\mathbf{s}$ is the conic locus $\mathbf{s}$ itself, and the relation between pole and polar is the same for the locus and the scroll. Thus, as in 4·57, a proper conic may be regarded either as a locus or as a scroll, and is determined either by a point matrix or by a line matrix.

11·15. Consider two conic loci with point matrices $\mathbf{l}$, $\mathbf{m}$ and a conic scroll with line matrix $\mathbf{N}$. The polars of a point P_1 relative to $\mathbf{l}$ and $\mathbf{m}$ are the lines $\mathbf{l}\mathbf{x}_1$ and $\mathbf{m}\mathbf{x}_1$, which are conjugate relative to $\mathbf{N}$ IF $\mathbf{x}_1'\mathbf{l}\mathbf{N}\mathbf{m}\mathbf{x}_1 = 0$, that is IF P_1 lies on the conic locus $\mathbf{x}'\mathbf{l}\mathbf{N}\mathbf{m}\mathbf{x} = 0$.‡ Thus:

* If $\mathbf{a}$ is skew-symmetric the scalar expression $\mathbf{x}'\mathbf{a}\mathbf{x}$ vanishes identically, for its transpose $\mathbf{x}'\mathbf{a}'\mathbf{x}$ is equal and opposite to itself.

† Since $\mathbf{x}_1'\mathbf{s}\mathbf{x}_2$ is a scalar it is equal to its transpose, namely $\mathbf{x}_2'\mathbf{s}'\mathbf{x}_1$, and $\mathbf{s}' = \mathbf{s}$, so that $\mathbf{x}_1'\mathbf{s}\mathbf{x}_2 = \mathbf{x}_2'\mathbf{s}\mathbf{x}_1$.

‡ It should be remarked that the matrix $\mathbf{l}\mathbf{N}\mathbf{m}$ is not necessarily symmetrical, although $\mathbf{l}$, $\mathbf{m}$ and $\mathbf{N}$ are symmetrical individually.

The locus of a point whose polar lines relative to two conic loci **l** *and* **m** *are conjugate relative to a conic scroll* **N** *is the conic locus*

$$\mathbf{x'lNmx} = 0.$$

In particular, taking $\mathbf{l} = \mathbf{m}$,

The reciprocal of a conic scroll **N** *relative to a conic locus* **m** *has point matrix* **mNm**.*

Dually:

The lines whose poles relative to two conic scrolls **L** *and* **M** *are conjugate relative to a conic locus* **n** *belong to the conic scroll* $\mathbf{X'LnMX} = 0$.

The reciprocal of a conic locus **n** *relative to a conic scroll* **M** *has line matrix* **MnM**.

11·16. If **m** and **n** are point matrices of two conic loci, the general conic locus of the pencil determined by them has point equation

$$\lambda\mathbf{x'mx} + \mu\mathbf{x'nx} = 0 \quad \text{or} \quad \mathbf{x'}(\lambda\mathbf{m} + \mu\mathbf{n})\mathbf{x} = 0,$$

and therefore has $\lambda\mathbf{m} + \mu\mathbf{n}$ as a point matrix. The line equation of this conic locus is

$$\mathbf{X'}(\lambda\mathbf{m} + \mu\mathbf{n})_a\mathbf{X} = 0.$$

But each coefficient in the matrix $(\lambda\mathbf{m} + \mu\mathbf{n})_a$ is a quadratic in (λ, μ) so the matrix may be expanded in the form

$$(\lambda\mathbf{m} + \mu\mathbf{n})_a = \mathbf{P}\lambda^2 + \mathbf{Q}\lambda\mu + \mathbf{R}\mu^2,$$

where **P**, **Q**, **R** are symmetric matrices. Evidently $\mathbf{P} = \mathbf{m}_a$ and $\mathbf{R} = \mathbf{n}_a$. The line equation

$$\mathbf{X'}(\lambda\mathbf{m} + \mu\mathbf{n})_a\mathbf{X} = 0$$

is thus

$$\lambda^2\mathbf{X'PX} + \lambda\mu\mathbf{X'QX} + \mu^2\mathbf{X'RX} = 0,$$

and it follows from the argument of 9·12 that the line equation of the harmonic envelope of **m** and **n** is $\mathbf{X'QX} = 0$. Thus:

If **m** *and* **n** *are point matrices of two conic loci and if*

$$(\lambda\mathbf{m} + \mu\mathbf{n})_a = \lambda^2\mathbf{m}_a + \lambda\mu\mathbf{Q} + \mu^2\mathbf{n}_a,$$

then **Q** *is a line matrix of the harmonic envelope of* **m** *and* **n**.

Dually, if **M** and **N** are line matrices of two conic scrolls, the general conic scroll of the tangential pencil determined by **M** and **N** has $\lambda\mathbf{M} + \mu\mathbf{N}$ as a line matrix. Also:

If $(\lambda\mathbf{M} + \mu\mathbf{N})_a = \lambda^2\mathbf{M}_a + \lambda\mu\mathbf{q} + \mu^2\mathbf{N}_a$, *then* **q** *is a point matrix of the harmonic locus of the conic scrolls* **M** *and* **N**.

* **mNm** is a symmetrical matrix.

11·2. Two proper conics. Harmonic locus and scroll

11·21. The main object of this chapter is to discuss the configuration determined by two proper conics. We shall denote their point matrices by $\mathbf{s}_1$ and $\mathbf{s}_2$,* and their line matrices $(\mathbf{s}_1)_a$ and $(\mathbf{s}_2)_a$ by $\mathbf{S}_1$ and $\mathbf{S}_2$. Let $\mathbf{s}_1 = (a_{ij})$, $\mathbf{s}_2 = (b_{ij})$, $\mathbf{S}_1 = (A_{ij})$, $\mathbf{S}_2 = (B_{ij})$, where $i, j = 1, 2, 3$. Then A_{ij} and B_{ij} are the co-factors of a_{ji} and b_{ji} in the determinants $|a_{ij}|$ and $|b_{ij}|$. The unit matrix $\mathbf{I}_3$ will be denoted by $\mathbf{I}$ and $|\mathbf{s}_1|$, $|\mathbf{s}_2|$, $|\mathbf{S}_1|$, $|\mathbf{S}_2|$ will be denoted by δ_1, δ_2, Δ_1, Δ_2. Since the conics are proper, none of δ_1, δ_2, Δ_1, Δ_2 is zero. Also $\Delta_1 = \delta_1^2$, $\Delta_2 = \delta_2^2$, $(\mathbf{S}_1)_a = \delta_1 \mathbf{s}_1$ and $(\mathbf{S}_2)_a = \delta_2 \mathbf{s}_2$. It is important to remember that

$$\mathbf{S}_1 \mathbf{s}_1 = \mathbf{s}_1 \mathbf{S}_1 = \delta_1 \mathbf{I} \qquad (\cdot 211)$$

and

$$\mathbf{S}_2 \mathbf{s}_2 = \mathbf{s}_2 \mathbf{S}_2 = \delta_2 \mathbf{I}. \qquad (\cdot 212)$$

The four expansions

$$|\lambda \mathbf{s}_1 + \mu \mathbf{s}_2| = \lambda^3 \delta_1 + \lambda^2 \mu \theta_1 + \lambda \mu^2 \theta_2 + \mu^3 \delta_2, \qquad (\cdot 213)$$

$$(\lambda \mathbf{s}_1 + \mu \mathbf{s}_2)_a = \lambda^2 \mathbf{S}_1 + \lambda \mu \mathbf{G} + \mu^2 \mathbf{S}_2, \qquad (\cdot 214)$$

$$|\lambda \mathbf{S}_1 + \mu \mathbf{S}_2| = \lambda^3 \Delta_1 + \lambda^2 \mu \Theta_1 + \lambda \mu^2 \Theta_2 + \mu^3 \Delta_2, \qquad (\cdot 215)$$

$$(\lambda \mathbf{S}_1 + \mu \mathbf{S}_2)_a = \lambda^2 (\mathbf{S}_1)_a + \lambda \mu \mathbf{f} + \mu^2 (\mathbf{S}_2)_a, \qquad (\cdot 216)$$

determine the scalars $\theta_1, \theta_2, \Theta_1, \Theta_2$ and the two symmetric matrices $\mathbf{G}$ and $\mathbf{f}$ as functions of the coefficients of $\mathbf{s}_1$ and $\mathbf{s}_2$. In fact

$$\theta_1 = A_{ij} b_{ij}, \quad \theta_2 = a_{ij} B_{ij},$$

where the summations are for $i, j = 1, 2, 3$, and $\mathbf{G}$ is the matrix

$$(g_{ij}), \; i, j = 1, 2, 3,$$

where

$$\begin{aligned} g_{11} &= a_{22} b_{33} + a_{33} b_{22} - 2 a_{23} b_{23}, \\ g_{22} &= a_{33} b_{11} + a_{11} b_{33} - 2 a_{31} b_{31}, \\ g_{33} &= a_{11} b_{22} + a_{22} b_{11} - 2 a_{12} b_{12}, \\ g_{23} = g_{32} &= a_{31} b_{12} + a_{12} b_{31} - a_{11} b_{23} - b_{11} a_{23}, \\ g_{31} = g_{13} &= a_{12} b_{23} + a_{23} b_{12} - a_{22} b_{31} - b_{22} a_{31}, \\ g_{12} = g_{21} &= a_{23} b_{31} + a_{31} b_{23} - a_{33} b_{12} - b_{33} a_{12}. \end{aligned}$$

The matrix $\mathbf{f}$ is similar to $\mathbf{G}$ but with A_{ij} and B_{ij} instead of a_{ij} and b_{ij}. Also since $(\mathbf{S}_1)_a = \delta_1 \mathbf{s}_1$ and $(\mathbf{S}_2)_a = \delta_2 \mathbf{s}_2$,

$$\Theta_1 = \delta_1 a_{ij} B_{ij} = \delta_1 \theta_2 \quad \text{and} \quad \Theta_2 = \delta_2 A_{ij} b_{ij} = \delta_2 \theta_1.$$

Thus:

$$|\lambda \mathbf{S}_1 + \mu \mathbf{S}_2| = \lambda^3 \delta_1^2 + \lambda^2 \mu \delta_1 \theta_2 + \lambda \mu^2 \delta_2 \theta_1 + \mu^3 \delta_2^2, \qquad (\cdot 217)$$

$$(\lambda \mathbf{S}_1 + \mu \mathbf{S}_2)_a = \lambda^2 \delta_1 \mathbf{s}_1 + \lambda \mu \mathbf{f} + \mu^2 \delta_2 \mathbf{s}_2. \qquad (\cdot 218)$$

* Cf. 9·11, p. 233.

It follows from 11·16 that $\mathbf{G}$ is a line matrix of the harmonic envelope of $\mathbf{s}_1$ and $\mathbf{s}_2$ and that $\mathbf{f}$ is a point matrix of the harmonic locus of $\mathbf{s}_1$ and $\mathbf{s}_2$.

11·22. Since

$$(\lambda\mathbf{s}_1+\mu\mathbf{s}_2)(\lambda\mathbf{s}_1+\mu\mathbf{s}_2)_a = (\lambda\mathbf{s}_1+\mu\mathbf{s}_2)_a(\lambda\mathbf{s}_1+\mu\mathbf{s}_2) = |\lambda\mathbf{s}_1+\mu\mathbf{s}_2|\,\mathbf{I},$$

and

$$(\lambda\mathbf{S}_1+\mu\mathbf{S}_2)(\lambda\mathbf{S}_1+\mu\mathbf{S}_2)_a = (\lambda\mathbf{S}_1+\mu\mathbf{S}_2)_a(\lambda\mathbf{S}_1+\mu\mathbf{S}_2) = |\lambda\mathbf{S}_1+\mu\mathbf{S}_2|\,\mathbf{I},$$

it follows that

$$\begin{aligned}(\lambda\mathbf{s}_1+\mu\mathbf{s}_2)(\lambda^2\mathbf{S}_1+\lambda\mu\mathbf{G}+\mu^2\mathbf{S}_2) &= (\lambda^2\mathbf{S}_1+\lambda\mu\mathbf{G}+\mu^2\mathbf{S}_2)(\lambda\mathbf{s}_1+\mu\mathbf{s}_2)\\ &= (\lambda^3\delta_1+\lambda^2\mu\theta_1+\lambda\mu^2\theta_2+\mu^3\delta_2)\,\mathbf{I},\end{aligned}$$

and

$$\begin{aligned}(\lambda\mathbf{S}_1+\mu\mathbf{S}_2)&(\lambda^2\delta_1\mathbf{s}_1+\lambda\mu\mathbf{f}+\mu^2\delta_2\mathbf{s}_2)\\ &= (\lambda^2\delta_1\mathbf{s}_1+\lambda\mu\mathbf{f}+\mu^2\delta_2\mathbf{s}_2)(\lambda\mathbf{S}_1+\mu\mathbf{S}_2)\\ &= (\lambda^3\delta_1^2+\lambda^2\mu\delta_1\theta_2+\lambda\mu^2\delta_2\theta_1+\mu^3\delta_2^2)\,\mathbf{I}.\end{aligned}$$

These equations are true for all values of λ and μ, so, by equating coefficients of $\lambda^2\mu$ and $\lambda\mu^2$, it follows that

$$\mathbf{s}_1\mathbf{G}+\mathbf{s}_2\mathbf{S}_1 = \mathbf{G}\mathbf{s}_1+\mathbf{S}_1\mathbf{s}_2 = \theta_1\mathbf{I}, \tag{·221}$$

$$\mathbf{s}_2\mathbf{G}+\mathbf{s}_1\mathbf{S}_2 = \mathbf{G}\mathbf{s}_2+\mathbf{S}_2\mathbf{s}_1 = \theta_2\mathbf{I}, \tag{·222}$$

$$\mathbf{S}_1\mathbf{f}+\delta_1\mathbf{S}_2\mathbf{s}_1 = \mathbf{f}\mathbf{S}_1+\delta_1\mathbf{s}_1\mathbf{S}_2 = \delta_1\theta_2\mathbf{I}, \tag{·223}$$

$$\mathbf{S}_2\mathbf{f}+\delta_2\mathbf{S}_1\mathbf{s}_2 = \mathbf{f}\mathbf{S}_2+\delta_2\mathbf{s}_2\mathbf{S}_1 = \delta_2\theta_1\mathbf{I}. \tag{·224}$$

From these four equations it is possible to deduce the following eighteen formulae, from which further formulae can be obtained by interchanging the suffixes 1 and 2.

11·23.

$$\mathbf{s}_1\mathbf{S}_2\mathbf{s}_1 = \theta_2\mathbf{s}_1-\mathbf{f}, \tag{·231}$$

$$\mathbf{s}_1\mathbf{S}_2\mathbf{f} = \delta_2(\theta_1\mathbf{s}_1-\delta_1\mathbf{s}_2), \tag{·232}$$

$$\mathbf{s}_1\mathbf{G}\mathbf{s}_1 = \theta_1\mathbf{s}_1-\delta_1\mathbf{s}_2, \tag{·233}$$

$$\mathbf{s}_1\mathbf{G}\mathbf{s}_2 = \mathbf{f}, \tag{·234}$$

$$\mathbf{s}_1\mathbf{G}\mathbf{f} = \theta_1\mathbf{f}-\delta_1(\theta_2\mathbf{s}_2-\delta_2\mathbf{s}_1), \tag{·235}$$

$$\mathbf{f}\mathbf{S}_1\mathbf{s}_2 = \delta_1(\theta_2\mathbf{s}_2-\delta_2\mathbf{s}_1), \tag{·236}$$

$$\mathbf{f}\mathbf{S}_1\mathbf{f} = \delta_1\{\theta_2\mathbf{f}-\delta_2(\theta_1\mathbf{s}_1-\delta_1\mathbf{s}_2)\}, \tag{·237}$$

$$\mathbf{f}\mathbf{G}\mathbf{s}_1 = \theta_1\mathbf{f}-\delta_1(\theta_2\mathbf{s}_2-\delta_2\mathbf{s}_1), \tag{·238}$$

$$\mathbf{f}\mathbf{G}\mathbf{f} = \delta_2(\theta_2\delta_1-\theta_1^2)\,\mathbf{s}_1+\delta_1(\theta_1\delta_2-\theta_2^2)\,\mathbf{s}_2+(\theta_1\theta_2+\delta_1\delta_2)\,\mathbf{f}. \tag{·239}$$

11·24.

$$\mathbf{S}_1\mathbf{s}_2\mathbf{S}_1 = \theta_1\mathbf{S}_1 - \delta_1\mathbf{G}, \qquad (\cdot 241)$$

$$\mathbf{S}_1\mathbf{s}_2\mathbf{G} = \theta_2\mathbf{S}_1 - \delta_1\mathbf{S}_2, \qquad (\cdot 242)$$

$$\mathbf{S}_1\mathbf{f}\mathbf{S}_1 = \delta_1(\theta_2\mathbf{S}_1 - \delta_1\mathbf{S}_2), \qquad (\cdot 243)$$

$$\mathbf{S}_1\mathbf{f}\mathbf{S}_2 = \delta_1\delta_2\mathbf{G}, \qquad (\cdot 244)$$

$$\mathbf{S}_1\mathbf{f}\mathbf{G} = \delta_1(\theta_2\mathbf{G} - \theta_1\mathbf{S}_2 + \delta_2\mathbf{S}_1), \qquad (\cdot 245)$$

$$\mathbf{G}\mathbf{s}_1\mathbf{S}_2 = \theta_1\mathbf{S}_2 - \delta_2\mathbf{S}_1, \qquad (\cdot 246)$$

$$\mathbf{G}\mathbf{s}_1\mathbf{G} = \theta_1\mathbf{G} - \theta_2\mathbf{S}_1 + \delta_1\mathbf{S}_2, \qquad (\cdot 247)$$

$$\mathbf{G}\mathbf{f}\mathbf{S}_1 = \delta_1(\theta_2\mathbf{G} - \theta_1\mathbf{S}_2 + \delta_2\mathbf{S}_1), \qquad (\cdot 248)$$

$$\mathbf{G}\mathbf{f}\mathbf{G} = (\theta_1\delta_2 - \theta_2^2)\,\mathbf{S}_1 + (\theta_2\delta_1 - \theta_1^2)\,\mathbf{S}_2 + (\theta_1\theta_2 + \delta_1\delta_2)\,\mathbf{G}. \qquad (\cdot 249)$$

11·25. The formulae (·231), (·232), (·233) and (·236) follow immediately from 11·22. For instance

$$\delta_1\mathbf{s}_1\mathbf{S}_2\mathbf{s}_1 = (\delta_1\theta_2\mathbf{I} - \mathbf{f}\mathbf{S}_1)\,\mathbf{s}_1, \quad \text{by } (\cdot 223),$$
$$= \delta_1(\theta_2\mathbf{s}_1 - \mathbf{f}), \quad \text{by } (\cdot 211),$$

which proves (·231).

Also
$$\mathbf{s}_1\mathbf{G}\mathbf{s}_2 = \mathbf{s}_1(\theta_2\mathbf{I} - \mathbf{S}_2\mathbf{s}_1), \quad \text{by } (\cdot 222),$$
$$= \theta_2\mathbf{s}_1 - \mathbf{s}_1\mathbf{S}_2\mathbf{s}$$
$$= \mathbf{f}, \quad \text{by } (\cdot 231),$$

which proves (·234).

Also
$$\mathbf{s}_1\mathbf{G}\mathbf{f} = (\theta_1\mathbf{I} - \mathbf{s}_2\mathbf{S}_1)\,\mathbf{f}, \quad \text{by } (\cdot 221),$$
$$= \theta_1\mathbf{f} - \delta_1(\theta_2\mathbf{s}_2 - \delta_2\mathbf{s}_1)$$

by interchanging the suffixes in (·232), which proves (·235).

Also
$$\mathbf{f}\mathbf{S}_1\mathbf{f} = \delta_1\mathbf{f}(\theta_2\mathbf{I} - \mathbf{S}_2\mathbf{s}_1), \quad \text{by } (\cdot 223),$$
$$= \delta_1\{\theta_2\mathbf{f} - \delta_2(\theta_1\mathbf{s}_1 - \delta_1\mathbf{s}_2)\}$$

by interchanging the suffixes in (·236), proving (·237).

Also
$$\mathbf{f}\mathbf{G}\mathbf{s}_1 = \mathbf{f}(\theta_1\mathbf{I} - \mathbf{S}_1\mathbf{s}_2), \quad \text{by } (\cdot 221),$$
$$= \theta_1\mathbf{f} - \delta_1(\theta_2\mathbf{s}_2 - \delta_2\mathbf{s}_1), \quad \text{by } (\cdot 236),$$

proving (·238).

Also
$$\delta_1\mathbf{f}\mathbf{G}\mathbf{f} = \mathbf{f}\mathbf{G}\mathbf{s}_1\mathbf{S}_1\mathbf{f}$$
$$= (\theta_1\mathbf{f} - \delta_1\theta_2\mathbf{s}_2 + \delta_1\delta_2\mathbf{s}_1)\,\mathbf{S}_1\mathbf{f}, \quad \text{by } (\cdot 238),$$

and (·239) follows from this and (·237).

Thus all the formulae of 11·23 are established. The proofs of the formulae of 11·24 are left to the reader.

It should be noticed that

$$\mathbf{s}_1 \mathbf{G} \mathbf{s}_2 = \mathbf{s}_2 \mathbf{G} \mathbf{s}_1, \tag{$\cdot 251$}$$

$$\mathbf{s}_1 \mathbf{G} \mathbf{f} = \mathbf{f} \mathbf{G} \mathbf{s}_1, \tag{$\cdot 252$}$$

$$\mathbf{s}_1 \mathbf{S}_2 \mathbf{f} = \mathbf{f} \mathbf{S}_2 \mathbf{s}_1, \tag{$\cdot 253$}$$

from 11·23, and similarly from 11·24

$$\mathbf{S}_1 \mathbf{f} \mathbf{S}_2 = \mathbf{S}_2 \mathbf{f} \mathbf{S}_1, \tag{$\cdot 254$}$$

$$\mathbf{S}_1 \mathbf{f} \mathbf{G} = \mathbf{G} \mathbf{f} \mathbf{S}_1, \tag{$\cdot 255$}$$

$$\mathbf{S}_1 \mathbf{s}_2 \mathbf{G} = \mathbf{G} \mathbf{s}_2 \mathbf{S}_1. \tag{$\cdot 256$}$$

11·26. The formulae of 11·23 and 11·24 may all be interpreted by means of the theorems of 11·15. For instance it follows from (·231) and (·241) that the reciprocal of $\mathbf{s}_2$ relative to $\mathbf{s}_1$ has point matrix $\theta_2 \mathbf{s}_1 - \mathbf{f}$ and line matrix $\theta_1 \mathbf{S}_1 - \delta_1 \mathbf{G}$. Similarly from (·233) and (·243) it follows that the reciprocal of $\mathbf{G}$ relative to $\mathbf{s}_1$ has point matrix $\theta_1 \mathbf{s}_1 - \delta_1 \mathbf{s}_2$, and that the reciprocal of $\mathbf{f}$ relative to $\mathbf{s}_1$ has line matrix $\theta_2 \mathbf{S}_1 - \delta_1 \mathbf{S}_2$. Also, from (·234) and (·244):

The conic $\mathbf{f}$ *is the locus of points whose polar lines relative to* $\mathbf{s}_1$ *and* $\mathbf{s}_2$ *are conjugate relative to* $\mathbf{G}$, *and dually the tangents of* $\mathbf{G}$ *are the lines whose poles relative to* $\mathbf{s}_1$ *and* $\mathbf{s}_2$ *are conjugate relative to* $\mathbf{f}$.

11·27. Consider triangles PQR which are self-polar relative to $\mathbf{s}_2$ and such that the vertices Q and R lie on $\mathbf{s}_1$. A point P is the vertex of such a triangle IF the polar of P relative to $\mathbf{s}_2$ meets $\mathbf{s}_1$ and $\mathbf{s}_2$ in apolar pairs of points, that is IF this polar is a tangent of $\mathbf{G}$. Thus the locus of P is the reciprocal of $\mathbf{G}$ relative to $\mathbf{s}_2$, which is a conic with point matrix $\theta_2 \mathbf{s}_2 - \delta_2 \mathbf{s}_1$. This conic belongs to the point pencil determined by $\mathbf{s}_1$ and $\mathbf{s}_2$ and meets $\mathbf{s}_1$ only in the base points of the pencil unless $\theta_2 = 0$, when the locus of P is the conic $\mathbf{s}_1$ itself. The theorem of 9·45 follows immediately, and $\mathbf{s}_1$ is outpolar to $\mathbf{s}_2$ IF $\theta_2 = 0$. Similarly $\mathbf{s}_2$ is outpolar to $\mathbf{s}_1$ IF $\theta_1 = 0$.

11·28. Dually, if p, q, r are the sides of a variable triangle self-polar relative to $\mathbf{s}_2$ and if q and r are tangents of $\mathbf{s}_1$, then the scroll formed by the lines p is the reciprocal of $\mathbf{f}$ relative to $\mathbf{s}_2$, which has line matrix $\theta_1 \mathbf{S}_2 - \delta_2 \mathbf{S}_1$. This scroll belongs to the line pencil determined by $\mathbf{s}_1$ and $\mathbf{s}_2$, and it follows that $\mathbf{s}_1$ is inpolar to $\mathbf{s}_2$ IF $\theta_1 = 0$. Similarly $\mathbf{s}_2$ is inpolar to $\mathbf{s}_1$ IF $\theta_2 = 0$.

11·3. Two proper conics (continued). A net of conics associated with two proper conics

11·31. Since $\mathbf{s}_1\mathbf{G}\mathbf{s}_1 = \theta_1\mathbf{s}_1 - \delta_1\mathbf{s}_2$ it follows that

$$\delta_1^2|\mathbf{G}| = |\mathbf{s}_1\mathbf{G}\mathbf{s}_1| = |\theta_1\mathbf{s}_1 - \delta_1\mathbf{s}_2| = \theta_1^3\delta_1 - \theta_1^2\delta_1\theta_1 + \theta_1\delta_1^2\theta_2 - \delta_1^3\delta_2,$$

by putting $\lambda = \theta_1$, $\mu = -\delta_1$, in (·213); hence

$$|\mathbf{G}| = \theta_1\theta_2 - \delta_1\delta_2. \tag{·311}$$

Thus $\mathbf{G}$ is a point-pair IF $\theta_1\theta_2 = \delta_1\delta_2$.

In metrical geometry, if $\mathbf{s}_1$ and $\mathbf{s}_2$ are two circles, it follows from 9·25 that this is a necessary condition for orthogonality.

Also, since $$\mathbf{f} = \mathbf{s}_1\mathbf{G}\mathbf{s}_2,$$

$$|\mathbf{f}| = \delta_1\delta_2(\theta_1\theta_2 - \delta_1\delta_2). \tag{·312}$$

Hence $\mathbf{f}$ is a line-pair IF $\mathbf{G}$ is a point-pair.

11·32. From the equation $\mathbf{s}_1\mathbf{G}\mathbf{s}_1 = \theta_1\mathbf{s}_1 - \delta_1\mathbf{s}_2$ it also follows that

$$\mathbf{s}_1(\lambda\mathbf{S}_1 + \nu\mathbf{G})\mathbf{s}_1 = (\lambda\delta_1 + \nu\theta_1)\mathbf{s}_1 - \nu\delta_1\mathbf{s}_2. \tag{·321}$$

Thus, by (·213),

$$\delta_1^2|\lambda\mathbf{S}_1 + \nu\mathbf{G}| = \delta_1(\lambda\delta_1 + \nu\theta_1)^3 - \theta_1\nu\delta_1(\lambda\delta_1 + \nu\theta_1)^2 + \theta_2\nu^2\delta_1^2(\lambda\delta_1 + \nu\theta_1) - \delta_2\nu^3\delta_1^3,$$

which gives

$$|\lambda\mathbf{S}_1 + \nu\mathbf{G}| = \lambda^3\delta_1^2 + 2\lambda^2\nu\delta_1\theta_1 + \lambda\nu^2(\theta_1^2 + \theta_2\delta_1) + \nu^3(\theta_1\theta_2 - \delta_1\delta_2). \tag{·322}$$

Similarly

$$|\mu\mathbf{S}_2 + \nu\mathbf{G}| = \mu^3\delta_2^2 + 2\mu^2\nu\delta_2\theta_2 + \mu\nu^2(\theta_2^2 + \theta_1\delta_2) + \nu^3(\theta_1\theta_2 - \delta_1\delta_2). \tag{·323}$$

Also from $$\mathbf{S}_1\mathbf{f}\mathbf{S}_1 = \delta_1(\theta_2\mathbf{S}_1 - \delta_1\mathbf{S}_2)$$

it follows that

$$\mathbf{S}_1(\lambda\mathbf{s}_1 + \nu\mathbf{f})\mathbf{S}_1 = \delta_1\{(\lambda + \nu\theta_2)\mathbf{S}_1 - \nu\delta_1\mathbf{S}_2\}, \tag{·324}$$

whence, by (·217),

$$|\lambda\mathbf{s}_1 + \nu\mathbf{f}| = \lambda^3\delta_1 + 2\lambda^2\nu\delta_1\theta_2 + \lambda\nu^2\delta_1(\theta_2^2 + \delta_2\theta_1) + \nu^3\delta_1\delta_2(\theta_1\theta_2 - \delta_1\delta_2), \tag{·325}$$

and similarly

$$|\mu\mathbf{s}_2 + \nu\mathbf{f}| = \mu^3\delta_2 + 2\mu^2\nu\delta_2\theta_1 + \mu\nu^2\delta_2(\theta_1^2 + \delta_1\theta_2) + \nu^3\delta_1\delta_2(\theta_1\theta_2 - \delta_1\delta_2). \tag{·326}$$

11·33. If $\mathbf{s}_1$ is outpolar to $\mathbf{s}_2$, then $\theta_2 = 0$, and it follows from (·231) and (·241) that the reciprocal of $\mathbf{s}_1$ relative to $\mathbf{s}_2$ is $\mathbf{G}$ and that the reciprocal of $\mathbf{s}_2$ relative to $\mathbf{s}_1$ is $\mathbf{f}$, as is indeed said in 11·27 and 11·28. It also follows from (·323) and (·325) that $\mathbf{s}_2$ is outpolar to $\mathbf{G}$ and that $\mathbf{s}_1$ is inpolar to $\mathbf{f}$.

Also from (·322) and (·326), $\theta_1^2+\theta_2\delta_1 = 0$ is the CONDITION for $\mathbf{s}_1$ to be inpolar to $\mathbf{G}$ and for $\mathbf{s}_2$ to be outpolar to $\mathbf{f}$. Similarly $\theta_2^2+\theta_1\delta_2 = 0$ is the CONDITION for $\mathbf{s}_2$ to be inpolar to $\mathbf{G}$ and for $\mathbf{s}_1$ to be outpolar to $\mathbf{f}$.

11·34. The adjoint of a product of matrices is the product of the adjoints in the reverse order, so from (·321) it follows that

$$\begin{aligned}\mathbf{S}_1(\lambda\mathbf{S}_1+\nu\mathbf{G})_a\mathbf{S}_1 &= \{(\lambda\delta_1+\nu\theta_1)\,\mathbf{s}_1-\nu\delta_1\,\mathbf{s}_2\}_a\\ &= (\lambda\delta_1+\nu\theta_1)^2\,\mathbf{S}_1-(\lambda\delta_1+\nu\theta_1)\,\nu\delta_1\,\mathbf{G}+\nu^2\delta_1^2\mathbf{S}_2,\end{aligned}$$

by (·214). Multiplying before and after by $\mathbf{s}_1$, and using (·233) and (·231),

$$\begin{aligned}\delta_1^2(\lambda\mathbf{S}_1+\nu\mathbf{G})_a = \delta_1(\lambda\delta_1+\nu\theta_1)^2\,\mathbf{s}_1\\ -\nu\delta_1(\lambda\delta_1+\nu\theta_1)\,(\theta_1\,\mathbf{s}_1-\delta_1\,\mathbf{s}_2)+\nu^2\delta_1^2(\theta_2\,\mathbf{s}_1-\mathbf{f}),\end{aligned}$$

which reduces to

$$(\lambda\mathbf{S}_1+\nu\mathbf{G})_a = \lambda^2\delta_1\,\mathbf{s}_1+\lambda\nu(\theta_1\,\mathbf{s}_1+\delta_1\,\mathbf{s}_2)+\nu^2(\theta_2\,\mathbf{s}_1+\theta_1\,\mathbf{s}_2-\mathbf{f}). \quad (·341)$$

Similarly

$$(\mu\mathbf{S}_2+\nu\mathbf{G})_a = \mu^2\delta_2\,\mathbf{s}_2+\mu\nu(\theta_2\,\mathbf{s}_2+\delta_2\,\mathbf{s}_1)+\nu^2(\theta_1\,\mathbf{s}_2+\theta_2\,\mathbf{s}_1-\mathbf{f}). \quad (·342)$$

Putting $\lambda = 0$ in (·341) gives

$$\mathbf{G}_a = \theta_2\,\mathbf{s}_1+\theta_1\,\mathbf{s}_2-\mathbf{f}, \quad (·343)$$

which might have been obtained more directly by taking the adjoint of each side of the equation

$$\mathbf{s}_1\,\mathbf{G}\mathbf{s}_1 = \theta_1\,\mathbf{s}_1-\delta_1\,\mathbf{s}_2.$$

Thus the point matrix of the harmonic envelope of $\mathbf{s}_1$ and $\mathbf{s}_2$ is $\theta_2\,\mathbf{s}_1+\theta_1\,\mathbf{s}_2-\mathbf{f}$.

The matrix $(\lambda\mathbf{S}_1+\mu\mathbf{S}_2+\nu\mathbf{G})_a$ can be expanded in the form

$$\lambda^2\mathbf{A}+\mu^2\mathbf{B}+\nu^2\mathbf{C}+\mu\nu\mathbf{P}+\nu\lambda\mathbf{Q}+\lambda\mu\mathbf{R},$$

where $\mathbf{A}$, $\mathbf{B}$, $\mathbf{C}$ are the adjoints of $\mathbf{S}_1$, $\mathbf{S}_2$, $\mathbf{G}$ and $\mathbf{P}$, $\mathbf{Q}$, $\mathbf{R}$ are the coefficients of $\mu\nu$, $\nu\lambda$, $\lambda\mu$ in the expansions of

$$(\mu\mathbf{S}_2+\nu\mathbf{G})_a, \quad (\lambda\mathbf{S}_1+\nu\mathbf{G})_a \quad \text{and} \quad (\lambda\mathbf{S}_1+\mu\mathbf{S}_2)_a.$$

Thus by (·341), (·342), (·343) and (·218):

$$\begin{aligned}(\lambda\mathbf{S}_1+\mu\mathbf{S}_2+\nu\mathbf{G})_a = \lambda^2\delta_1\,\mathbf{s}_1+\mu^2\delta_2\,\mathbf{s}_2+\nu^2(\theta_2\,\mathbf{s}_1+\theta_1\,\mathbf{s}_2-\mathbf{f})\\ +\mu\nu(\theta_2\,\mathbf{s}_2+\delta_2\,\mathbf{s}_1)+\nu\lambda(\theta_1\,\mathbf{s}_1+\delta_1\,\mathbf{s}_2)+\lambda\mu\mathbf{f}. \quad (·344)\end{aligned}$$

By a similar argument it follows that

$$(\lambda \mathbf{s}_1 + \nu \mathbf{f})_a = \lambda^2 \mathbf{S}_1 + \lambda\nu(\theta_2 \mathbf{S}_1 + \delta_1 \mathbf{S}_2) + \nu^2(\theta_1\delta_2 \mathbf{S}_1 + \theta_2\delta_1 \mathbf{S}_2 - \delta_1\delta_2 \mathbf{G}), \quad (\cdot 345)$$

$$(\mu \mathbf{s}_2 + \nu \mathbf{f})_a = \mu^2 \mathbf{S}_2 + \mu\nu(\theta_1 \mathbf{S}_2 + \delta_2 \mathbf{S}_1) + \nu^2(\theta_2\delta_1 \mathbf{S}_2 + \theta_1\delta_2 \mathbf{S}_1 - \delta_1\delta_2 \mathbf{G}), \quad (\cdot 346)$$

and

$$\mathbf{f}_a = \theta_1\delta_2 \mathbf{S}_1 + \theta_2\delta_1 \mathbf{S}_2 - \delta_1\delta_2 \mathbf{G}, \quad (\cdot 347)$$

$$(\lambda \mathbf{s}_1 + \mu \mathbf{s}_2 + \nu \mathbf{f})_a = \lambda^2 \mathbf{S}_1 + \mu^2 \mathbf{S}_2 + \nu^2(\theta_1\delta_2 \mathbf{S}_1 + \theta_2\delta_1 \mathbf{S}_2 - \delta_1\delta_2 \mathbf{G})$$
$$+ \mu\nu(\delta_2 \mathbf{S}_1 + \theta_1 \mathbf{S}_2) + \nu\lambda(\theta_2 \mathbf{S}_1 + \delta_1 \mathbf{S}_2) + \lambda\mu \mathbf{G}. \quad (\cdot 348)$$

11·35. The point matrix of the harmonic locus of the conics $\mathbf{s}_1$ and $\mathbf{G}$ is the coefficient of $\lambda\nu$ in $(\lambda \mathbf{S}_1 + \nu \mathbf{G})_a$, which by (·341) is $\theta_1 \mathbf{s}_1 + \delta_1 \mathbf{s}_2$. Thus:

The harmonic locus of $\mathbf{s}_1$ *and* $\mathbf{G}$ *belongs to the point pencil determined by the conics* $\mathbf{s}_1$ *and* $\mathbf{s}_2$.*

Similarly the harmonic envelope of $\mathbf{s}_1$ and $\mathbf{f}$ has line matrix $\theta_2 \mathbf{S}_1 + \delta_1 \mathbf{S}_2$, so:

The harmonic envelope of $\mathbf{s}_1$ *and* $\mathbf{f}$ *belongs to the line pencil determined by the conics* $\mathbf{s}_1$ *and* $\mathbf{s}_2$.†

11·36. The harmonic locus of $\mathbf{s}_1$ and $\mathbf{s}_2$ cannot coincide with $\mathbf{s}_1$ or with $\mathbf{s}_2$. For if $\mathbf{f} = \kappa \mathbf{s}_1$, then by (·223)

$$\mathbf{s}_1 \mathbf{S}_2 = (\theta_2 - \kappa)\mathbf{I},$$

and therefore

$$\delta_2 \mathbf{s}_1 = (\theta_2 - \kappa)\mathbf{s}_2,$$

which implies that the conics $\mathbf{s}_1$ and $\mathbf{s}_2$ coincide. Similarly $\mathbf{G}$ cannot coincide with either of the conics $\mathbf{s}_1$ and $\mathbf{s}_2$.

11·37. The harmonic locus belongs to the pencil determined by the conic loci $\mathbf{s}_1$ and $\mathbf{s}_2$ IF $\mathbf{s}_1$ and $\mathbf{s}_2$ have double contact.‡ For, by (·231), $\mathbf{f}$ is of the form $\kappa_1 \mathbf{s}_1 + \kappa_2 \mathbf{s}_2$ IF $\mathbf{s}_1 \mathbf{S}_2 \mathbf{s}_1$ is of the form $\lambda \mathbf{s}_1 + \mu \mathbf{s}_2$. But $\mathbf{s}_1 \mathbf{S}_2 \mathbf{s}_1$ is a point matrix of the reciprocal of $\mathbf{s}_2$ relative to $\mathbf{s}_1$, and, if this conic contains the common points of $\mathbf{s}_1$ and $\mathbf{s}_2$, then the

* It may be noted for future reference (in 11·41) that this theorem is true even if $\mathbf{s}_2$ is not proper. For both $\mathbf{G}$ and the harmonic locus of $\mathbf{s}_1$ and $\mathbf{G}$ are definable and unique in this case; the theorem that the point matrix of this harmonic locus is $\theta_1 \mathbf{s}_1 + \delta_1 \mathbf{s}_2$ is clearly ultimately equivalent to a number of algebraic identities in the coefficients of $\mathbf{s}_1$ and $\mathbf{s}_2$, and these will remain true if δ_2 vanishes.

† Strictly we should say 'the line pencil determined by the line conics $\mathbf{S}_1$ and $\mathbf{S}_2$'. Stated in this form the theorem is true even if $\mathbf{S}_2$ is a point pair or a repeated point.

‡ Four-point contact is regarded as a particular case of double contact.

tangent to $\mathbf{s}_1$ at each common point is also a tangent to $\mathbf{s}_2$, so $\mathbf{s}_1$ and $\mathbf{s}_2$ touch wherever they meet. Conversely it was shown in 8·22 that when $\mathbf{s}_1$ and $\mathbf{s}_2$ have double contact the reciprocal of $\mathbf{s}_2$ relative to $\mathbf{s}_1$ belongs to the pencil determined by $\mathbf{s}_1$ and $\mathbf{s}_2$, so $\mathbf{s}_1\mathbf{S}_2\mathbf{s}_1$ is of the form $\lambda\mathbf{s}_1+\mu\mathbf{s}_2$ and $\mathbf{f}$ belongs to the pencil also.

Dually $\mathbf{G}$ belongs to the line pencil determined by $\mathbf{s}_1$ and $\mathbf{s}_2$ IF $\mathbf{s}_1$ and $\mathbf{s}_2$ have double contact.

11·38. Thus, provided $\mathbf{s}_1$ and $\mathbf{s}_2$ do not have double contact, the conic locus $\mathbf{f}$ is not dependent on the conic loci $\mathbf{s}_1$ and $\mathbf{s}_2$, and the conic scroll $\mathbf{G}$ is not dependent on the scrolls $\mathbf{S}_1$ and $\mathbf{S}_2$. Consequently the conic loci with point matrices

$$\lambda\mathbf{s}_1+\mu\mathbf{s}_2+\nu\mathbf{f} \qquad (\cdot 381)$$

form a linear ∞^2 system, or net, while the conic scrolls with line matrices

$$\lambda\mathbf{S}_1+\mu\mathbf{S}_2+\nu\mathbf{G} \qquad (\cdot 382)$$

form a linear ∞^2 system, or line net. The formulae of 11·23 and 11·24 express certain products of the matrices $\mathbf{s}_1$, $\mathbf{s}_2$, $\mathbf{f}$, $\mathbf{S}_1$, $\mathbf{S}_2$, $\mathbf{G}$ in one of these forms. Also (·344) and (·348) show that the conic scrolls of the line net have point matrices of the form (·381), and that the conic loci of the net have line matrices of the form (·382). Thus:

If $\mathbf{s}_1$ *and* $\mathbf{s}_2$ *do not have double contact, the proper conics of the net determined by* $\mathbf{s}_1$, $\mathbf{s}_2$, $\mathbf{f}$ *are also the proper conics of the line net determined by* $\mathbf{S}_1$, $\mathbf{S}_2$, $\mathbf{G}$.

If $\mathbf{s}_1$ *and* $\mathbf{s}_2$ *have double contact, the same is true, with 'pencil' instead of 'net'.*

The system (·381) of conic loci and the system (·382) of conic scrolls will both be denoted by χ. The improper conic loci of (·381) have line matrices of the form (·382), but cannot be regarded as conic scrolls.

When the four common points of $\mathbf{s}_1$ and $\mathbf{s}_2$ are distinct, the two conics have one common self-polar triangle, and χ is the system of conics relative to which this triangle is self-polar.

11·39. Any two conic loci of χ have point matrices

$$a_1\mathbf{s}_1+b_1\mathbf{s}_2+c_1\mathbf{f} \quad \text{and} \quad a_2\mathbf{s}_1+b_2\mathbf{s}_2+c_2\mathbf{f},$$

and the line matrix of their harmonic envelope is the coefficient of $\lambda\mu$ in the adjoint of the matrix

$$(\lambda a_1+\mu a_2)\,\mathbf{s}_1+(\lambda b_1+\mu b_2)\,\mathbf{s}_2+(\lambda c_1+\mu c_2)\,\mathbf{f}.$$

This adjoint may be written down by (·348), and the coefficient of $\lambda\mu$ is evidently of the form (·382). Similarly for the harmonic locus. Thus:

The harmonic envelope and locus of any two conics of χ also belong to χ.

11·4. Two proper conics (continued). The ν and W-conics

11·41. It was shown in 9·33 that there is a unique line conic, W_1, such that $\mathbf{s}_2$ is the harmonic locus of $\mathbf{s}_1$ and W_1. This also follows from the work of the present chapter. For, if such a line conic W_1 exists, then by applying 11·35 to $\mathbf{S}_1$ and W_1, we see that the harmonic envelope of $\mathbf{s}_2$ and $\mathbf{s}_1$ belongs to the line pencil determined by $\mathbf{s}_1$ and W_1.* Thus the line matrix $\mathbf{W}_1$ of W_1 must be of the form

$$\mathbf{W}_1 = \rho\mathbf{S}_1 + \sigma\mathbf{G}.$$

But the point matrix of the harmonic locus of $\mathbf{W}_1$ and $\mathbf{s}_1$ is the coefficient of $\lambda\mu$ in the adjoint of

$$\lambda(\rho\mathbf{S}_1 + \sigma\mathbf{G}) + \mu\mathbf{S}_1,$$

which by (·341) is

$$(\lambda\rho+\mu)^2\delta_1\mathbf{s}_1 + (\lambda\rho+\mu)\,\sigma\lambda(\theta_1\mathbf{s}_1 + \delta_1\mathbf{s}_2) + \lambda^2\sigma^2(\theta_2\mathbf{s}_1 + \theta_1\mathbf{s}_2 - \mathbf{f}).$$

The coefficient of $\lambda\mu$ is

$$(2\rho\delta_1 + \sigma\theta_1)\,\mathbf{s}_1 + \sigma\delta_1\mathbf{s}_2,$$

which gives $\mathbf{s}_2$ IF $\qquad 2\rho\delta_1 + \sigma\theta_1 = 0.$

Thus the conic W_1 exists and is unique, and its line matrix may be taken as

$$\mathbf{W}_1 = \theta_1\mathbf{S}_1 - 2\delta_1\mathbf{G}. \qquad (\cdot 411)$$

Similarly there is ONE conic W_2 such that $\mathbf{s}_1$ is the harmonic locus of $\mathbf{s}_2$ and W_2, and the line matrix of W_2 is

$$\mathbf{W}_2 = \theta_2\mathbf{S}_2 - 2\delta_2\mathbf{G}. \qquad (\cdot 412)$$

Dually, if there is a conic v_1 such that $\mathbf{s}_2$ is the harmonic envelope of $\mathbf{s}_1$ and v_1, then by 11·35 v_1 must have a point matrix of the form

$$\mathbf{v}_1 = \rho\mathbf{s}_1 + \sigma\mathbf{f}.$$

By (·345) the adjoint of $\lambda(\rho\mathbf{s}_1 + \sigma\mathbf{f}) + \mu\mathbf{s}_1$ is

$$(\lambda\rho+\mu)^2\mathbf{S}_1 + (\lambda\rho+\mu)\,\lambda\sigma(\theta_2\mathbf{S}_1 + \delta_1\mathbf{S}_2) + \lambda^2\sigma^2(\theta_1\delta_2\mathbf{S}_1 + \theta_2\delta_2\mathbf{S}_2 - \delta_1\delta_2 G),$$

and the coefficient of $\lambda\mu$ in this is

$$(2\rho + \sigma\theta_2)\,\mathbf{S}_1 + \sigma\delta_1\mathbf{S}_2,$$

which reduces to $\mathbf{S}_2$ IF $\qquad 2\rho + \sigma\theta_2 = 0.$

* W_1 may be a degenerate line conic. See the footnotes to 11·35, p. 306.

Thus there is ONE conic v_1, and its point matrix may be taken as

$$\mathbf{v}_1 = \theta_2 \mathbf{s}_1 - 2\mathbf{f}. \qquad (\cdot 413)$$

Similarly there is ONE conic v_2 such that $\mathbf{s}_1$ is the harmonic envelope of $\mathbf{s}_2$ and v_2, and the point matrix of v_2 is

$$\mathbf{v}_2 = \theta_1 \mathbf{s}_2 - 2\mathbf{f}. \qquad (\cdot 414)$$

11·42. From (·411) it follows that

$$\mathbf{s}_1 \mathbf{W}_1 \mathbf{s}_1 = \theta_1 \delta_1 \mathbf{s}_1 - 2\delta_1(\theta_1 \mathbf{s}_1 - \delta_1 \mathbf{s}_2),$$

or

$$\mathbf{s}_1 \mathbf{W}_1 \mathbf{s}_1 = \delta_1(2\delta_1 \mathbf{s}_2 - \theta_1 \mathbf{s}_1). \qquad (\cdot 421)$$

Thus by 11·15 the reciprocal of $\mathbf{W}_1$ relative to $\mathbf{s}_1$ has point matrix $2\delta_1 \mathbf{s}_2 - \theta_1 \mathbf{s}_1$. Similarly from (·413) it follows that

$$\mathbf{S}_1 \mathbf{v}_1 \mathbf{S}_1 = \delta_1(2\delta_1 \mathbf{S}_2 - \theta_2 \mathbf{S}_1). \qquad (\cdot 422)$$

From these equations or from (·322) and (·325) it follows that

$$|\mathbf{W}_1| = \delta_1^2(8\delta_1^2\delta_2 - 4\delta_1\theta_1\theta_2 + \theta_1^3), \qquad (\cdot 423)$$

$$|\mathbf{v}_1| = \delta_1(8\delta_2^2\delta_1 - 4\delta_2\theta_1\theta_2 + \theta_2^3), \qquad (\cdot 424)$$

with similar expressions for $|\mathbf{W}_2|$ and $|\mathbf{v}_2|$, obtained by interchanging the suffices 1 and 2. From (·341) we have

$$(\mathbf{W}_1)_a = \theta_1^2\delta_1 \mathbf{s}_1 - 2\theta_1\delta_1(\theta_1 \mathbf{s}_1 + \delta_1 \mathbf{s}_2) + 4\delta_1^2(\theta_2 \mathbf{s}_1 + \theta_1 \mathbf{s}_2 - \mathbf{f})$$

Thus

$$\left.\begin{aligned}(\mathbf{W}_1)_a &= \delta_1(4\delta_1\theta_2 - \theta_1^2)\,\mathbf{s}_1 + 2\delta_1^2\theta_1 \mathbf{s}_2 - 4\delta_1^2\mathbf{f}\\ &= \delta_1(4\delta_1\theta_2 - \theta_1^2)\,\mathbf{s}_1 + 2\delta_1^2 \mathbf{v}_2,\end{aligned}\right\} \qquad (\cdot 425)$$

by (·414)

Similarly, from (·345),

$$\left.\begin{aligned}(\mathbf{v}_1)_a &= (4\delta_2\theta_1 - \theta_2^2)\,\mathbf{S}_1 + 2\delta_1\theta_2 \mathbf{S}_2 - 4\delta_1\delta_2 G\\ &= (4\delta_2\theta_1 - \theta_2^2)\,\mathbf{S}_1 + 2\delta_1 \mathbf{W}_2\end{aligned}\right\} \qquad (\cdot 426)$$

We have thus found expressions for the point matrix of $\mathbf{W}_1$ and the line matrix of $\mathbf{v}_1$. Similar expressions can be written down for $(\mathbf{W}_2)_a$ and $(\mathbf{v}_2)_a$.

11·43. Since $\mathbf{s}_2$ is the harmonic locus of $\mathbf{s}_1$ and $\mathbf{W}_1$, the tangents to $\mathbf{s}_1$ from any point of $\mathbf{s}_2$ are conjugate relative to $\mathbf{W}_1$. Thus, if P, Q, R are the vertices of a triangle whose sides p, q, r touch $\mathbf{s}_1$ and if Q and R lie on $\mathbf{s}_2$, then the pairs of lines $|p, q|$ and $|p, r|$ are conjugate relative to $\mathbf{W}_1$, so P is the pole of p relative to $\mathbf{W}_1$. Conversely, if p is any tangent of $\mathbf{s}_1$ meeting $\mathbf{s}_2$ in Q and R, and if P is the pole of p relative to $\mathbf{W}_1$, then QP and RP are both conjugate to p relative to $\mathbf{W}_1$ and therefore touch $\mathbf{s}_1$. Thus the locus of the vertex P of a triangle having the above property is the reciprocal of $\mathbf{s}_1$ relative to $\mathbf{W}_1$, which by 11·15 has line matrix $\mathbf{W}_1\mathbf{s}_1\mathbf{W}_1$. But

$$\begin{aligned}\mathbf{W}_1\mathbf{s}_1\mathbf{W}_1 &= (\theta_1\mathbf{S}_1-2\delta_1\mathbf{G})\mathbf{S}_1(\theta_1\mathbf{S}_1-2\delta_1\mathbf{G})\\ &= \theta_1^2\delta_1\mathbf{S}_1-4\theta_1\delta_1^2\mathbf{G}+4\delta_1^2\mathbf{G}\mathbf{s}_1\mathbf{G}\\ &= \theta_1^2\delta_1\mathbf{S}_1-4\theta_1\delta_1^2\mathbf{G}+4\delta_1^2(\theta_1\mathbf{G}-\theta_2\mathbf{S}_1+\delta_1\mathbf{S}_2) \text{ by } (\cdot 247)\\ &= \delta_1(\theta_1^2-4\delta_1\theta_2)\,\mathbf{S}_1+4\delta_1^3\mathbf{S}_2.\end{aligned}$$

The locus therefore belongs to the line pencil determined by $\mathbf{s}_1$ and $\mathbf{s}_2$, and coincides with $\mathbf{s}_2$ IF $\theta_1^2-4\delta_1\theta_2=0$. Thus:

The conic $\mathbf{s}_1$ *is triangularly inscribed in* $\mathbf{s}_2$ IF $\theta_1^2-4\delta_1\theta_2=0$. *Similarly* $\mathbf{s}_2$ *is triangularly inscribed in* $\mathbf{s}_1$ IF $\theta_2^2-4\delta_2\theta_1=0$.

11·44. Dually the points of intersection of $\mathbf{s}_1$ with any tangent to $\mathbf{s}_2$ are conjugate relative to $\mathbf{v}_1$. It follows that if P, Q, R are the vertices of a variable triangle inscribed in $\mathbf{s}_1$ whose sides PQ and PR touch $\mathbf{s}_2$, then the scroll formed by the remaining sides QR is the reciprocal of $\mathbf{s}_1$ relative to $\mathbf{v}_1$. The point matrix of this scroll is $\mathbf{v}_1\mathbf{S}_1\mathbf{v}_1$, or

$$(\theta_2\mathbf{s}_1-2\mathbf{f})\,\mathbf{S}_1(\theta_2 s_1-2\mathbf{f}),$$

which reduces to

$$\delta_1(\theta_2^2-4\theta_1\delta_2)\,\mathbf{s}_1-4\delta_1^2\delta_2\,\mathbf{s}_2.$$

Thus, as in 11·43, it follows that $\mathbf{s}_2$ is triangularly inscribed in $\mathbf{s}_1$ IF $\theta_2^2-4\theta_1\delta_2=0$.

11·45. If $\mathbf{s}_1$ is triangularly inscribed in $\mathbf{s}_2$, $\theta_1^2-4\theta_1\delta_2=0$, so by (·425) the conics $\mathbf{W}_1$ and $\mathbf{v}_2$ are identical. Similarly if $\mathbf{s}_2$ is triangularly inscribed in $\mathbf{s}_1$ the conics $\mathbf{W}_2$ and $\mathbf{v}_1$ coincide.*

11·46. If $\mathbf{W}_1$ is a point-pair $\|H, K\|$, then $\mathbf{s}_2$ is the locus of points P such that the lines PH, PK are conjugate relative to $\mathbf{s}_1$. It follows from the argument of 9·35 that a CONDITION for the existence of ∞^1 sets of four points A, B, C, D on $\mathbf{s}_2$ such that the lines AB, BC, CD, DA touch $\mathbf{s}_1$ is $|\mathbf{W}_1|=0$, or

$$8\delta_1^2\delta_2-4\delta_1\theta_1\theta_2+\theta_1^3=0.$$

Similarly a CONDITION for the existence of ∞^1 sets of four points A, B, C, D on $\mathbf{s}_1$ such that the lines AB, BC, CD, DA touch $\mathbf{s}_2$ is

$$8\delta_2^2\delta_1-4\delta_2\theta_2\theta_1+\theta_2^3=0.$$

It should be noticed that $|\mathbf{W}_1|=0$ implies $|\mathbf{v}_2|=0$ and $|\mathbf{W}_2|=0$ implies $|\mathbf{v}_1|=0$, in agreement with 9·35.

* Cf. 9·44, p. 244.

11·47. $\mathbf{s}_1\mathbf{W}_1\mathbf{s}_2 = \mathbf{s}_1(\theta_1\mathbf{S}_1 - 2\delta_1\mathbf{G})\mathbf{s}_2 = \delta_1\theta_1\mathbf{s}_2 - 2\delta_1\mathbf{f}$ by (·234),

so $$\mathbf{s}_1\mathbf{W}_1\mathbf{s}_2 = \mathbf{s}_2\mathbf{W}_1\mathbf{s}_1 = \delta_1\mathbf{v}_2.$$

Similarly $$\mathbf{S}_1\mathbf{v}_1\mathbf{S}_2 = \mathbf{S}_2\mathbf{v}_1\mathbf{S}_1 = \delta_1\mathbf{W}_2.$$

Thus $\mathbf{v}_2$ is the locus of points whose polar lines relative to $\mathbf{s}_1$ and $\mathbf{s}_2$ are conjugate relative to $\mathbf{W}_1$, and $\mathbf{W}_2$ is the scroll formed by lines whose poles relative to $\mathbf{s}_1$ and $\mathbf{s}_2$ are conjugate relative to $\mathbf{v}_1$. The same applies to $\mathbf{v}_1$ and $\mathbf{W}_2$ and to $\mathbf{W}_1$ and $\mathbf{v}_2$.

11·48. The CONDITIONS for $\mathbf{s}_1$ to be inpolar, outpolar, triangularly inscribed and triangularly circumscribed to $\mathbf{s}_2$ are respectively

$$\theta_1 = 0, \quad \theta_2 = 0, \quad \theta_1^2 - 4\delta_1\theta_2 = 0, \quad \text{and} \quad \theta_2^2 - 4\delta_2\theta_1 = 0.$$

If the last two CONDITIONS are satisfied it does not follow that $\theta_1 = \theta_2 = 0$, but if any other pair of the four CONDITIONS are satisfied it does follow that $\theta_1 = \theta_2 = 0$, and that all the CONDITIONS are satisfied. (It is assumed that δ_1 and δ_2 are not zero.)

The properties of a pair of mutually apolar conics have already been discussed in 9·69, but it also follows from the present chapter that when $\theta_1 = \theta_2 = 0$ the conics $\mathbf{f}$, $\mathbf{G}$, $\mathbf{v}_1$, $\mathbf{v}_2$, $\mathbf{W}_1$ and $\mathbf{W}_2$ all coincide in a conic $\mathbf{s}_3$, and that the three conics form a symmetrical set, each being the reciprocal of either of the others relative to the third, and each being the harmonic envelope and the harmonic locus of the other two. Any two of the three conics are mutually apolar and also triangularly inscribed and circumscribed to each other. Such a set of three conics is known as a *Hessian triad.*

11·49. The two conics $\mathbf{s}_1$ and $\mathbf{s}_2$ will be triangularly inscribed and circumscribed to each other IF

$$\theta_1^2 - 4\delta_1\theta_2 = 0 \quad \text{and} \quad \theta_2^2 - 4\delta_2\theta_1 = 0.$$

If θ_1 and θ_2 are not zero these equations imply

$$\theta_1^3 = 64\delta_1\delta_2^3, \quad \theta_2^3 = 64\delta_2\delta_1^3.$$

To obtain an example of two such conics we may consider the conics

$$s_1 \equiv x_1^2 + x_2^2 + x_3^2 = 0, \quad s_2 \equiv a_1x_1^2 + a_2x_2^2 + a_3x_3^2 = 0,$$

where $a_1a_2a_3 = 1$. Then

$$\delta_1 = \delta_2 = 1, \quad \theta_1 = a_1 + a_2 + a_3, \quad \text{and} \quad \theta_2 = a_2a_3 + a_3a_1 + a_1a_2.$$

Then the CONDITIONS are satisfied if $\theta_1 = \theta_2 = 4$, when a_1, a_2, a_3 are the roots of the equation

$$x^3 - 4x^2 + 4x - 1 = 0,$$

namely
$$1, \quad \frac{3+\sqrt{5}}{2} \quad \text{and} \quad \frac{3-\sqrt{5}}{2}.$$

11·5. Conics which reciprocate one conic into another

11·51. It was shown in Chapter VIII that it is always possible to find a conic relative to which $\mathbf{s}_1$ and $\mathbf{s}_2$ are reciprocals. Let $\mathbf{C}$ be the line matrix of such a conic. With respect to $\mathbf{C}$ the conic loci $\mathbf{s}_1$, $\mathbf{s}_2$ and $\mathbf{f}$ reciprocate into the conic scrolls $\mathbf{S}_2$, $\mathbf{S}_1$ and $\mathbf{G}$; hence reciprocation with respect to $\mathbf{C}$ transforms the conic loci of χ into the conic scrolls of χ.* We prove that:

Provided $\mathbf{s}_1$ *and* $\mathbf{s}_2$ *do not have double contact the conic* $\mathbf{C}$ *must belong to* χ.

For there are ∞^2 conics of χ with point matrices of the form $\lambda\mathbf{s}_1 + \mu\mathbf{s}_2 + \nu\mathbf{f}$, and it is possible to choose a conic $\mathbf{C}_1$ of χ which is both outpolar and inpolar to $\mathbf{C}$, since this implies choosing $\lambda:\mu:\nu$ to satisfy a linear and a quadratic equation. Suppose that $\mathbf{C}$ does not belong to χ, so that $\mathbf{C}_1$ is distinct from $\mathbf{C}$. Then by 11·48 the harmonic locus and envelope of $\mathbf{C}$ and $\mathbf{C}_1$ coincide in a conic $\mathbf{C}_2$, distinct from $\mathbf{C}_1$, which is the reciprocal of $\mathbf{C}_1$ relative to $\mathbf{C}$ and therefore belongs to χ. But by 11·48 $\mathbf{C}$ is the harmonic envelope of $\mathbf{C}_1$ and $\mathbf{C}_2$, and therefore, by 11·39, $\mathbf{C}$ belongs to χ, which is contrary to hypothesis.

11·52. Thus, provided $\mathbf{s}_1$ and $\mathbf{s}_2$ do not have double contact, the line matrix of $\mathbf{C}$ must be of the form

$$\mathbf{C} = \lambda\mathbf{S}_1 + \mu\mathbf{S}_2 + \nu\mathbf{G}.$$

The line matrix of the reciprocal of $\mathbf{s}_1$ relative to $\mathbf{C}$ is

$$(\lambda\mathbf{S}_1 + \mu\mathbf{S}_2 + \nu\mathbf{G})\,\mathbf{s}_1(\lambda\mathbf{S}_1 + \mu\mathbf{S}_2 + \nu\mathbf{G}),$$

which by use of the formulae of 11·24 reduces to

$$(\lambda^2\delta_1 - 2\mu\nu\delta_2 - \nu^2\theta_2)\,\mathbf{S}_1 + (2\lambda\mu\delta_1 + 2\mu\nu\theta_1 + \mu^2\theta_2 + \nu^2\delta_1)\,\mathbf{S}_2 + (\nu^2\theta_1 + 2\lambda\nu\delta_1 - \mu^2\delta_2)\,\mathbf{G}.$$

This is the line matrix of $\mathbf{S}_2$ IF

$$\theta_2\nu^2 + 2\delta_2\mu\nu - \delta_1\lambda^2 = 0 \quad \text{and} \quad \theta_1\nu^2 + 2\delta_1\lambda\nu - \delta_2\mu^2 = 0.$$

* See 11·38, p. 307.

These equations give

$$\nu^2 : 2\nu : 1 = (\delta_1^2\lambda^3 - \delta_2^2\mu^3) : -(\delta_1\theta_1\lambda^2 - \delta_2\theta_2\mu^2) : (\delta_1\theta_2\lambda - \delta_2\theta_1\mu).$$

Thus the conics **C** are obtained by solving the quartic equation

$$(\delta_1\theta_1\lambda^2 - \delta_2\theta_2\mu^2)^2 - 4(\delta_1^2\lambda^3 - \delta_2^2\mu^3)(\delta_1\theta_2\lambda - \delta_2\theta_1\mu) = 0.$$

11·53. If $\theta_2^2 - 4\theta_1\delta_2 = 0$, the coefficient of μ^4 in the quartic equation of 11·52 vanishes, so one of the conics **C** is given by

$$\lambda = 0, \quad \theta_2\nu + 2\delta_2\mu = 0,$$

and is therefore the conic $\mathbf{W}_2$, which in this case coincides with $\mathbf{v}_1$. This agrees with 11·43.

CHAPTER XII

INVARIANTS AND COVARIANTS

12·1. Invariants

12·11. Consider the non-singular linear transformation T between the variables $x_1, x_2, \ldots, x_n$ and $y_1, y_2, \ldots, y_n$ given by

$$\mathbf{x} = \mathbf{my}, \qquad (\cdot 111)$$

where $\mathbf{x}$, $\mathbf{y}$ are the column matrices $(x_1, x_2, \ldots, x_n)'$, $(y_1, y_2, \ldots, y_n)'$ and $\mathbf{m}$ is a non-singular matrix (m_{ij}), $(i, j = 1, 2, \ldots, n)$. A homogeneous polynomial of degree h in $x_1, x_2, \ldots, x_n$ is called a form of degree h in (x), and will be denoted by $f(x)$, or $\phi(x), \ldots$. On substituting for $x_1, x_2, \ldots, x_n$ the linear forms in (y) to which they are equal in (·111), a form $f(x)$ of degree h in (x) becomes a form $g(y)$ of degree h in y. This is expressed symbolically by writing

$$Tf(x) = g(y). \qquad (\cdot 112)$$

The coefficient of a term $x_1^{p_1} x_2^{p_2} \ldots x_n^{p_n}$ in $f(x)$ is said to correspond to the coefficient of $y_1^{p_1} y_2^{p_2} \ldots y_n^{p_n}$ in $g(y)$.

Suppose that $f_1(x), f_2(x), \ldots, f_r(x)$ are a given set of forms in (x), with coefficients $a_1, a_2, \ldots, a_s$. Let $g_1(y), g_2(y), \ldots, g_r(y)$ be the forms in (y) obtained by applying T, and let $b_1, b_2, \ldots, b_s$ be the corresponding coefficients. Then a polynomial function $I(a)$ of the coefficients $a_1, a_2, \ldots, a_s$ is said to be a *projective invariant* of the given set of forms IF

$$I(b) = \phi(m_{ij})\, I(a), \qquad (\cdot 113)$$

where $\phi(m_{ij})$ is a function of the coefficients of m, and does not depend on the coefficients $a_1, a_2, \ldots, a_s$. It can in fact be shown,* though a proof is unnecessary for the purposes of this chapter, that $\phi(m_{ij})$ is always an integral power of m, the determinant of the matrix $\mathbf{m}$. If $\phi(m_{ij}) = m^w$, $I(a)$ is said to be a projective invariant of weight w, and in particular, if $w = 0$, $I(b) = I(a)$ and $I(a)$ is said to be an absolute projective invariant. In this chapter the word invariant will always mean projective invariant.

* H. W. Turnbull, *The Theory of Determinants, Matrices, and Invariants* (Glasgow, 1929), pp. 169–70.

12·12. As an example consider the quadratic form in two variables

$$f(x) = a_1 x_1^2 + 2a_2 x_1 x_2 + a_3 x_2^2.$$

On applying the transformation

$$\begin{pmatrix} x_1 \\ x_2 \end{pmatrix} = \begin{pmatrix} m_{11} & m_{12} \\ m_{21} & m_{22} \end{pmatrix} \begin{pmatrix} y_1 \\ y_2 \end{pmatrix}$$

the form

$$g(y) = b_1 y_1^2 + 2b_2 y_1 y_2 + b_3 y_2^2$$

is obtained, where

$$\begin{aligned} b_1 &= a_1 m_{11}^2 + 2a_2 m_{11} m_{21} + a_3 m_{21}^2, \\ b_2 &= a_1 m_{11} m_{12} + a_2(m_{11} m_{22} + m_{21} m_{12}) + a_3 m_{21} m_{22}, \\ b_3 &= a_1 m_{12}^2 + 2a_2 m_{12} m_{22} + a_3 m_{22}^2. \end{aligned}$$

It is easily seen that

$$\begin{vmatrix} b_1 & b_2 \\ b_2 & b_3 \end{vmatrix} = m^2 \begin{vmatrix} a_1 & a_2 \\ a_2 & a_3 \end{vmatrix},$$

where

$$m = \begin{vmatrix} m_{11} & m_{12} \\ m_{21} & m_{22} \end{vmatrix},$$

so the determinant $\begin{vmatrix} a_1 & a_2 \\ a_2 & a_3 \end{vmatrix}$ is an invariant of weight two of the quadratic form $f(x)$.

12·13. More generally consider a quadratic form, in n variables $x_1, x_2, \ldots, x_n$,

$$f(x) = \mathbf{x}'\mathbf{a}\mathbf{x},$$

where there is no loss of generality in supposing that the matrix $a \equiv (a_{ij})$, $(i, j = 1, 2, \ldots, n)$, is symmetric. On applying the linear transformation T given by

$$\mathbf{x} = \mathbf{m}\mathbf{y},$$

we have

$$f(x) = \mathbf{y}'\mathbf{m}'\mathbf{a}\mathbf{m}\mathbf{y} = \mathbf{y}'\mathbf{b}\mathbf{y},$$

where $\mathbf{b}$ is the symmetric matrix $\mathbf{m}'\mathbf{a}\mathbf{m}$. In fact

$$Tf(x) = g(y),$$

where

$$g(y) = \mathbf{y}'\mathbf{b}\mathbf{y}.$$

But $|\mathbf{b}| = m^2 |\mathbf{a}|$, so $|\mathbf{a}|$ is an invariant of $f(x)$ of weight two.

12·14. Consider now two quadratic forms

$$f_1(x) = \mathbf{x}'\mathbf{a}_1\mathbf{x}, \quad f_2(x) = \mathbf{x}'\mathbf{a}_2\mathbf{x}$$

in $x_1, \ldots, x_n$, where $\mathbf{a}_1$ and $\mathbf{a}_2$ are symmetric. Then

$$\begin{aligned}\lambda f_1(x) + \mu f_2(x) &= \mathbf{x}'(\lambda\mathbf{a}_1 + \mu\mathbf{a}_2)\mathbf{x} \\ &= \mathbf{y}'\mathbf{m}'(\lambda\mathbf{a}_1 + \mu\mathbf{a}_2)\,\mathbf{my} \\ &= \mathbf{y}'(\lambda\mathbf{b}_1 + \mu\mathbf{b}_2)\,\mathbf{y} \\ &= \lambda g_1(y) + \mu g_2(y),\end{aligned}$$

where $\mathbf{b}_1$, $\mathbf{b}_2$ are the symmetric matrices $\mathbf{m}'\mathbf{a}_1\mathbf{m}$, $\mathbf{m}'\mathbf{a}_2\mathbf{m}$, and $g_1(y) = Tf_1(x)$, $g_2(y) = Tf_2(x)$. But the determinant $|\lambda\mathbf{a}_1 + \mu\mathbf{a}_2|$ may be expanded in the form

$$|\lambda\mathbf{a}_1 + \mu\mathbf{a}_2| = \theta_0\lambda^n + \theta_1\lambda^{n-1}\mu + \ldots + \theta_n\mu^n,$$

where $\theta_0, \theta_1, \ldots, \theta_n$ are functions of the coefficients of $f_1(x)$ and $f_2(x)$. In particular $\theta_0 = |\mathbf{a}_1|$ and $\theta_n = |\mathbf{a}_2|$. Similarly

$$|\lambda\mathbf{b}_1 + \mu\mathbf{b}_2| = \phi_0\lambda^n + \phi_1\lambda^{n-1}\mu + \ldots + \phi_n\mu^n.$$

But
$$\lambda\mathbf{b}_1 + \mu\mathbf{b}_2 = \mathbf{m}'(\lambda\mathbf{a}_1 + \mu\mathbf{a}_2)\,\mathbf{m},$$

so
$$|\lambda\mathbf{b}_1 + \mu\mathbf{b}_2| = m^2\,|\lambda\mathbf{a}_1 + \mu\mathbf{a}_2|,$$

and this is true for all values of λ and μ, so that

$$\phi_r = m^2\theta_r,$$

for $r = 0, 1, \ldots, n$. In fact $\theta_0, \theta_1, \ldots, \theta_n$ are invariants of weight two of the two forms $f_1(x)$ and $f_2(x)$.

The invariant θ_i is a polynomial of degree $n-i$ in the coefficients of $\mathbf{a}_1$ and of degree i in the coefficients of $\mathbf{a}_2$. Also, if $F(\theta)$ is a homogeneous polynomial in $\theta_0, \ldots, \theta_n$, then the equation $F(\theta) = 0$ implies $F(\phi) = 0$.

Consider the case $n = 2$ and regard x_1, x_2 as co-ordinates of points of a line.

The quadratic forms $f_1(x), f_2(x)$ may be written

$$f_1(x) = p_1x_1^2 + 2q_1x_1x_2 + r_1x_2^2,$$
$$f_2(x) = p_2x_2^2 + 2q_2x_1x_2 + r_2x_2^2,$$

where
$$|\mathbf{a}_1| = \theta_0 = p_1r_1 - q_1^2,$$
$$\theta_1 = p_1r_2 + p_2r_1 - 2q_1q_2,$$
$$|\mathbf{a}_2| = \theta_2 = p_2r_2 - q_2^2.$$

The equations $f_1(x) = 0$, $f_2(x) = 0$ give two pairs of points $|P_1, Q_1|$, $|P_2, Q_2|$, and it is known that

(a) P_1, Q_1 coincide IF $\theta_0 = 0$.

(b) P_2, Q_2 coincide IF $\theta_2 = 0$.

(c) $|P_1, Q_1|$, $|P_2, Q_2|$ are apolar IF $\theta_1 = 0$.

(d) $|P_1, Q_1|$, $|P_2, Q_2|$ have a common point IF $\theta_1^2 = 4\theta_0\theta_2$.

Similarly in the case $n = 3$, regarding x_1, x_2, x_3 as co-ordinates in a plane, the equations $f_1(x) = 0, f_2(x) = 0$ give two conic loci s_1, s_2 with point matrices $\mathbf{a}_1$, $\mathbf{a}_2$. In Chapters IV, IX and XI, when a slightly different notation was used and θ_0, θ_3 were denoted by δ_1, δ_2, it was shown that

(i) s_1 is a line-pair IF $\theta_0 = 0$.
(ii) s_2 is a line-pair IF $\theta_3 = 0$.
(iii) s_1 is outpolar to s_2 IF $\theta_2 = 0$.
(iv) s_2 is outpolar to s_1 IF $\theta_1 = 0$.
(v) The G conic of s_1, s_2 is a point-pair IF $\theta_1\theta_2 = \theta_0\theta_3$.
(vi) The f conic of s_1, s_2 is a line-pair IF $\theta_0\theta_3(\theta_1\theta_2 - \theta_0\theta_3) = 0$.
(vii) s_1 is triangularly inscribed in s_2 IF $\theta_1^2 = 4\theta_0\theta_2$.
(viii) s_2 is triangularly inscribed in s_1 IF $\theta_2^2 = 4\theta_1\theta_3$.
(ix) There are ordered quadrangles inscribed in s_1 and circumscribed to s_2 IF $\theta_2^3 - 4\theta_1\theta_2\theta_3 + 8\theta_0\theta_3^2 = 0$.
(x) There are ordered quadrangles inscribed in s_2 and circumscribed to s_1 IF $\theta_1^3 - 4\theta_0\theta_1\theta_2 + 8\theta_0^2\theta_3 = 0$.*

Thus a projective property of two conics, or of two pairs of points on a line may be equivalent to a relation between the invariants of two quadratic forms whose vanishing gives the conics or pairs of points.

12·15. A projective property of two conics s_1, s_2 is independent of the choice of a co-ordinate system, so if a polynomial equation $P(\theta) = 0$ in the invariants θ_0, θ_1, θ_2, θ_3 of two quadratic forms $f_1(x)$ and $f_2(x)$, whose vanishing gives s_1 and s_2, is to be a CONDITION for some projective property, then the truth of the equation $P(\theta) = 0$ must be independent of the method of writing down the equations of s_1 and s_2.

* For (i) and (ii) see 4·11, p. 100, for (iii) and (iv) 9·64, p. 250, for (v) and (vi) 11·31, p. 304, for (vii) and (viii) 11·43, p. 310, and for (ix) and (x) 11·46, p. 310.

Even when a co-ordinate system is assigned, the values of the invariants are not determined by the conics, for s_1 and s_2 are also given by the equations $c_1 f_1(x) = 0$, $c_2 f_2(x) = 0$, where c_1, c_2 are any non-zero constants, and the invariants of the forms $c_1 f_1(x)$, $c_2 f_2(x)$ are θ_0', θ_1', θ_2', θ_3', where

$$\theta_0' = c_1^3\theta_0, \quad \theta_1' = c_1^2 c_2 \theta_1, \quad \theta_2' = c_1 c_2^2 \theta_2, \quad \theta_3' = c_2^3\theta_3.$$

It is therefore essential that $P(\theta) = 0$ shall imply $P(\theta') = 0$. This will be the case if all the terms of $P(\theta)$ are of the same degree p in the coefficients of $f_1(x)$ and of the same degree q in the coefficients of $f_2(x)$. For then, if $h\theta_0^\alpha \theta_1^\beta \theta_2^\gamma \theta_3^\delta$ is a general term of $P(\theta)$,

$$3\alpha + 2\beta + \gamma = p, \qquad (\cdot 151)$$

$$\beta + 2\gamma + 3\delta = q, \qquad (\cdot 152)$$

and consequently $$\alpha + \beta + \gamma + \delta = \frac{p+q}{3} = r. \qquad (\cdot 153)$$

Thus the corresponding term $h\theta_0'^\alpha \theta_1'^\beta \theta_2'^\gamma \theta_3'^\delta$ in $P(\theta')$ is equal to $hc_1^p c_2^q \theta_0^\alpha \theta_1^\beta \theta_2^\gamma \theta_3^\delta$, so

$$P(\theta') = c_1^p c_2^q P(\theta).$$

It follows also from ·153 that the polynomial $P(\theta)$ is homogeneous in θ_0, θ_1, θ_2, θ_3, and it is convenient to make the following definition:

If two conics s_1, s_2 are given by the equations $f_1(x) = 0$, $f_2(x) = 0$, a polynomial equation $P(\theta) = 0$ is said to be an invariant equation of the two conics IF *all the terms of $P(\theta)$ are of the same degree p in the coefficients of $f_1(x)$ and of the same degree q in the coefficients of $f_2(x)$. The polynomial $P(\theta)$ is then homogeneous in θ_0, θ_1, θ_2, θ_3.**

It has already been shown that the truth of the equation $P(\theta) = 0$ is unaffected by multiplying the forms $f_1(x)$, $f_2(x)$ by non-zero numbers, and it will now be shown that it is also unaffected by a change of co-ordinates. In fact:

The truth of an invariant equation of two conics is independent of the method of writing down the equations of the conics.

For any co-ordinate system y_1, y_2, y_3 is connected with x_1, x_2, x_3 by a transformation T, and if $g_1(y)$, $g_2(y)$ are the transforms of $f_1(x)$, $f_2(x)$, the invariants ϕ_0, ϕ_1, ϕ_2, ϕ_3 of $g_1(y)$, $g_2(y)$ are given by

$$\phi_0 = m^2\theta_0, \quad \phi_1 = m^2\theta_1, \quad \phi_2 = m^2\theta_2, \quad \phi_3 = m^2\theta_3.$$

* For a proof that *all* projective polynomial invariant equations are of this form, see H. W. Turnbull, loc. cit., pp. 304–306.

Since an invariant equation $P(\theta) = 0$ is homogeneous in $\theta_0, \theta_1, \theta_2, \theta_3$ it follows that $P(\theta) = 0$ implies $P(\phi) = 0$. But the equations of the conics in the co-ordinates y_1, y_2, y_3 can only be of the form

$$d_1 g_1(y) = 0, \quad d_2 g_2(y),$$

where $d_1 d_2 \neq 0$; it has already been shown that the truth of $P(\phi) = 0$ is unaffected by replacing $g_1(y), g_2(y)$ by $d_1 g_1(y)$ and $d_2 g_2(y)$, and so the theorem is established.

12·16. Two particular invariant equations are

$$\theta_1^2 - k_1\theta_0\theta_2 = 0, \quad \theta_2^2 - k_2\theta_1\theta_3 = 0,$$

where k_1, k_2 are constants. It thus follows that the values of the expressions

$$Q_1 = \frac{\theta_1^2}{\theta_0\theta_2}, \quad Q_2 = \frac{\theta_2^2}{\theta_1\theta_3}$$

are independent of the method of writing down the equations of the two conics. Q_1, Q_2 are accordingly said to be absolute invariants of the two conics. This has already been proved in 9·62.

An invariant equation $P(\theta) = 0$ may in general be expressed as an equation in Q_1 and Q_2. For if $h\theta_0^\alpha\theta_1^\beta\theta_2^\gamma\theta_3^\delta$ is a general term of $P(\theta)$, then the equations (·151), (·152), (·153) follow, where p, q, r are the same for all terms of $P(\theta)$. On substituting for θ_0, θ_3 the expressions

$$\theta_0 = \frac{\theta_1^2}{\theta_2 Q_1}, \quad \theta_3 = \frac{\theta_2^2}{\theta_1 Q_2},$$

we have

$$\begin{aligned} P(\theta) &= \Sigma h\theta_0^\alpha\theta_1^\beta\theta_2^\gamma\theta_3^\delta \\ &= \Sigma h Q_1^{-\alpha} Q_2^{-\delta}\,\theta_1^{2\alpha+\beta-\delta}\,\theta_2^{-\alpha+\gamma+2\delta} \\ &= \theta_1^{p-r}\,\theta_2^{q-r}\,\Sigma h Q_1^{-\alpha} Q_2^{-\delta}. \end{aligned}$$

So $P(\theta) = 0$ is equivalent to a polynomial equation $f(Q_1, Q_2) = 0$, which is not usually homogeneous.

12·17. The theory of invariant equations may be used to establish the properties (i), (ii), ..., (x) of 12·14. Indeed the fact that $\theta_1 = 0$ and $\theta_2 = 0$ are invariant equations has already been established and used in 9·63 and 9·64. Consider for example the theorem that, if there is one triangle inscribed in s_1 and circumscribed to s_2, there are ∞^1 such triangles, and that a CONDITION for this to happen is $\theta_2^2 = 4\theta_1\theta_3$. Suppose first that there is one such triangle, and take it as triangle of reference. Then the equations may be written

$$s_1 \equiv 2yz + 2zx + 2xy = 0,$$

$$s_2 \equiv a^2x^2 + b^2y^2 + c^2z^2 - 2bcyz - 2cazx - 2abxy = 0,$$

when

$$\theta_0 = 2, \quad \theta_1 = -(a+b+c)^2, \quad \theta_2 = 4abc(a+b+c), \quad \theta_3 = -4a^2b^2c^2,$$

whence
$$\phi \equiv \theta_2^2 - 4\theta_1\theta_3 = 0.$$

But the terms θ_2^2 and $4\theta_1\theta_3$ of ϕ are both of degree two in the coefficients of s_1 and of degree four in the coefficients of s_2; hence $\phi = 0$ is an invariant equation, and is therefore true for all methods of writing down the equation of the conics.

Suppose then that a general tangent of s_2 meets s_1 in Y and Z. Let the other tangents to s_2 through Y and Z meet in X. With X, Y, Z as triangle of reference the equation may be written in the form

$$s_1 \equiv \lambda x^2 + 2yz + 2zx + 2xy = 0,$$
$$s_2 \equiv a^2x^2 + b^2y^2 + c^2z^2 - 2bcyz - 2cazx - 2abxy = 0,$$

where $abc \neq 0$. Then

$$\theta_3 = -4a^2b^2c^2, \quad \theta_2 = 4abc(a+b+c), \quad \theta_1 = 2bc\lambda - (a+b+c)^2.$$

Since $\phi = \theta_2^2 - 4\theta_1\theta_3 = 0$, it follows that $\lambda = 0$, so X lies on s_1. There are thus ∞^1 triangles, and the necessity and sufficiency of the condition $\theta_2^2 = 4\theta_1\theta_3$ have been established.

This example serves to illustrate the method and further examples will be found in Sommerville, *Analytical Conics*, Chapter XX.*

12·2. Covariants

12·21. A projective invariant of a set of forms in n variables was defined in 12·11. Using the same notation, suppose that $C(a,x)$ is a form in the variables $x_1, x_2, \ldots, x_n$ whose coefficients $a_1, a_2, \ldots, a_s$ are polynomials in the coefficients of $f_1(x), f_2(x), \ldots, f_r(x)$. Let $C(b,y)$ be the form similarly obtained from the forms $g_1(y), g_2(y), \ldots, g_r(y)$. Then $C(a,x)$ is said to be a *projective covariant* of the given set of forms IF

$$C(b,y) \equiv \phi(m_{ij})\, C(a,x), \qquad (\cdot 211)$$

where $\phi(m_{ij})$ is a function of the coefficients of $\mathbf{m}$ and does not depend on the coefficients $a_1, a_2, \ldots, a_s$. The equation (·211) means that if the substitution $\mathbf{x} = \mathbf{my}$ is made in the form $\phi(m_{ij})\, C(a,x)$, the form $C(b,y)$ is obtained. In particular, if $\phi(m_{ij}) = m^w$, $C(a,x)$ is

* London: G. Bell and Sons, Ltd. 1924.

said to be a covariant of weight w. It should be noticed that an invariant is a number, whereas a covariant is a form. As we shall only be concerned with linear transformations the term 'covariant' is to be understood in the following as meaning 'projective covariant'.

12·22. *Hessian of a form in two variables*

As an example some covariants of forms in two variables will be considered, and for convenience the variables will be denoted by x, y instead of x_1, x_2. Let $f(x,y)$ be a form of degree n, which will be denoted by f. Thus:

$$f \equiv f(x,y) \equiv a_0 x^n + a_1 x^{n-1} y + \dots + a_n y^n.$$

When a non-singular transformation

$$\begin{pmatrix} x \\ y \end{pmatrix} = \mathbf{m} \begin{pmatrix} x' \\ y' \end{pmatrix} = \begin{pmatrix} m_{11} & m_{12} \\ m_{21} & m_{22} \end{pmatrix} \begin{pmatrix} x' \\ y' \end{pmatrix}$$

is applied, the form $f(x,y)$ becomes a form $u(x',y')$ in the new variables x', y'. Thus:

$$f \equiv f(x,y) \equiv u(x',y') \equiv u.$$

Now
$$\frac{\partial}{\partial x'} = m_{11} \frac{\partial}{\partial x} + m_{21} \frac{\partial}{\partial y},$$

$$\frac{\partial}{\partial y'} = m_{12} \frac{\partial}{\partial x} + m_{22} \frac{\partial}{\partial y},$$

so
$$\begin{pmatrix} \dfrac{\partial}{\partial x'} \\ \dfrac{\partial}{\partial y'} \end{pmatrix} = \mathbf{m}' \begin{pmatrix} \dfrac{\partial}{\partial x} \\ \dfrac{\partial}{\partial y} \end{pmatrix}.$$

Consider the form
$$H(f) \equiv \begin{vmatrix} \dfrac{\partial^2 f}{\partial x^2} & \dfrac{\partial^2 f}{\partial x \, \partial y} \\ \dfrac{\partial^2 f}{\partial y \, \partial x} & \dfrac{\partial^2 f}{\partial y^2} \end{vmatrix}$$

and the corresponding form

$$H(u) \equiv \begin{vmatrix} \dfrac{\partial^2 u}{\partial x'^2} & \dfrac{\partial^2 u}{\partial x' \, \partial y'} \\ \dfrac{\partial^2 u}{\partial y' \, \partial x'} & \dfrac{\partial^2 u}{\partial y'^2} \end{vmatrix}.$$

Now
$$\begin{pmatrix} \dfrac{\partial^2 u}{\partial x'^2} & \dfrac{\partial^2 u}{\partial x'\,\partial y'} \\ \dfrac{\partial^2 u}{\partial y'\,\partial x'} & \dfrac{\partial^2 u}{\partial y'^2} \end{pmatrix} = \begin{pmatrix} \dfrac{\partial}{\partial x'} \\ \dfrac{\partial}{\partial y'} \end{pmatrix} \left(\frac{\partial}{\partial x'}, \frac{\partial}{\partial y'}\right) u$$

$$= \mathbf{m}' \begin{pmatrix} \dfrac{\partial}{\partial x} \\ \dfrac{\partial}{\partial y} \end{pmatrix} \left(\frac{\partial}{\partial x}, \frac{\partial}{\partial y}\right) \mathbf{m} f$$

$$= \mathbf{m}' \begin{pmatrix} \dfrac{\partial^2 f}{\partial x^2} & \dfrac{\partial^2 f}{\partial x\,\partial y} \\ \dfrac{\partial^2 f}{\partial y\,\partial x} & \dfrac{\partial^2 f}{\partial y^2} \end{pmatrix} \mathbf{m}.$$

Taking determinants of both sides,

$$H(u) = m^2 H(f).$$

Thus:

The Hessian $H(f)$ of the form $f(x,y)$ is a covariant of weight two.

This argument applies readily to a form in any number of variables.

12·23. *Jacobian of two forms in two variables*

Let $f(x,y)$, $g(x,y)$ be two forms, and suppose that under the transformation

$$\begin{pmatrix} x \\ y \end{pmatrix} = \mathbf{m} \begin{pmatrix} x' \\ y' \end{pmatrix}$$

they become $u(x',y')$, $v(x',y')$. Thus:

$$f \equiv f(x,y) \equiv u(x',y') \equiv u,$$
$$g \equiv g(x,y) \equiv v(x',y') \equiv v.$$

Consider the form
$$J(f,g) \equiv \begin{vmatrix} \dfrac{\partial f}{\partial x} & \dfrac{\partial g}{\partial x} \\ \dfrac{\partial f}{\partial y} & \dfrac{\partial g}{\partial y} \end{vmatrix}$$

and the corresponding form

$$J(u,v) \equiv \begin{vmatrix} \dfrac{\partial u}{\partial x'} & \dfrac{\partial v}{\partial x'} \\ \dfrac{\partial u}{\partial y'} & \dfrac{\partial v}{\partial y'} \end{vmatrix}.$$

Since

$$\begin{pmatrix} \frac{\partial u}{\partial x'} & \frac{\partial v}{\partial x'} \\ \frac{\partial u}{\partial y'} & \frac{\partial v}{\partial y'} \end{pmatrix} = \begin{pmatrix} \frac{\partial}{\partial x'} \\ \frac{\partial}{\partial y'} \end{pmatrix} (u, v) = \mathbf{m}' \begin{pmatrix} \frac{\partial}{\partial x} \\ \frac{\partial}{\partial y} \end{pmatrix} (f, g) = \mathbf{m}' \begin{pmatrix} \frac{\partial f}{\partial x}, & \frac{\partial g}{\partial x} \\ \frac{\partial f}{\partial y}, & \frac{\partial g}{\partial y} \end{pmatrix},$$

it follows that $$J(u, v) = mJ(f, g),$$
so:

The Jacobian, $J(f, g)$, of two forms $f(x, y)$, $g(x, y)$ is a covariant of weight one.

The argument extends to n forms in n variables.

12·24. *Hessian of a set of elements of an $R\infty^1$*

The ratio variable (x, y) may be regarded as a ratio parameter of an $R\infty^1$. For convenience of language we shall consider the $R\infty^1$ as the points of a line, though what follows is applicable to any $R\infty^1$.

Consider then a line p and a set of n points $(A_1, A_2, \ldots, A_n)$ on p. If a ratio parameter (x, y) is chosen, the points (A) are given by an equation $$f(x, y) \equiv a_0 x^n + a_1 x^{n-1} y + \ldots + a_n y^n = 0,$$

in which the ratio set $(a_0, a_1, \ldots, a_n)$ is uniquely determined. The equation $$H(f) = 0$$

then determines a unique set of $2(n-2)$ points $(H_1, H_2, \ldots, H_{2(n-2)})$, or (H), because the coefficients in $H(f)$ are homogeneous in $a_0, a_1, \ldots, a_n$.

Any other ratio parameter (x', y') of p is connected with (x, y) by a non-singular transformation

$$\begin{pmatrix} x \\ y \end{pmatrix} = \mathbf{m} \begin{pmatrix} x' \\ y' \end{pmatrix}, \quad m = |\mathbf{m}| \neq 0,$$

so in terms of (x', y') the set (A) is given by the equation

$$u(x', y') = 0,$$

where $u(x', y')$ is the transform of $f(x, y)$. The equation

$$H(u) = 0$$

gives a set of $2(n-2)$ points, which is actually the set (H), for it has been shown that $$H(u) \equiv m^2 H(f).$$

Since $m \neq 0$ it follows that, if P_1 is a point of p given by the values (x_1, y_1), (x_1', y_1') of the ratio parameters (x, y), (x', y'), then $H(u_1) = 0$ implies $H(f_1) = 0$ and $H(f_1) = 0$ implies $H(u_1) = 0$, where

$$f_1 = f(x_1, y_1) \quad \text{and} \quad u_1 = u(x_1', y_1').$$

In fact, whatever ratio parameter is used, the same set (H) is obtained. Thus:

A set of n points $(A_1, A_2, \ldots, A_n)$ on a line p determines a set of $2(n-2)$ points, called the Hessian set of $(A_1, A_2, \ldots, A_n)$. If (x, y) is any ratio parameter of the line, and if the set of points $(A_1, A_2, \ldots, A_n)$ is given by $f(x, y) = 0$, then the Hessian set is given by $H(f) = 0$.

Further, if (x, y), (x', y') are ratio parameters of two lines p, p', any proper $(1, 1)$ correspondence between p and p' is given by equations of the form

$$\begin{pmatrix} x \\ y \end{pmatrix} = \mathbf{m} \begin{pmatrix} x' \\ y' \end{pmatrix},$$

where $m = |\mathbf{m}| \neq 0$. It follows that:

If in a proper $T(1, 1)$ between two lines p and p' the points $A_1, A_2, \ldots, A_n$ of p correspond to the points $A_1', A_2', \ldots, A_n'$ of p', then the Hessian set of $A_1, A_2, \ldots, A_n$ corresponds to the Hessian set of $A_1', A_2', \ldots, A_n'$.

12·25. *Polar sets*

By a similar argument it follows that, if $A_1, A_2, \ldots, A_n$ and $B_1, B_2, \ldots, B_m$ are two sets of points on p given by

$$f(x, y) \equiv a_0 x^n + a_1 x^{n-1} y + \ldots + a_n y^n = 0,$$
$$g(x, y) \equiv b_0 x^m + b_1 x^{m-1} y + \ldots + b_m y^m = 0,$$

then the set of $m + n - 2$ points given by

$$J(f, g) = 0$$

is independent of the co-ordinate system, and is therefore determined by the two given sets of points.

In particular let

$$g(x, y) \equiv -y_1 x + x_1 y,$$

so that the set given by $g(x, y) = 0$ consists of the single point P_1. Then

$$J(f, g) = x_1 \frac{\partial f}{\partial x} + y_1 \frac{\partial f}{\partial y},$$

so the set of $n-1$ points given by

$$\left(x_1 \frac{\partial}{\partial x}+y_1 \frac{\partial}{\partial y}\right)f(x,y) = 0$$

is determined by P_1 and $(A_1, A_2, \ldots, A_n)$. This set is called the *first polar set* of P_1 relative to $(A_1, A_2, \ldots, A_n)$.* The first polar set of P_1 relative to the first polar set of P_1 relative to $(A_1, A_2, \ldots, A_n)$ is given by

$$\left(x_1 \frac{\partial}{\partial x}+y_1 \frac{\partial}{\partial y}\right)^2 f(x,y) = 0,$$

and is called the *second polar set* of P_1 relative to $(A_1 \ldots A_n)$.

Similarly the set of $n-r$ points given by

$$\left(x_1 \frac{\partial}{\partial x}+y_1 \frac{\partial}{\partial y}\right)^r f(x,y) = 0$$

is independent of the co-ordinate system, and is called the rth polar set of P_1 relative to $(A_1, A_2, \ldots, A_n)$.

As for the Hessian set it follows that, if in a homography between p and another line p' the points $P_1, A_1, A_2, \ldots, A_n$ correspond to $P'_1, A'_1, A'_2, \ldots, A'_n$, then the polar sets of P_1 relative to $(A_1, A_2, \ldots, A_n)$ correspond to the polar sets of P'_1 relative to $(A'_1, A'_2, \ldots, A'_n)$.

The first polar sets are connected with the Hessian set of $(A_1, A_2, \ldots, A_n)$ as follows. The first polar set of P_1 will in general consist of $n-1$ distinct points, but the point P_2 counts twice in the polar set IF (x_2, y_2) is a double root of

$$\left(x_1 \frac{\partial}{\partial x}+y_1 \frac{\partial}{\partial y}\right)f(x,y) = 0,$$

that is IF

$$\frac{\partial}{\partial x_2}\left(x_1 \frac{\partial}{\partial x_2}+y_1 \frac{\partial}{\partial y_2}\right)f(x_2,y_2) = \frac{\partial}{\partial y_2}\left(x_1 \frac{\partial}{\partial x_2}+y_1 \frac{\partial}{\partial y_2}\right)f(x_2,y_2) = 0.$$

Eliminating x_1, y_1 we see that the points P_2 which count twice in some first polar set are just the points of the Hessian set given by

$$\begin{vmatrix} \dfrac{\partial^2}{\partial x^2}f(x,y), & \dfrac{\partial^2}{\partial x\,\partial y}f(x,y) \\ \dfrac{\partial^2}{\partial y\,\partial x}f(x,y), & \dfrac{\partial^2}{\partial y^2}f(x,y) \end{vmatrix} = 0.$$

* Cf. 7·7 and 9·89, pp. 206, 265.

12·26. By an extension of the argument of 12·23 it is easily shown that the Jacobian of three forms $f(x,y,z)$, $g(x,y,z)$, $h(x,y,z)$ in three variables, defined by

$$J(f,g,h) = \begin{vmatrix} \frac{\partial f}{\partial x} & \frac{\partial g}{\partial x} & \frac{\partial h}{\partial x} \\ \frac{\partial f}{\partial y} & \frac{\partial g}{\partial y} & \frac{\partial h}{\partial y} \\ \frac{\partial f}{\partial z} & \frac{\partial g}{\partial z} & \frac{\partial h}{\partial z} \end{vmatrix}$$

is a covariant of weight one of the three forms. Consider the geometrical interpretations of the equation $J(f,g,h) = 0$, when the forms are linear or quadratic. Let

$$s_i \equiv (a_i b_i c_i f_i g_i h_i \between x,y,z)^2,$$
$$p_i \equiv X_i x + Y_i y + Z_i z.$$

Then $J(p_1,p_2,p_3) = 0$ is the CONDITION *for the lines p_1, p_2, p_3 to be concurrent. $J(s_1,p_1,p_2) = 0$ is the polar line of the point $p_1 p_2$ relative to the conic s_1. $J(s_1,s_2,p_1) = 0$ is the eleven-point conic of p_1 and the point pencil determined by s_1 and s_2. $J(s_1,s_2,s_3) = 0$ is the locus of vertices of line-pairs of the net of conics $\lambda s_1 + \mu s_2 + \nu s_3 = 0$, and also the locus of points whose polars relative to the conics of this net are concurrent.*

The first two of these results are obvious. The polars of P_1 relative to s_1, s_2 are

$$x_1 \frac{\partial s_1}{\partial x} + y_1 \frac{\partial s_1}{\partial y} + z_1 \frac{\partial s_1}{\partial z} = 0,$$

$$x_1 \frac{\partial s_2}{\partial x} + y_1 \frac{\partial s_2}{\partial y} + z_1 \frac{\partial s_2}{\partial z} = 0,$$

and meet in the point P_1' which is conjugate to P_1 relative to all conics $\lambda s_1 + \mu s_2 = 0$. Also P_1 lies on p_1 IF

$$x_1 \frac{\partial p_1}{\partial x} + y_1 \frac{\partial p_1}{\partial y} + z_1 \frac{\partial p_1}{\partial z} = 0,$$

so, eliminating x_1, y_1, z_1, the locus of P_1' as P_1 moves on p_1 is the conic $J(s_1,s_2,p_1) = 0$.

Also the polar of (x, y, z) relative to $\lambda s_1 + \mu s_2 = 0$ is p_1 if

$$\lambda \frac{\partial s_1}{\partial x} + \mu \frac{\partial s_2}{\partial x} = \rho \frac{\partial p_1}{\partial x},$$

$$\lambda \frac{\partial s_2}{\partial y} + \mu \frac{\partial s_2}{\partial y} = \rho \frac{\partial p_1}{\partial y},$$

$$\lambda \frac{\partial s_1}{\partial z} + \mu \frac{\partial s_2}{\partial z} = \rho \frac{\partial p_1}{\partial z}.$$

So, eliminating λ, μ, ρ, the conic $J(s_1, s_2, p) = 0$ is also the locus of poles of p_1 relative to the conics $\lambda s_1 + \mu s_2 = 0$.

For the third result it is only necessary to remark that the conic $\lambda s_1 + \mu s_2 + \nu s_3 = 0$ is a line-pair with vertex (x, y, z) if

$$\lambda \frac{\partial s_1}{\partial x} + \mu \frac{\partial s_2}{\partial x} + \nu \frac{\partial s_3}{\partial x} = 0,$$

$$\lambda \frac{\partial s_1}{\partial y} + \mu \frac{\partial s_2}{\partial y} + \nu \frac{\partial s_3}{\partial y} = 0,$$

$$\lambda \frac{\partial s_1}{\partial z} + \mu \frac{\partial s_2}{\partial z} + \nu \frac{\partial s_3}{\partial z} = 0,$$

and the rest is obvious.

12·3. Covariants and contravariants of two conics

12·31. *The covariant f of two quadratic forms in three variables*

In 11·21 it was shown that the harmonic locus of the conics $\mathbf{x}'\mathbf{s}_1\mathbf{x} = 0$, $\mathbf{x}'\mathbf{s}_2\mathbf{x} = 0$ is given by the equation $\mathbf{x}'\mathbf{f}\mathbf{x} = 0$, where the symmetric matrix $\mathbf{f}$ is determined by the identity

$$(\lambda \mathbf{S}_1 + \mu \mathbf{S}_2)_a \equiv \lambda^2 \delta_1 \mathbf{s}_1 + \lambda\mu \mathbf{f} + \mu^2 \delta_2 \mathbf{s}_2.$$

We shall use the notation of Chapter XI, using co-ordinates x_1, x_2, x_3.

It will now be shown that the quadratic form

$$f(x_1, x_2, x_3) \equiv \mathbf{x}'\mathbf{f}\mathbf{x}$$

is a covariant of weight two of the two quadratic forms

$$s_1(x_1, x_2, x_3) \equiv \mathbf{x}'\mathbf{s}_1\mathbf{x}, \quad s_2(x_1, x_2, x_3) \equiv \mathbf{x}'\mathbf{s}_2\mathbf{x}.$$

Consider a change of co-ordinates from x_1, x_2, x_3 to y_1, y_2, y_3 given by a non-singular transformation

$$\mathbf{x} = \mathbf{m}\mathbf{y}.$$

Since $$\mathbf{x}'\mathbf{s}_1\mathbf{x} = \mathbf{y}'\mathbf{m}'\mathbf{s}_1\mathbf{m}\mathbf{y},$$
the quadratic form $s_1(x_1, x_2, x_3)$ becomes the quadratic form
$$\bar{s}_1(y_1, y_2, y_3) \equiv \mathbf{y}'\bar{\mathbf{s}}_1\mathbf{y},$$
where $\bar{\mathbf{s}}_1 = \mathbf{m}'\mathbf{s}_1\mathbf{m}$, and similarly for $s_2(x_1, x_2, x_3)$. Also
$$\bar{\mathbf{S}}_1 = (\bar{\mathbf{s}}_1)_a = \mathbf{m}_a\mathbf{S}_1\mathbf{m}'_a,$$
and similarly $$\bar{\mathbf{S}}_2 = \mathbf{m}_a\mathbf{S}_2\mathbf{m}'_a.$$

Scalars $\bar{\delta}_1$, $\bar{\theta}_1$, $\bar{\theta}_2$, $\bar{\delta}_2$ and a matrix $\bar{\mathbf{f}}$ are determined by $\bar{\mathbf{s}}_1$ and $\bar{\mathbf{s}}_2$ just as δ_1, θ_1, θ_2, δ_2 and $\mathbf{f}$ are determined by $\mathbf{s}_1$ and $\mathbf{s}_2$. It has already been shown in 12·14, where a different notation was used, that
$$(\bar{\delta}_1, \bar{\theta}_1, \bar{\theta}_2, \bar{\delta}_2) = m^2(\delta_1, \theta_1, \theta_2, \delta_2),$$
so that δ_1, θ_1, θ_2, δ_2 are invariants of weight two of the quadratic forms
$$s_1(x_1, x_2, x_3) \quad \text{and} \quad s_2(x_1, x_2, x_3).$$
Now $$\lambda\bar{\mathbf{S}}_1 + \mu\bar{\mathbf{S}}_2 = \mathbf{m}_a(\lambda\mathbf{S}_1 + \mu\mathbf{S}_2)\,\mathbf{m}'_a,$$
so the identity which defines $\bar{\mathbf{f}}$ is
$$\begin{aligned}\lambda^2\bar{\delta}_1\bar{\mathbf{s}}_1 + \lambda\mu\bar{\mathbf{f}} + \mu^2\bar{\delta}_2\bar{\mathbf{s}}_2 &= (\lambda\bar{\mathbf{S}}_1 + \mu\bar{\mathbf{S}}_2)_a \\ &= m^2\mathbf{m}'(\lambda\mathbf{S}_1 + \mu\mathbf{S}_2)_a\mathbf{m} \\ &= m^2\mathbf{m}'(\lambda^2\delta_1\mathbf{s}_1 + \lambda\mu\mathbf{f} + \mu^2\delta_2\mathbf{s}_2)\,\mathbf{m}.\end{aligned}$$
Thus, equating coefficients of $\lambda\mu$,
$$\bar{\mathbf{f}} = m^2\mathbf{m}'\mathbf{f}\mathbf{m},$$
so $$\mathbf{y}'\bar{\mathbf{f}}\mathbf{y} \equiv m^2\mathbf{x}'\mathbf{f}\mathbf{x}.$$
This proves that the quadratic form $\mathbf{x}'\mathbf{f}\mathbf{x}$ is a covariant of weight two of the forms $\mathbf{x}'\mathbf{s}_1\mathbf{x}$ and $\mathbf{x}'\mathbf{s}_2\mathbf{x}$.

12·32. *Covariant equations of two conics*

Let two conics γ_1, γ_2 be given. With a particular co-ordinate system x_1, x_2, x_3 let the equations of the conics be
$$s_1 \equiv \mathbf{x}'\mathbf{s}_1\mathbf{x} = 0, \quad s_2 \equiv \mathbf{x}'\mathbf{s}_2\mathbf{x} = 0,$$
and denote the form $\mathbf{x}'\mathbf{f}\mathbf{x}$ by f. We shall consider the conic given by
$$\theta_2 s_1 - f = 0. \tag{·321}$$
It was shown in 11·26 that this conic is the reciprocal of γ_2 relative to γ_1, so the equation $\theta_2 s_1 - f = 0$ determines the same locus of points however the equations of the conics are written down. This fact may also be deduced from the properties that have just been

established. For any other method of writing down the equations of γ_1, γ_2 may be obtained by a combination of the three operations

(i) Multiplying the coefficients of s_1 by a constant.

(ii) Multiplying the coefficients of s_2 by a constant.

(iii) Applying a change of variables given by a non-singular transformation $\mathbf{x} = \mathbf{my}$.

None of these operations changes the locus given by (·321), because

(i) The coefficients of $\theta_2 s_1$ and f are all quadratic in the coefficients of s_1.

(ii) The coefficients of $\theta_2 s_1$ and f are all quadratic in the coefficients of s_2.

(iii) In the notation of 12·31

$$\bar{s}_1 \equiv s_1, \quad \bar{\theta}_2 = m^2\theta_2, \quad \bar{f} \equiv m^2 f,$$

and consequently $\qquad \bar{\theta}_2\bar{s}_1 - \bar{f} \equiv m^2(\theta_2 s_1 - f).$

Hence any point whose co-ordinates (x_1, x_2, x_3) satisfy (·321) is such that its co-ordinates (y_1, y_2, y_3) satisfy $\bar{\theta}_2\bar{s}_1 - \bar{f} = 0$.

In a similar way it is easily shown that, however the equations $s_1 = 0$, $s_2 = 0$ of γ_1, γ_2 are written down, the equations

$$\theta_2 s_1 + \theta_1 s_2 - f = 0,$$
$$\delta_1(4\delta_1\theta_2 - \theta_1^2)\, s_1 + 2\delta_1^2\theta_1 s_2 - 4\delta_1^2 f = 0,$$
$$f^2 - 4\delta_1\delta_2 s_1 s_2 = 0$$

give the same loci. These loci are actually the G-conic, the W_1-conic, and the common tangents of γ_1 and γ_2.

More generally:

A polynomial equation $\psi(\delta_1, \theta_1, \theta_2, \delta_2, s_1, s_2, f) = 0$ *is said to be a covariant equation of the two conics* γ_1, γ_2 *if the polynomial* ψ *is* (*a*) *homogeneous in the coefficients of* s_1, (*b*) *homogeneous in the coefficients of* s_2, (*c*) *homogeneous in the co-ordinates* x_1, x_2, x_3.

The locus given by a covariant equation is independent of the method of writing down the equations $s_1 = 0$, $s_2 = 0$ *of* γ_1, γ_2.*

* It is not, however, true that any covariant equation of γ_1, γ_2 is expressible in the above form. It has been shown, though the proof is beyond the scope of this book, that any covariant equation of γ_1, γ_2 is expressible as a polynomial equation of the form

$$\psi\{\delta_1, \theta_1, \theta_2, \delta_2, s_1, s_2, f, J(s_1, s_2, f)\} = 0$$

in which conditions (*a*), (*b*) and (*c*) are satisfied. See, e.g. Grace and Young, *Algebra of Invariants* (Cambridge, 1903), Ch. XIII.

For suppose that a general term of ψ is

$$h\delta_1^{\xi}\theta_1^{\eta}\theta_2^{\zeta}\delta_2^{\tau}s_1^{\lambda}s_2^{\mu}f^{\nu}.$$

Then, from the properties (*a*), (*b*), (*c*),

$$3\xi+2\eta+\zeta \quad +\lambda+2\nu = 3p, \qquad (\cdot 322)$$

$$\eta+2\zeta+3\tau+\mu+2\nu = 3q, \qquad (\cdot 323)$$

$$\lambda+\mu+\nu \quad = 3r, \qquad (\cdot 324)$$

where p, q, r are the same for all terms of ψ. From these equations it follows that

$$\xi+\eta+\zeta+\tau+\nu = p+q-r. \qquad (\cdot 325)$$

Since $\bar{s}_1 \equiv s_1, \quad \bar{s}_2 \equiv s_2, \quad \bar{f} \equiv m^2 f,$

and $(\bar{\delta}_1, \bar{\theta}_1, \bar{\theta}_2, \bar{\delta}_2) = m^2(\delta_1, \theta_1, \theta_2, \delta_2),$

it follows that

$$\psi(\bar{\delta}_1, \bar{\theta}_1, \bar{\theta}_2, \bar{\delta}_2, \bar{s}_1, \bar{s}_2, \bar{f}) \equiv m^{2(p+q-r)}\, \psi(\delta_1, \theta_1, \theta_2, \delta_2, s_1, s_2, f),$$

so that the equation

$$\psi(\bar{\delta}_1, \bar{\theta}_1, \ldots, \bar{f}) = 0$$

gives the same locus as

$$\psi(\delta_1, \theta_1, \ldots, f) = 0.$$

Also the properties (*a*), (*b*) ensure that this locus is unchanged when the coefficients of s_1 or of s_2 are multiplied by a constant, so the property of covariant equations is established.

12·33. This property of covariant equations may be used to prove many of the results of Chapter XI. For example, suppose it is desired to prove that the reciprocal of the conic $s_2 = 0$ relative to the conic $s_1 = 0$ is the conic $\theta_2 s_1 - f = 0$.

Suppose first that the conics $s_1 = 0$, $s_2 = 0$ are general, so that their equations may be taken in the form

$$s_1 \equiv ax^2+by^2+cz^2 = 0, \quad s_2 \equiv x^2+y^2+z^2 = 0.$$

Then $f \equiv a(b+c)\,x^2+b(c+a)\,y^2+c(a+b)\,z^2$

and $\theta_2 = a+b+c.$

It is easily seen that the reciprocal of s_2 relative to s_1 is given by the equation

$$a^2x^2+b^2y^2+c^2z^2 = 0,$$

which may be written $\theta_2 s_1 - f = 0.$

This satisfies the three conditions for a covariant equation, so it follows that however the equations $s_1 = 0$, $s_2 = 0$ of the two given conics are written down, the equation $\theta_2 s_1 - f = 0$ gives the same locus, which is the reciprocal of s_2 relative to s_1.

Thus the theorem is established for the general case. To deal with the various particular cases we can either use the standard forms of equations obtained in Chapter VIII, or we can use the following limiting argument.

If it is not possible to find a common self-polar triangle of the conics γ_1, γ_2, choose a conic γ_3 such that γ_1 and γ_3 have a common self-polar triangle. Let the equations be

$$s_1 = 0, \quad s_2 = 0, \quad s_3 = 0.$$

Then for general values of λ the conics $s_1 = 0$ and $s_\lambda \equiv s_2 + \lambda s_3 = 0$ have a common self-polar triangle, and consequently the reciprocal of $s_\lambda = 0$ relative to $s_1 = 0$ is

$$\theta_\lambda s_1 - f_\lambda = 0,$$

where θ_λ and f_λ are obtained from s_1 and s_λ just as θ_2 and f are obtained from s_1 and s_2. But as $\lambda \to 0$ the conic $s_\lambda \to$ the conic s_2, the reciprocal of s_λ relative to $s_1 \to$ the reciprocal of s_2 relative to s_1, and $\theta_\lambda s_1 - f_\lambda \to \theta_2 s_1 - f$. So the theorem is established.

In view of this difficulty with the particular cases, the direct method of Chapter XI is to be preferred.

12·34. *The contragredient transformation*

When the point co-ordinates of a plane are changed by the non-singular transformation

$$\mathbf{x} = \mathbf{m}\mathbf{y},$$

the equation
$$X_1 x_1 + X_2 x_2 + X_3 x_3 = 0$$
becomes
$$Y_1 y_1 + Y_2 y_2 + Y_3 y_3 = 0,$$
where
$$\mathbf{Y} = \mathbf{m}'\mathbf{X},$$
since
$$\begin{aligned} X_1 x_1 + X_2 x_2 + X_3 x_3 = \mathbf{x}'\mathbf{X} &= \mathbf{y}'\mathbf{m}'\mathbf{X} \\ &= \mathbf{y}'\mathbf{Y} \\ &= Y_1 y_1 + Y_2 y_2 + Y_3 y_3. \end{aligned}$$

Thus the line co-ordinates are changed by the transformation

$$\mathbf{X} = (\mathbf{m}')^{-1}\mathbf{Y},$$

which is known as the *contragredient transformation*.*

* Cf. 1·6, p. 23.

Just as $\delta_1, \theta_1, \theta_2, \delta_2$ are invariants of weight two of the forms s_1 and s_2 for linear transformations of point co-ordinates, so, dually, $\Delta_1, \Theta_1, \Theta_2, \Delta_2$ are invariants of weight two of the forms S_1 and S_2 for linear transformations of the line co-ordinates.* Since

$$(\Delta_1, \Theta_1, \Theta_2, \Delta_2) = (\delta_1^2, \delta_1\theta_2, \delta_2\theta_1, \delta_2^2)$$

it follows that if an invariant equation

$$\phi(\delta_1, \theta_1, \theta_2, \delta_2) = 0$$

is a CONDITION for the two conics to have some property, then the equation

$$\phi(\delta_1^2, \delta_1\theta_2, \delta_2\theta_1, \delta_2^2) = 0,$$

which is also an invariant equation, is a CONDITION for the two conics to have the dual property.

When the transformation

$$\mathbf{x} = \mathbf{my}$$

is regarded as fundamental, a form in X_1, X_2, X_3 which is a covariant for the contragredient transformation

$$\mathbf{X} = (\mathbf{m}')^{-1}\,\mathbf{Y}$$

is said to be a *contravariant* for the transformation $\mathbf{x} = \mathbf{my}$.

Now, for the two quadratic forms s_1 and s_2, since

$$\bar{\mathbf{s}}_1 = \mathbf{m}'\mathbf{s}_1\mathbf{m},$$

it follows that $$\bar{\mathbf{S}}_1 = \mathbf{m}_a\mathbf{S}_1\mathbf{m}'_a,$$

so $$\bar{S}_1 \equiv \mathbf{Y}'\bar{\mathbf{S}}_1\mathbf{Y} \equiv \mathbf{X}'\mathbf{m}\mathbf{m}_a\mathbf{S}_1\mathbf{m}'_a\mathbf{m}'\mathbf{X} \equiv m^2 S_1$$

and similarly $$\bar{\mathbf{S}}_2 \equiv m^2\mathbf{S}_2.$$

Also
$$\begin{aligned}\lambda^2\bar{\mathbf{S}}_1 + \lambda\mu\bar{\mathbf{G}} + \mu^2\bar{\mathbf{S}}_2 &= (\lambda\bar{\mathbf{s}}_1 + \mu\bar{\mathbf{s}}_2)_a \\ &= \{\mathbf{m}'(\lambda\mathbf{s}_1 + \mu\mathbf{s}_2)\,\mathbf{m}\}_a \\ &= \mathbf{m}_a(\lambda^2\mathbf{S}_1 + \lambda\mu\mathbf{G} + \mu^2\mathbf{S}_2)\,\mathbf{m}'_a,\end{aligned}$$

so $$\bar{\mathbf{G}} = \mathbf{m}_a\mathbf{G}\mathbf{m}'_a,$$

and therefore $$\bar{G} \equiv \mathbf{Y}'\bar{\mathbf{G}}\mathbf{Y} \equiv m^2\mathbf{X}'\mathbf{G}\mathbf{X} \equiv m^2 G.$$

Thus the forms S_1, S_2, G in X_1, X_2, X_3 are contravariants of weight two of the forms s_1, s_2 in x_1, x_2, x_3.

A polynomial equation $\psi(\delta_1, \theta_1, \theta_2, \delta_2, S_1, S_2, G) = 0$ *is said to be a contravariant equation of the two conic loci* γ_1, γ_2 *with point equations* $s_1 = 0$, $s_2 = 0$ IF ψ *is homogeneous* (i) *in the coefficients of* s_1, (ii) *in the coefficients of* s_2, (iii) *in* X_1, X_2, X_3.

* Cf. 9·65, 9·66, pp. 250, 251.

*The scroll given by a contravariant equation is independent of the method of writing down the point equations $s_1 = 0$, $s_2 = 0$ of γ_1, γ_2.**

For, if a general term of ψ is

$$h\delta_1^\xi \theta_1^\eta \theta_2^\zeta \delta_2^\tau S_1^\lambda S_2^\mu G^\nu,$$

the properties (i), (ii), (iii) show that

$$\begin{aligned} 3\xi + 2\eta + \zeta \quad + 2\lambda + \nu &= 3p, \\ \eta + 2\zeta + 3\tau + 2\mu + \nu &= 3q, \\ \lambda + \mu \quad + \nu &= 3r, \end{aligned}$$

where p, q, r are the same for all terms of ψ. It follows that

$$\xi + \eta + \zeta + \tau + \lambda + \mu + \nu = p + q + r,$$

whence

$$\psi(\bar{\delta}_1, \bar{\theta}_1, \bar{\theta}_2, \bar{\delta}_2, \bar{S}_1, \bar{S}_2, \bar{G}) \equiv m^{2(p+q+r)} \psi(\delta_1, \theta_1, \theta_2, \delta_2, S_1, S_2, G).$$

Thus the equation $\psi = 0$ is unaffected by the transformation $\mathbf{x} = m\mathbf{y}$, and the properties (i), (ii) show that it is unaffected by multiplying the coefficients of s_1 or of s_2 by a constant.

12·4. Four points on a conic

12·41. The polar sets and Hessian set of a triad of points on a conic were discussed in 7·7. Consider now a proper conic γ, with parametric equation

$$x : y : z = \lambda^2 : 2\lambda : 1, \tag{·411}$$

and on it a set of four points A_1, A_2, A_3, A_4 given by the equation

$$a_0\lambda^4 + 4a_1\lambda^3 + 6a_2\lambda^2 + 4a_3\lambda + a_4 = 0, \tag{·412}$$

and denoted by (A). The points λ, λ_i of γ will be denoted by L, L_i. The conics through A_1, A_2, A_3, A_4 form a point pencil which will be denoted by σ.

The first, second and third polars of L_1 relative to (A) are given by

$$\lambda_1(a_0\lambda^3 + 3a_1\lambda^2 + 3a_2\lambda + a_3) + (a_1\lambda^3 + 3a_2\lambda^2 + 3a_3\lambda + a_4) = 0, \tag{·413}$$

$$\lambda_1^2(a_0\lambda^2 + 2a_1\lambda + a_2) + 2\lambda_1(a_1\lambda^2 + 2a_2\lambda + a_3) + (a_2\lambda^2 + 2a_3\lambda + a_4) = 0, \tag{·414}$$

$$\lambda_1^3(a_0\lambda + a_1) + 3\lambda_1^2(a_1\lambda + a_2) + 3\lambda_1(a_2\lambda + a_3) + (a_3\lambda + a_4) = 0. \tag{·415}$$

* Not all contravariant equations are of the above form, but any contravariant equation can be expressed as a polynomial equation

$$\psi\{\delta_1, \theta_1, \theta_2, \delta_2, S_1, S_2, G, J(S_1, S_2, G)\} = 0.$$

Cf. the footnote to 12·32.

The Hessian of (A) is given by

$$\begin{vmatrix} a_0\lambda^2+2a_1\lambda+a_2, & a_1\lambda^2+2a_2\lambda+a_3 \\ a_1\lambda^2+2a_2\lambda+a_3, & a_2\lambda^2+2a_3\lambda+a_4 \end{vmatrix} = 0. \qquad (\cdot 416)$$

12·42. *The pencil σ contains* ONE *conic outpolar to γ, namely the conic*

$$s_a \equiv a_0x^2+a_2y^2+a_4z^2+2a_3yz+2a_2zx+2a_1xy = 0.$$

For this conic is outpolar to γ and meets γ in the set (A). Also σ cannot contain another conic outpolar to γ, or all conics of σ, including γ itself, would be outpolar to γ.

Alternatively the conic

$$ax^2+by^2+cz^2+2fyz+2gzx+2hxy = 0 \qquad (\cdot 421)$$

meets γ in the four points given by

$$a\lambda^4+4h\lambda^3+(4b+2g)\lambda^2+4f\lambda+c = 0. \qquad (\cdot 422)$$

Comparing this with (·412) it follows that the general conic of σ is

$$a_0x^2+uy^2+a_4z^2+2a_3yz+2vzx+2a_1xy = 0, \qquad (\cdot 423)$$

where u, v may have any values satisfying

$$4u+2v = 6a_2. \qquad (\cdot 424)$$

But the line equation of γ is

$$Y^2-ZX = 0,$$

so the conic (·423) is outpolar to γ IF

$$u-v = 0,$$

and with (·424) this gives

$$u = v = a_2.$$

12·43. *The first polar set of a point L_1 of γ relative to (A) is the intersection with γ other than L_1 of the eleven-point conic of the tangent to γ at L_1 and the pencil σ.**

For by 12·26 this eleven-point conic is given by the vanishing of the Jacobian of the forms s_a, y^2-4zx, and $x-\lambda_1y+\lambda_1^2z$. Its equation is therefore

$$\begin{vmatrix} a_0x+a_1y+a_2z, & a_1x+a_2y+a_3z, & a_2x+a_3y+a_4z \\ 2z & -y & 2x \\ 1 & -\lambda_1 & \lambda_1^2 \end{vmatrix} = 0.$$

* The author is indebted to P. W. Wood for several results, including this one.

The intersections of this eleven-point conic with γ are given by

$$\begin{vmatrix} a_0\lambda^2+2a_1\lambda+a_2, & a_1\lambda^2+2a_2\lambda+a_3, & a_2\lambda^2+2a_3\lambda+a_4 \\ 1 & -\lambda & \lambda^2 \\ 1 & -\lambda_1 & \lambda_1^2 \end{vmatrix} = 0$$

which reduces to

$$(\lambda-\lambda_1)\{\lambda_1(a_0\lambda^3+3a_1\lambda^2+3a_2\lambda+a_3) \\ +(a_1\lambda^3+3a_2\lambda^2+3a_3\lambda+a_4)\} = 0.$$

12·44. *The second polar set of L_1 relative to (A) is the intersection with γ of the polar line of L_1 relative to s_a.*

For the polar line of L_1 relative to s_a is

$$\lambda_1^2(a_0x+a_1y+a_2z)+2\lambda_1(a_1x+a_2y+a_3z)+(a_2x+a_3y+a_4z) = 0,$$

and the intersections of this line with γ are given by (·414).

Since the second polar set of L_1 relative to (A) is its first polar set relative to its first polar set relative to (A), it follows from 7·78 that the polar line of L_1 relative to s_a is also the polar line of L_1 relative to the triangle whose vertices are the first polar set of L_1 relative to (A).

12·45. *The third polar of L_1 relative to (A) is the other intersection with γ of the line joining L_1 to the pole relative to γ of the polar line of L_1 relative to s_a.*

For this point is the first polar of L_1 relative to the second polar of L_1 relative to (A).

Alternatively it follows from (·413) and (·415) that L_2 is the third polar of L_1 relative to (A) IF L_1 belongs to the first polar of L_2 relative to (A), that is if L_1 is the pole of the tangent to γ at L_2 relative to some conic of σ other than γ, that is IF the tangent to γ at L_2 passes through the point of concurrence of the polar lines of L_1 relative to the conics of σ, whence the result follows.

12·46. Denote by $S_a(2,2)$ the symmetric $(2,2)$ correspondence between points of γ which are conjugate relative to s_a. The equation of $S_a(2,2)$ is

$$\lambda'^2(a_0\lambda^2+2a_1\lambda+a_2)+2\lambda'(a_1\lambda^2+2a_2\lambda+a_3) \\ +(a_2\lambda^2+2a_3\lambda+a_4) = 0.$$

Comparing with (·414):

The second polar set of L_1 relative to (A) is the pair of points of γ which correspond to L_1 in $S_a(2,2)$.

The self-corresponding points of $S_a(2,2)$ are the points (A).

$S_a(2,2)$ is not a general $S(2,2)$ because s_a is outpolar to γ.

12·47. *The Hessian set of (A) is formed by the contacts with γ of the common tangents of γ and s_a.*

For the equation (·416) which gives the Hessian set also gives the branch points of $S_a(2, 2)$. By 10·21 these branch points are the intersection of γ with the reciprocal of γ relative to s_a, which is also the harmonic locus of γ and s_a since s_a is outpolar to γ.* Also the harmonic locus of γ and s_a contains the contacts of the common tangents of γ and s_a.

12·48. *The invariants I and J.*

The conic $s_a + \rho(y^2 - 4zx) = 0$ of σ is a line-pair IF

$$\begin{vmatrix} a_0 & a_1 & a_2 - 2\rho \\ a_1 & a_2 + \rho & a_3 \\ a_2 - 2\rho & a_3 & a_4 \end{vmatrix} = 0,$$

or
$$4\rho^3 - I\rho - J = 0, \qquad (\cdot 481)$$
where
$$I \equiv a_0 a_4 - 4a_1 a_3 + 3a_2^2$$
and
$$J \equiv \begin{vmatrix} a_0 & a_1 & a_2 \\ a_1 & a_2 & a_3 \\ a_2 & a_3 & a_4 \end{vmatrix}.$$

The coefficient of ρ^2 in (·481) vanishes because s_a is outpolar to γ. $I = 0$ IF s_a is also inpolar to γ, and $J = 0$ IF s_a is a line-pair. I and J are actually invariants of the quartic form (·412), namely

$$a_0\lambda^4 + 4a_1\lambda^3 + 6a_2\lambda^2 + 4a_3\lambda + a_4, \qquad (\cdot 482)$$

and the equations $I = 0$, $J = 0$ correspond to projective properties of the four roots of (·412), namely that they form respectively an equianharmonic or a harmonic set. Expressed as theorems for the points (A) of γ, the results to be proved are:

(a) *The set (A) is harmonic on γ* IF $J = 0$.
(b) *The set (A) is equianharmonic on γ* IF $I = 0$.

Suppose that the tangent at A_1 to the conic $s_a + \rho(y^2 - 4zx) = 0$ of σ meets γ again in the point whose parameter is λ. This establishes a (1, 1) correspondence between λ and ρ. The 'tangents' to the line-pairs of σ meet γ in A_2, A_3, A_4, and the particular value $\rho = \infty$ corresponds to the parameter λ_1 of A_1. Thus

$$(A_1A_2A_3A_4) = (\lambda_1\lambda_2\lambda_3\lambda_4) = (\infty\rho_1\rho_2\rho_3) = \frac{\rho_3 - \rho_2}{\rho_1 - \rho_2},$$

* Cf. 11·33, p. 304.

where ρ_1, ρ_2, ρ_3 are the roots of (·481) in some order. Since

$$\rho_1+\rho_2+\rho_3 = 0$$

it follows that (A) is harmonic IF (·481) has a zero root, i.e. IF $J = 0$. More simply, $J = 0$ is a CONDITION for s_a to be a line-pair, i.e. for the joins of two points of (A) to be conjugate relative to γ to the join of the other two points of (A), i.e. for (A) to be a harmonic set on γ. Again, by 7·64, four numbers form an equianharmonic set if each is one of the Hessian numbers of the other three. The Hessian numbers for the roots ρ_1, ρ_2, ρ_3 of (·481) are given by

$$\begin{vmatrix} 12\rho & -I \\ -I & -I\rho-3J \end{vmatrix} = 0.$$

This has an infinite root IF $I = 0$, which is therefore a CONDITION for (A) to be equianharmonic.

It should be noticed that $J = 0$ is also the CONDITION for $S_a(2,2)$ to be compounded of an involution,* and that $I = 0$, the CONDITION for s_a to be inpolar to γ, is also the CONDITION for $S_a(2,2)$ to be compounded of triads.†

12·49. *Solution of a quartic equation*

The equation (·481), which is usually called Euler's *reducing cubic*, arises in Ferrari's method of solution of the quartic equation

$$a_0\lambda^4+4a_1\lambda^3+6a_2\lambda^2+4a_3\lambda+a_4 = 0.$$

In this we multiply the left-hand side by a_0 and express the equation in the form

$$(a_0\lambda^2+2a_1\lambda+a_2+2\rho)^2-(2p\lambda+q)^2 = 0,$$

where, by equating coefficients,

$$p^2 = a_1^2-a_0a_2+a_0\rho,\ pq = a_1a_2-a_0a_3+2a_1\rho,\ q^2 = (a_2+2\rho)^2-a_0a_4.$$

Eliminating p and q we find

$$4\rho^3-I\rho-J = 0.$$

If ρ is a root of this it determines p and q uniquely, apart from an immaterial sign, and the solution of the quartic is reduced to the solution of the two quadratics

$$(a_0\lambda^2+2a_1\lambda+a_2+2\rho)\pm(2px+q) = 0.$$

It was shown in 12·48 that $(\lambda_1\lambda_2\lambda_3\lambda_4) = (\infty\rho_1\rho_2\rho_3)$; since (·481) has no infinite root it follows that the quartic equation (·412) has a

* Cf. 10·41, 10·42, p. 289. † Cf. 10·24, p. 278.

repeated root IF the reducing cubic (·481) has a repeated root. The CONDITION for this is

$$\Delta \equiv I^3 - 27J^2 = 0.$$

The expression Δ is called the *discriminant* of the quartic, and it may be shown† that

$$a_0^6 \prod_{i \neq j} (\lambda_i - \lambda_j)^2 = a_0^6 . 64^2 \prod_{i \neq j} (\rho_i - \rho_j)^2 = 256\Delta.$$

The expressions corresponding to I and J for the quartic form

$$\sigma(a_0\lambda^4 + 4a_1\lambda^3 + 6a_2\lambda^2 + 4a_3\lambda + a_4)$$

are I^* and J^*, where $I^* = \sigma^2 I$, $J^* = \sigma^3 J$; the corresponding reducing cubic is

$$4\rho^3 - \sigma^2 I\rho - \sigma^3 J = 0, \qquad (·491)$$

whose roots are σ times those of (·481). In fact the equation $(\lambda_1\lambda_2\lambda_3\lambda_4) = (\infty\rho_1\rho_2\rho_3)$, together with $\rho_1 + \rho_2 + \rho_3 = 0$, determines ρ_1, ρ_2, ρ_3, apart from a common factor, as the roots of (·491), in which σ is arbitrary. Let λ' be a parameter connected with λ by a non-singular projectivity

$$\alpha\lambda\lambda' + \beta\lambda + \gamma\lambda' + \delta = 0 \quad (\alpha\delta \neq \beta\gamma).$$

On substituting
$$\lambda = -\frac{\gamma\lambda' + \delta}{\alpha\lambda' + \beta}$$
in the quartic form (·482) and multiplying by $(\alpha\lambda' + \beta)^4$ we obtain a projectively equivalent quartic form in λ', say

$$a_0'\lambda'^4 + 4a_1'\lambda'^3 + 6a_2'\lambda'^2 + 4a_3'\lambda' + a_4'.$$

If $\lambda_1', \lambda_2', \lambda_3', \lambda_4'$ are the values of λ' for which this vanishes, and if I' and J' are the expressions obtained from I and J by replacing a_0, a_1, a_2, a_3, a_4 by $a_0', a_1', a_2', a_3', a_4'$, it follows, from

$$(\lambda_1'\lambda_2'\lambda_3'\lambda_4') = (\lambda_1\lambda_2\lambda_3\lambda_4) = (\infty\rho_1\rho_2\rho_3),$$

that ρ_1, ρ_2, ρ_3 are the roots of a cubic

$$4\rho^3 - \sigma'^2 I'\rho - \sigma'^3 J' = 0, \qquad (·492)$$

in which σ' is arbitrary. Comparing (·491) and (·492) we see that $I^3/J^2 = I'^3/J'^2$. The expression I^3/J^2 is thus an absolute projective invariant of the quartic form (·482).

12·5. A general point pencil of conics. Combinant conic

12·51. *The base points A_1, A_2, A_3, A_4 of a general point pencil σ of conics determine a Hessian set of four points on each conic of σ. There*

† See, for example, H. W. Turnbull, *Theory of Equations* (Edinburgh, 1946), p. 133.

is a conic which contains all these Hessian sets, and this conic is called the combinant conic of σ.

For, taking the diagonal triangle of A_1, A_2, A_3, A_4 as triangle of reference and A_1 as unit point, the general conic of σ is

$$s \equiv \lambda x^2 + \mu y^2 + \nu z^2 = 0,$$

where
$$\lambda + \mu + \nu = 0.$$

The conic of σ which is outpolar to s is

$$\lambda(\mu - \nu)\, x^2 + \mu(\nu - \lambda)\, y^2 + \nu(\lambda - \mu)\, z^2 = 0,$$

and the f-conic of this and s is

$$c \equiv \lambda(\mu - \nu)^2 x^2 + \mu(\nu - \lambda)^2 y^2 + \nu(\lambda - \mu)^2 z^2 = 0.$$

By 12·47 the Hessian set of (A) on s is the section of s by this f-conic. But, since $\lambda + \mu + \nu = 0$,

$$\mu^2 + \mu\nu + \nu^2 = \nu^2 + \nu\lambda + \lambda^2 = \lambda^2 + \lambda\mu + \mu^2 = -(\mu\nu + \nu\lambda + \lambda\mu),$$

so
$$c + (\mu\nu + \nu\lambda + \lambda\mu)\, s + 3\lambda\mu\nu(x^2 + y^2 + z^2) \equiv 0.$$

Thus the Hessian sets of (A) on all the conics s of σ lie on the conic

$$x^2 + y^2 + z^2 = 0,$$

which is therefore the combinant conic of σ.

12·52. *If* $s_1 = 0$, $s_2 = 0$ *are the point equations of two conics of a general point pencil* σ, *the combinant conic of* σ *is*

$$\theta_2 s_1 + \theta_1 s_2 - 3f = 0.$$

For the equations of the two conics may be taken as

$$s_1 \equiv \lambda_1 x^2 + \mu_1 y^2 + \nu_1 z^2 = 0,$$
$$s_2 \equiv \lambda_2 x^2 + \mu_2 y^2 + \nu_2 z^2 = 0,$$

where
$$\lambda_1 + \mu_1 + \nu_1 = 0,$$
$$\lambda_2 + \mu_2 + \nu_2 = 0,$$

and consequently

$$\mu_1 \nu_2 - \mu_2 \nu_1 = \nu_1 \lambda_2 - \nu_2 \lambda_1 = \lambda_1 \mu_2 - \lambda_2 \mu_1 = \kappa.$$

Then it is easily verified that

$$\theta_2 s_1 + \theta_1 s_2 - 3f \equiv -\kappa^2(x^2 + y^2 + z^2),$$

and it has just been shown that the conic

$$x^2 + y^2 + z^2 = 0$$

is the combinant conic of σ.

Thus, when the equations of the two given conics are written down in this particular way, the combinant conic is given by

$$\theta_2 s_1 + \theta_1 s_2 - 3f = 0.$$

But this is a covariant equation, and is therefore the equation of the combinant conic however the equations $s_1 = 0$, $s_2 = 0$ of the conics are written down.

It should be noticed that $\theta_2 s_1 + \theta_1 s_2 - 3f = 0$ is more than a covariant equation, for it still gives the same conic if the two given conics are replaced by any other two conics of σ.

12·53. *A general point pencil σ of conics contains* ONE *pair of conics which are mutually apolar. If the conics of σ are*

$$\lambda x^2 + \mu y^2 + \nu z^2 = 0, \quad \lambda + \mu + \nu = 0,$$

the mutually apolar conics are

$$x^2 + \omega y^2 + \omega^2 z^2 = 0, \quad x^2 + \omega^2 y^2 + \omega z^2 = 0,$$

where $\omega^3 = 1$, $\omega \neq 1$.

For the conic of σ outpolar to

$$s \equiv \lambda x^2 + \mu y^2 + \nu z^2 = 0$$

is

$$\lambda(\mu - \nu) x^2 + \mu(\nu - \lambda) y^2 + \nu(\lambda - \mu) z^2 = 0,$$

and this is also inpolar to $s = 0$ IF

$$\frac{1}{\mu - \nu} + \frac{1}{\nu - \lambda} + \frac{1}{\lambda - \mu} = 0.$$

With

$$\lambda + \mu + \nu = 0,$$

this reduces to

$$\mu\nu + \nu\lambda + \lambda\mu = 0,$$

so λ, μ, ν must be proportional to the roots of the equation $t^3 = 1$.

12·54. *The combinant conic and the mutually apolar conics of σ form a Hessian triad.**

For it is easily verified from the equations just obtained that the f-conic and G-conic of the mutually apolar conics coincide in the combinant conic.

12·55. *The combinant conic of σ is the* ONE *conic which is inpolar to all the conics of σ and has the diagonal triangle of $A_1A_2A_3A_4$ as a self-polar triangle.*

For the only conic of the form

$$px^2 + qy^2 + rz^2 = 0,$$

* See 11·48, p. 311.

which is inpolar to all conics of σ is evidently

$$x^2+y^2+z^2 = 0.$$

The conic scrolls apolar to all the conic loci of σ form a linear ∞^3 system, so we should expect ONE of them to have a given triangle as self-polar triangle.

12·56. *The mutually apolar conics of σ are the Hessian pair of the three line-pairs of σ in the $R\infty^1$ formed by the conics of σ.*

For the conics of σ are

$$\xi(x^2+\omega y^2+\omega^2 z^2)+\eta(x^2+\omega^2 y^2+\omega z^2) = 0,$$

where (ξ, η) is a ratio parameter of the $R\infty^1$. The line-pairs of σ are given by

$$(\xi+\eta)(\omega\xi+\omega^2\eta)(\omega^2\xi+\omega\eta) = 0,$$

or

$$\xi^3+\eta^3 = 0,$$

and the Hessian pair of the roots of this equation are given by

$$\xi\eta = 0.$$

12·57. *A conic s of σ is one of the mutually apolar conics* IF *the base points A_1, A_2, A_3, A_4 form an equianharmonic set on s.*

For there is a homographic relation between the $R\infty^1$ formed by the conics of σ and the $R\infty^1$ formed by the pencil of lines through A_1 in which each conic s of σ corresponds to the tangent a_1 to s at A_1. The line-pairs of σ correspond to the lines A_1A_2, A_1A_3, A_1A_4, so s is one of the mutually apolar conics of σ IF a_1 is one of the Hessian pairs of the lines A_1A_2, A_1A_3, A_1A_4 in the pencil of lines with vertex A_1. By 7·64 this happens IF the lines a_1, A_1A_2, A_1A_3, A_1A_4 form an equianharmonic subpencil, that is IF A_1, A_2, A_3, A_4 form an equianharmonic set on s. This argument is essentially the same as that of 12·48.

12·6. Polarity relative to a triangle

12·61. *Polarity relative to a curve of order n*

We have in 12·25 defined the first, second, ... rth ... polar sets of a point of a line p with respect to a set of n points on p and shown that they are independent of the co-ordinate system. In a similar way we define the *first polar curve* of $P_1(x_1, y_1, z_1)$ with respect to the plane curve f whose equation is

$$f(x, y, z) = 0,$$

where $f(x, y, z)$ is a homogeneous polynomial of degree n, as the curve of order $n-1$ whose equation is

$$\left(x_1 \frac{\partial}{\partial x} + y_1 \frac{\partial}{\partial y} + z_1 \frac{\partial}{\partial z}\right) f = 0.$$

The *second polar curve* of P_1 relative to f is of order $n-2$ and its equation is

$$\left(x_1 \frac{\partial}{\partial x} + y_1 \frac{\partial}{\partial y} + z_1 \frac{\partial}{\partial z}\right)^2 f = 0,$$

and this is the first polar curve of P_1 relative to its first polar curve relative to f. Generally, the rth polar curve of P_1 relative to f is of order $n-r$ and its equation is

$$\left(x_1 \frac{\partial}{\partial x} + y_1 \frac{\partial}{\partial y} + z_1 \frac{\partial}{\partial z}\right)^r f = 0.$$

The argument of 12·25 is readily extended to show that these polar curves depend only on P_1 and the curve f, and not on the co-ordinate system; a proof, based on the Clebsch-Aronhold symbolism, has indeed already been sketched in 9·88.

If P_1, P_2 are two points of the plane, a general point of the line joining them is $\lambda P_1 + \mu P_2$, and the intersections of the line with f are given by

$$\phi(\lambda, \mu) \equiv f(\lambda x_1 + \mu x_1, \lambda y_1 + \mu y_2, \lambda z_1 + \mu z_2) = 0.$$

If we write

$$f(x_1, y_1, z_1) = f_1, f(x_2, y_2, z_2) = f_2,$$

$$x_1 \frac{\partial}{\partial x_2} + y_1 \frac{\partial}{\partial y_2} + z_1 \frac{\partial}{\partial z_2} = \Delta_{12}, x_2 \frac{\partial}{\partial x_1} + y_2 \frac{\partial}{\partial y_1} + z_2 \frac{\partial}{\partial z_1} = \Delta_{21},$$

this can be expanded in either of the forms

$$\phi(\lambda, \mu) \equiv \lambda^n f_1 + \binom{n}{1} \lambda^{n-1} \mu \Delta_{21} f_1 + \dots + \binom{n}{r} \lambda^{n-r} \mu^r \Delta_{21}^r f_1 + \dots + \mu^n \Delta_{21}^n f_1 = 0,$$

or

$$\phi(\lambda, \mu) \equiv \lambda^n \Delta_{12}^n f_2 + \binom{n}{1} \lambda^{n-1} \mu \Delta_{12}^{n-1} f_2 + \dots + \binom{n}{r} \lambda^{n-r} \mu^r \Delta_{12}^{n-r} f_2 + \dots + \mu^n f_2 = 0.$$

Equating coefficients,

$$\Delta_{21}^r f_1 = \Delta_{12}^{n-r} f_2 \quad (r = 0, 1, 2, \dots, n).$$

The equation $$\Delta_{21}^r f_1 = 0$$

is the CONDITION that P_1 should lie on the rth polar of P_2. Thus:

If the rth polar of P_2 passes through P_1 then the $(n-r)$th polar of P_1 passes through P_2.

We next prove:

The rth polar curve of P_1 relative to f is the locus of the rth polar sets of P_1 relative to the sets of points in which f is met by variable lines p through P_1.

In fact, since the polar curves depend only on P_1 and the curve f, there is no loss of generality if we take any particular line p as $z = 0$, and P_1 as $(x_1, y_1, 0)$. The rth polar curve of P_1 is then

$$\left(x_1\frac{\partial}{\partial x}+y_1\frac{\partial}{\partial y}\right)^r f(x,y,z) = 0$$

and it meets p where

$$\left(x_1\frac{\partial}{\partial x}+y_1\frac{\partial}{\partial y}\right)^r f(x,y,0) = 0.$$

By 12·25 this is the rth polar set of P_1 relative to the n points, given by $f(x,y,0) = 0$, in which f meets p.

12·62. *Polarity relative to a triangle*

If the curve f of 12·61 is of order three, a general point P_1 has relative to f a first and a second polar curve, known respectively as the *polar conic* and the *polar line* of P_1 relative to f. If the polar conic of a point P_1 contains P_2, then the polar line of P_2 contains P_1.

We are concerned here with the particular case when f degenerates into three lines forming a triangle T. If we take T as the triangle of reference the equation of f is

$$xyz = 0.$$

The polar conic of a point P_1 relative to T will be denoted by γ_1, and its equation is

$$\frac{x_1}{x}+\frac{y_1}{y}+\frac{z_1}{z} = 0.$$

The polar line of P_1 will be denoted by p_1, and its equation is

$$\frac{x}{x_1}+\frac{y}{y_1}+\frac{z}{z_1} = 0.$$

It thus appears that the polar line of P_1 relative to T, as defined here, is identical with the polar line for which a construction was given in 5·5.

More generally, if the sides of T are the three non-concurrent lines $[a_i, b_i, c_i]$, $(i = 1, 2, 3)$, the polar conic of P_1 is

$$\frac{a_1x_1+b_1y_1+c_1z_1}{a_1x+b_1y+c_1z}+\frac{a_2x_1+b_2y_1+c_2z_1}{a_2x+b_2y+c_2z}+\frac{a_3x_1+b_3y_1+c_3z_1}{a_3x+b_3y+c_3z}=0,$$

and its polar line is

$$\frac{a_1x+b_1y+c_1z}{a_1x_1+b_1y_1+c_1z_1}+\frac{a_2x+b_2y+c_2z}{a_2x_1+b_2y_1+c_2z_1}+\frac{a_3x+b_3y+c_3z}{a_3x_1+b_3y_1+c_3z_1}=0.$$

A general line is the polar relative to T of a unique point, which is called the pole of the line relative to T.

A point P_2 is said to be *conjugate* to P_1 relative to T IF P_2 lies on the polar line p_1 of P_1 relative to T. When T is taken as triangle of reference, this CONDITION is

$$\frac{x_2}{x_1}+\frac{y_2}{y_1}+\frac{z_2}{z_1}=0.$$

The polar line of P_1 is indeterminate IF P_1 is X, Y or Z. If P_1 is $(0, y_1 z_1)$, p_1 is $x = 0$. The pole P_1 of a line p_1 is indeterminate IF p_1 is YZ, ZX or XY. If p_1 is $[0, Y_1, Z_1]$, P_1 is X. The polar conic of $(0, y_1, z_1)$ is $x(z_1y+y_1z) = 0$, and the polar conic of X is $yz = 0$.

12·63. *The tangent at P_1 to the conic P_1P_2XYZ contains the pole of the line P_1P_2 relative to T IF P_2 is conjugate to P_1 relative to T.*

For the conic P_1P_2XYZ is

$$x_1x_2(y_1z_2-y_2z_1)\,yz+y_1y_2(z_1x_2-z_2x_1)\,zx+z_1z_2(x_1y_2-x_2y_1)\,xy=0,$$

and the tangent to this at P_1 is

$$x_2y_1z_1(y_1z_2-y_2z_1)\,x+y_2z_1x_1(z_1x_2-z_2x_1)\,y$$
$$+z_2x_1y_1(x_1y_2-x_2y_1)\,z=0.$$

The pole of P_1P_2 relative to T is

$$\left(\frac{1}{y_1z_2-y_2z_1},\quad \frac{1}{z_1x_2-z_2x_1},\quad \frac{1}{x_1y_2-x_2y_1}\right),$$

which lies on the tangent line IF

$$x_2y_1z_1+y_2z_1x_1+z_2x_1y_1=0,$$

i.e. IF P_2 is conjugate to P_1 relative to T.

12·64. *Self-polar triangles*

If P_2 is conjugate to P_1 relative to T it does not follow that P_1 is conjugate to P_2. Two points are said to be *mutually conjugate* with respect to T IF each lies on the polar line of the other. A triangle is said to be *self-polar* relative to T IF each side is the polar line of the opposite vertex relative to T, i.e. IF each pair of vertices is mutually conjugate.

The points which are mutually conjugate to P_1 are given by

$$\frac{x}{x_1}+\frac{y}{y_1}+\frac{z}{z_1} = \frac{x_1}{x}+\frac{y_1}{y}+\frac{z_1}{z} = 0,$$

and are therefore the two points Q_1, R_1 with co-ordinates

$$(x_1, \omega y_1, \omega^2 z_1), \quad (x_1, \omega^2 y_1, \omega z_1).$$

But Q_1 and R_1 are themselves mutually conjugate, and so $P_1 Q_1 R_1$ is a self-polar triangle relative to T. Thus:

A general point (x_1, y_1, z_1) is a vertex of ONE *triangle which is self-polar relative to T, the other vertices being $(x_1, \omega y_1, \omega^2 z_1)$ and*

$$(x_1, \omega^2 y_1, \omega z_1).$$

The polar conic γ_1 of P_1 contains Q_1 and R_1 and touches $P_1 Q_1$ and $P_1 R_1$; similarly the polar conic of Q_1 touches $Q_1 P_1$, $Q_1 R_1$ at P_1, R_1 and the polar conic of R_1 touches $R_1 P_1$, $R_1 Q_1$ at P_1, Q_1.

The sides of the triangle $P_1 Q_1 R_1$ are the polar lines

$$\frac{x}{x_1}+\frac{y}{y_1}+\frac{z}{z_1} = 0, \quad \frac{x}{x_1}+\frac{\omega^2 y}{y_1}+\frac{\omega z}{z_1} = 0, \quad \frac{x}{x_1}+\frac{\omega y}{y_1}+\frac{\omega^2 z}{z_1} = 0$$

of the vertices P_1, Q_1, R_1.

The polar line of X relative to the triangle $P_1 Q_1 R_1$ is, by 12·62,

$$x_1\left(\frac{x}{x_1}+\frac{y}{y_1}+\frac{z}{z_1}\right)+x_1\left(\frac{x}{x_1}+\frac{\omega^2 y}{y_1}+\frac{\omega z}{z_1}\right)+x_1\left(\frac{x}{x_1}+\frac{\omega y}{y_1}+\frac{\omega^2 z}{z_1}\right) = 0,$$

i.e. the line $x = 0$. Thus:

If a triangle PQR is self-polar relative to a triangle XYZ, then XYZ is self-polar relative to PQR.

12·65. *A general line p contains two points Q, R which are mutually conjugate relative to T; these are the Hessian pair of the intersections L, M, N of p with YZ, ZX, XY.*

In fact, if P is the pole of p, Q and R are the other two vertices of the self-polar triangle determined by P. Alternatively, by 12·61, the polar conic and polar line of a point P_1 of p meet p in the first polar pair and the second polar point of P_1 relative to the triad L, M, N. If Q and R are mutually conjugate points on p, the second polar of Q relative to the triad must be R, and the first polar pair of

Q must contain R and, in fact, consist of R counted twice, since the second polar of Q is the harmonic conjugate of Q relative to the first polar pair. Thus, by 7·76, Q and R are the Hessian points of L, M, N.

12·7. Three mutually apolar conics

12·71 Consider the three conics

$$s_\lambda \equiv \lambda x^2 + 2yz = 0, \quad s_\mu \equiv \mu y^2 + 2zx = 0, \quad s_\nu \equiv \nu z^2 + 2xy = 0,$$

each two of which are mutually apolar for all non-zero values of λ, μ, ν.* If s_λ contains P_1, then s_λ also contains the points Q_1, R_1 which form with P_1 a self-polar triangle relative to the triangle of reference XYZ. For the points P_1, Q_1, R_1 are

$$(x_1, y_1, z_1), \quad (x_1, wy_1, w^2 z_1), \quad (x_1, w^2 y_1, wz_1),$$

and each of them lies on s_λ IF

$$\lambda x_1^2 + 2y_1 z_1 = 0.$$

The same is true for s_μ and s_ν. Thus if the three conics have one common point P, they also have two other common points Q, R. The triangle PQR is self-polar relative to XYZ, so XYZ is self-polar relative to PQR, and s_λ, s_μ, s_ν are the polar conics of X, Y, Z relative to PQR.

But at the common points of s_μ and s_ν

$$\mu\nu y^2 z^2 = 4x^2 yz,$$

so either $yz = 0$ or $4x^2 = \mu\nu yz$. Thus the three common points of s_μ and s_ν other than X lie on the conic $4x^2 = \mu\nu yz$, which is s_λ IF $\lambda\mu\nu + 8 = 0$. In fact

The conics s_λ, s_μ, s_ν only have a common point if $\lambda\mu\nu + 8 = 0$. If $\lambda\mu\nu + 8 = 0$ the three conics have three common points P, Q, R; the triangle PQR is self-polar relative to XYZ, and the conics are the polar conics of X, Y and Z relative to PQR.

12·72. The triangle $P_1 Q_1 R_1$ is self-polar relative to s_λ IF

$$\lambda x_1^2 - y_1 z_1 = 0,$$

i.e. IF P_1 (and consequently Q_1 and R_1) lies on the conic

$$\lambda x^2 - yz = 0.$$

* This is easily verified algebraically, but it also follows immediately from the end of 10·28, p. 285.

Similarly $P_1Q_1R_1$ is self-polar relative to s_μ if P_1 lies on

$$\mu y^2 - zx = 0,$$

and relative to s_ν IF P_1 lies on

$$\nu z^2 - xy = 0.$$

But by 12·71 the common points of the last two conics, other than X, form a triangle self-polar relative to XYZ, which is therefore the unique common self-polar triangle of s_μ and s_ν. Further, by 12·71, the conics

$$\lambda x^2 - yz = 0, \quad \mu y^2 - zx = 0, \quad \nu z^2 - xy = 0,$$

have three common points IF $\lambda\mu\nu = 1$, so s_λ, s_μ, s_ν have a common self-polar triangle IF $\lambda\mu\nu = 1$.

The harmonic locus and scroll of s_μ, s_ν coincide in

$$x^2 + 2\mu\nu yz = 0,$$

which is s_λ IF $\lambda\mu\nu = 1$, so

$\lambda\mu\nu = 1$ *is both a* CONDITION *for* s_λ, s_μ, s_ν *to have a common self-polar triangle and a* CONDITION *for the three conics to form a Hessian triad.*

12·8. A general point pencil of conics

12·81. *Representation of a net of conics*

The polar theory of a triangle could be used to prove theorems about a general point pencil of conic loci, whose equations may be taken in the form

$$\lambda x^2 + \mu y^2 + \nu z^2 = 0,$$

where the ratio set (λ, μ, ν) takes all the ∞^1 values satisfying a linear equation

$$a\lambda + b\mu + c\nu = 0.$$

To do this we consider all the ∞^2 conics s given by equations of the form

$$s \equiv \lambda x^2 + \mu y^2 + \nu z^2 = 0,$$

which form a net Λ, and we regard the coefficients λ, μ, ν as coordinates of a point P' of another plane π'. Let $X'Y'Z'$ be the triangle of reference in π' and denote it by T'. Then each conic s of Λ determines ONE point P' of π', and each point P' of π' determines ONE conic s of Λ. In fact the ∞^2 conics of Λ are represented by the ∞^2 points of π'.

The line-pairs of Λ are given by $\lambda\mu\nu = 0$, and therefore correspond to the points of the sides of T'. In particular the repeated lines of Λ are

$$x^2 = 0, \quad y^2 = 0, \quad z^2 = 0,$$

and correspond to the points X', Y', Z'.

12·82. *Representation of apolarity*

A particular conic of Λ

$$s_i \equiv \lambda_i x^2 + \mu_i y^2 + \nu_i z^2 = 0,$$

corresponds to the point $(\lambda_i, \mu_i, \nu_i)$ of π', which will be denoted by P'_i. A conic s_2 of Λ is outpolar to a conic s_1 of Λ IF

$$\frac{\lambda_2}{\lambda_1} + \frac{\mu_2}{\mu_1} + \frac{\nu_2}{\nu_1} = 0;$$

i.e. IF P'_2 is conjugate to P'_1 relative to T'.

Also the harmonic locus of two conics s_1 and s_2 of Λ is

$$f \equiv \lambda_1\lambda_2(\mu_1\nu_2 + \mu_2\nu_1)\,x^2 + \mu_1\mu_2(\nu_1\lambda_2 + \nu_2\lambda_1)\,y^2 + \nu_1\nu_2(\lambda_1\mu_2 + \lambda_2\mu_1)\,z^2 = 0,$$

which belongs to Λ and corresponds to the point P'_f of π' with co-ordinates

$$\{\lambda_1\lambda_2(\mu_1\nu_2 + \mu_2\nu_1), \quad \mu_1\mu_2(\nu_1\lambda_2 + \nu_2\lambda_1), \quad \nu_1\nu_2(\lambda_1\mu_2 + \lambda_2\mu_1)\}.$$

This point is easily seen to be the pole of $P'_1 P'_2$ relative to the conic of π' through X', Y', Z', P'_1 and P'_2.

12·83. *Point pencils of* Λ

Consider a general point pencil σ of Λ, given by the values of λ, μ, ν which satisfy the linear equation

$$a\lambda + b\mu + c\nu = 0,$$

where $abc \neq 0$. The conics of σ correspond to the points of the line p' in π' given by

$$ax' + by' + cz' = 0,$$

where x', y', z' are the co-ordinates in π'. There is in fact a (1, 1) correspondence between the $R\infty^1$ formed by the conics of σ and the points of p'. Conversely the points of any line in π' correspond to the conics of a point pencil in Λ.

The line-pairs of σ correspond to the intersections L', M', N' of p' with the sides of T'.

12·84. *Combinant conic of σ*

The line p' has a unique pole P' relative to T', so Λ contains a unique conic s_p which is inpolar to all conics of σ. P' is $\left(\frac{1}{a}, \frac{1}{b}, \frac{1}{c}\right)$, so s_p is

$$\frac{x^2}{a}+\frac{y^2}{b}+\frac{z^2}{c} = 0.$$

A general conic s_1 of σ corresponds to a general point P_1' of p'. The polar line of P_1' relative to T' meets p' in the ONE point P_2' which lies on p' and is conjugate to P_1' relative to T'. This point P_2' corresponds to the ONE conic s_2 of σ which is outpolar to s_1. By 12·47 the Hessian set on s_1 of the base points of σ is the intersection of s_1 with the harmonic locus f of s_1 and s_2. But by 12·82 f corresponds to the pole P_f' of p' relative to the conic $P_1'P_2'X'Y'Z'$, and consequently P_f' lies on the tangent at P_1' to this conic. Also by 12·63, since P_2' is conjugate to P_1' relative to T', this tangent contains P'. Thus the points P_1', P', P_f' are collinear, and consequently the conics s_1, s_p, f belong to a point pencil. It follows that the Hessian set on s_1 of the base points of σ is the intersection of s_1 with s_p, so the theorems of 12·51 and 12·55 are established.

12·85. *Mutually apolar conics of σ*

By 12·65 the line p' contains ONE pair of points Q' and R' which are mutually conjugate relative to T', these being the Hessian pair of L', M', N' on p'. It follows that σ contains ONE pair of mutually apolar conics s_q and s_r, and that these are the Hessian pair of the line-pairs of σ in the $R\,\infty^1$ formed by the conics of σ, as was shown in 12·53 and 12·56. Also the triangle $P'Q'R'$ is self-polar relative to T', so P' is the pole of $Q'R'$ relative to the conic $X'Y'Z'Q'R'$, which is the polar conic of P' relative to T'. Thus s_p is the harmonic locus of s_q and s_r, and similarly each of these three conics is the harmonic locus of the other two. Thus s_p, s_q, s_r form a Hessian triad, as was shown in 12·54. In fact the vertices of any triangle in π' which is self-polar relative to T' correspond to a Hessian triad of conics in Λ. Also each conic of Λ is the combinant conic of ONE point pencil of Λ.

BOOKS REFERRED TO IN THE TEXT

BAKER, H. F. *Principles of Geometry*, I and II (Cambridge, 1922).

BERTINI, E. *Geometria Proiettiva degli Iperspazi* (Messina, 1923).

BÔCHER, M. *Introduction to Higher Algebra* (New York, 1936).

CAYLEY, A. *Collected Mathematical Papers*, IV and VII (Cambridge, 1891 and 1894).

DURELL, C. V. and ROBSON, A. *Advanced Algebra*, II and III (London, 1937).

GRACE, J. H. and YOUNG, A. *Algebra of Invariants* (Cambridge, 1903).

HARDY, G. H. *Pure Mathematics* (Cambridge, 1938).

HILTON, H. J. *Plane Algebraic Curves* (Oxford, 1920).

ROBSON, A. *An Introduction to Analytical Geometry*, I and II (Cambridge, 1940 and 1947); referred to in the text $R(1)$ and as $R(2)$

SOMMERVILLE, D. M. Y. *Analytical Conics* (London, 1924).

TURNBULL, H. W. *The Theory of Determinants, Matrices and Invariants* (Glasgow, 1929).

TURNBULL, H. W. *Theory of Equations* (Edinburgh, 1946).

Printed in the United States
By Bookmasters